Finite Mathematics

with Computer-Supported Applications

Finite Mathematics

with Computer-Supported Applications

James Calvert and William Voxman
Department of Mathematics and Statistics
University of Idaho

McGraw-Hill, Inc.
New York St. Louis San Francisco Auckland Bogotá Caracas
Lisbon London Madrid Mexico City Milan Montreal New Delhi
San Juan Singapore Sydney Tokyo Toronto

Finite Mathematics
with Computer-Supported Applications

This book is printed on acid-free, recycled paper containing a minimum of 50% total recycled fiber with 10% postconsumer de-inked fiber.

1 2 3 4 5 6 7 8 9 0 DOC DOC 9 0 9 8 7 6 5 4 3

P/N 009669-4

The editor was Michael Johnson;
the production supervisor was Friederich W. Schulte.
R. R. Donnelley & Sons Company was printer and binder.

This book was prepared with LaTeX and PC TeX using Computer Modern fonts. It was published from camera-ready pages printed on a laser printer. TeX is a trademark of the American Mathematical Society. PC TeX is a trademark of Personal TeX, Inc.

Library of Congress Cataloging-in-Publication Data

Calvert, James (James E.)
Finite mathematics with computer-supported applications / James Calvert and William Voxman.
p. cm.
Includes bibliographical references and index.
P/N 009669-4
1. Mathematics. I. Voxman, William L. II. Title.
QA39.2.C315 1994
510—dc20 93–4626

International Edition

When ordering this title, use ISBN 0-07-113248 -1

Contents

Instructions for CALIPSO

This list will help you locate features of CALIPSO when you need to review an operation. Some of the references are to subsections that give instructions on solving problems with CALIPSO and others are to examples that may be useful.

Preface

Finite Mathematics with Computer-Supported Applications provides a computer-assisted approach to learning finite mathematics. Although the text can be used successfully without recourse to computers, the software package CALIPSO will enable students to solidify their understanding of finite mathematics and increase their problem solving skills. Remarkably easy to use, CALIPSO eliminates much of the tedium in carrying out routine arithmetic and algebraic manipulations and allows the student to focus on basic principles and problem-solving techniques. CALIPSO incorporates pedagogic devices that permit one to make the basic decisions for carrying out various procedures without having to perform the accompanying hand-calculations.

The first three chapters deal with basic matrix operations and their applications to solving systems of equations and finding inverses. A special feature is the treatment of LU decompositions, a technique frequently used to deal with large systems of linear equations.

Chapters 4, 5 and 6 provide an introductory but relatively thorough look at linear programming. CALIPSO permits the student to carry out the simplex method and its variations without suffering through long and unedifying hand calculations. It also enables the inclusion of more realistic examples that demonstrate the power and utility of linear programming. Chapter 6 contains a number of topics (duality, sensitivity analysis, etc.) not commonly presented in depth in a beginning finite mathematics course. The student who takes courses in operations management will find considerable use for this chapter and CALIPSO's modeling and report generation capabilities. [1]

Problems in counting, probability, and statistics form the core of Chapters 7, 8, and 9. These chapters serve to enhance problem solving skills. The student can use the calculating features of CALIPSO to perform numerical calculations of considerable magnitude. The student can also use CALIPSO to create histograms, probability distributions, and areas under normal curves. In addition, we have included a section on linear regression, a subject of fundamental interest in dealing with empirical data.

---

[1] The text often refers the student seeking a deeper understanding of linear programming to [CV], which is a text by Calvert and Voxman titled *Linear Programming* and published by Harcourt-Brace-Jovanovich in 1989.

Chapter 10 treats the mathematics of finance. CALIPSO allows one to make the needed calculations to solve problems involving interest rates, annuities, and amortization. We conclude the text with some of the basic principles and manipulations underlying elementary logic.

Answers to many problems are provided at the end of the text. The symbol § following the number of a problem indicates that its answer is given in the back of the book. Some sections of the text are marked with a ⋆ to indicate that they are somewhat more difficult than other sections and can be omitted without loss of continuity.

FOR THE INSTRUCTOR

Instead of the usual solutions manual we give instructors an Instructor's Answer Disk (IAD) containing answers that can be read or processed by CALIPSO. The answers to problems not given in the text can be read from the IAD on the computer screen or printed on paper. We solved all of the matrix algebra and linear programming problems and stored the solutions on the IAD. The instructor can read these solutions directly into the CALIPSO program to see how we expect that a student will solve the problems with CALIPSO.

The IAD also contains a program called GAUSS that can be used for classroom demonstrations of Gauss elimination, solving systems of equations, inverting matrices, and LU factorization. The IAD contains a file named README that tells how to use these demonstrations.

Your McGraw-Hill representative can provide you with a computer-generated test system. This test system enables instructors to choose questions either manually or randomly by section of the text, level of difficulty, and need for calculator or computer support. This system is available for IBM compatible and MacIntosh computers. A printed version of the questions in the computerized test bank is also available.

The authors thank the McGraw-Hill staff, including especially Michael Johnson, the mathematics editor, Margery Luhrs, the editing supervisor, and Bruce Marcus of the software division. We also thank the following reviewers who offered many valuable suggestions: August Zarcone, College of Du Page; Barbara Williams, North Hennepin Community College; Dean Brown, Youngstown State University; Mike Ecker, Pennsylvania State, Wilkes-Barre; Robert Keicher, Delta College; Jerry Farlow, University of Maine; and Ken Meerdink, University of Idaho.

Finite Mathematics

with Computer-Supported Applications

0 Sets and Linear Equations

This chapter provides a brief review of elementary set theory as well as an introduction to linear equations. Sets will play an especially important role in Chapter 7. Linear equations occur throughout the text and are particularly emphasized in Chapters 2, 4, 5, 6, and 9.

0.1 SETS

For purposes of this text we view a *set* as a collection of objects. For instance, the collection of students taking finite mathematics forms a set. The collection of prime numbers lying between 8 and 18 forms another set. The objects that belong to a set are called *elements* or *members* of the set. The prime number 11 is a member of the set of primes between 8 and 18, while the number 23 does not belong to this set.

Set Notation

Brackets { } are commonly used to denote sets. We can denote, for example, the set of prime numbers between 8 and 18 in the following ways:

- {All prime numbers between 8 and 18}
- $\{x : x \text{ is a prime number and } 8 < x < 18\}$ (read the colon ":" as "such that")
- $\{11, 13, 17\}$

We frequently use capital letters $A, B, C, \ldots$ to indicate sets and we use the symbol $\in$ to indicate membership in a set. Thus, for example, the expression

$$x \in A$$

means that x is a *member* of the set A. If x is a member of A we can also say x is an *element* of A or x *belongs to* A.

Example 1: Set Notation

If A is the set of all outcomes obtained when tossing a coin twice, then we can write

$$A = \{HH, HT, TH, TT\}$$

where, for instance, HT indicates that the first toss results in heads and the second toss results in tails.

If $B = \{z : z \text{ is an even integer and } 3 < z < 100\}$, then we can write $8 \in B$ to indicate that 8 belongs to (is an element or member of) the set B. To indicate that 31 is not a member of B, we write $31 \notin B$.

Subsets

If A is set, then a set B is a *subset* of A if each member of B is also a member of A. We write $B \subset A$ to indicate that B is a subset of A.

Example 2: Subsets

If A is the set of all even integers and $B = \{4, 8, 10, 14\}$, then $B \subset A$, since each member of B is also a member of A. The set $C = \{6, 8, 10, 11\}$ is not a subset of A, since 11 belongs to C but 11 is not a member of A.

Unions

If A and B are sets, then the *union* of A and B is the collection of all members of A or B (or possibly both). We write $A \cup B$ to denote the union of A and B.

Example 3: Unions

If $A = \{3, 6, 9, 12\}$ and $B = \{b : b \text{ is an odd integer and } 4 < b < 17\}$, then

$$A \cup B = \{3, 5, 6, 7, 9, 11, 12, 13, 15\}$$

Note that although the integer 9 is a member of both A and B, we include it once, not twice, in $A \cup B$.

Intersections

If A and B are sets, then the *intersection* of A and B is the collection of all elements that belong to both A and B. We write $A \cap B$ to denote the intersection of A and B.

Example 4: Intersections

If $A = \{-4, 7, 10, 1/2, 52, 0\}$ and $B = \{3, 7, 12, 52, 6\}$, then $A \cap B = \{7, 52\}$.

If $C = \{x : x \text{ is an odd integer }\}$ and $D = \{-8, 2/3, 16, 22\}$, then there are no elements that belong to both C and D. In this case we say that the intersection of C and D is *empty* or that C and D are *disjoint*, and we write $C \cap D = \emptyset$, where $\emptyset$ denotes the empty set.

Complements

If A and B are sets, then the *complement* of B in A is the set of all elements that are in A but not in B. We write $A - B$ to denote the complement of B in A; that is,

$$A - B = \{x : x \in A \text{ and } x \notin B\}$$

Example 5: Complements

If $A = \{1, 3, 5, 7, 9, 11, 13\}$ and $B = \{3, 4, 5, 6, 7\}$, then $A - B = \{1, 9, 11, 13\}$.

All of the concepts discussed thus far can be extended to three or more sets. For instance $A \cap B \cap C$ is the set of all elements that belong to all of the sets A, B, and C. The set $X - (E \cup F)$ indicates the elements that belong to X but are not members of E or F.

0.1A Exercises : Set Operations

These problems refer to the sets $A = \{-2, 0, 1/2, 7, 8, 9, 12\}, B = \{w : 4 < w < 14 \text{ and } w \text{ is an integer}\}, C = \{y : y = 3x \text{ and } x \text{ is an integer with absolute value less than } 16\}$, $D = \{-3, 0, 4, \%, *, 12, \#\}$.

1. § Find $B \cup C$, $B \cap C$, and $B - C$.
2. Find three subsets of D.
3. § Find $(A \cup B) \cap D$.
4. Find $(A \cap B) \cup (C \cap D)$.
5. § Find $D \cap (B \cup C)$.
6. Find $D \cup (B \cap C)$.

Venn Diagrams

It is frequently useful to use *Venn diagrams* to visualize various set relationships. A Venn diagram usually consists of a rectangle representing a set X and circles representing subsets of X. Unions, intersections, and complements of various subsets of X are represented by shading appropriate portions of the circles. In Figures 1, 2, 3, 4 we have shaded in the areas representing $A \cap B$, $A \cup B$, $A - B$, and $X - A$, respectively.

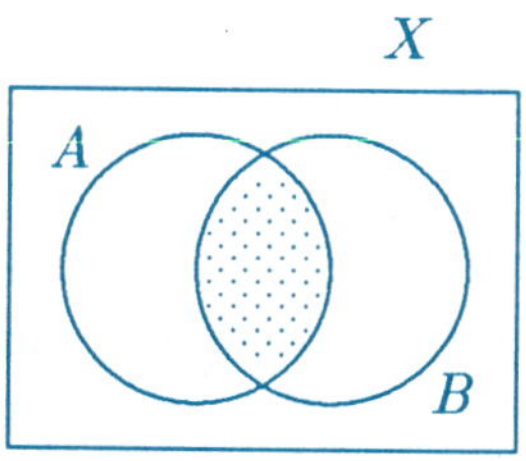

Figure 1

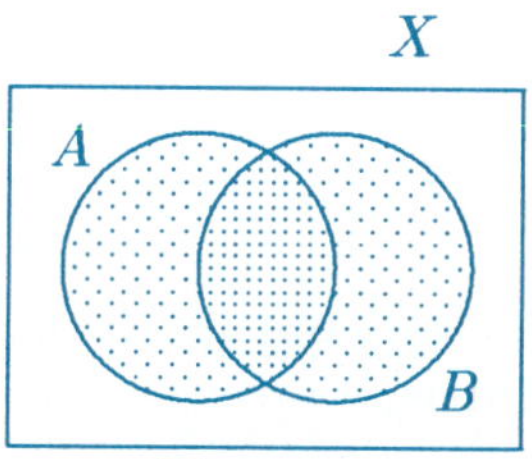

Figure 2

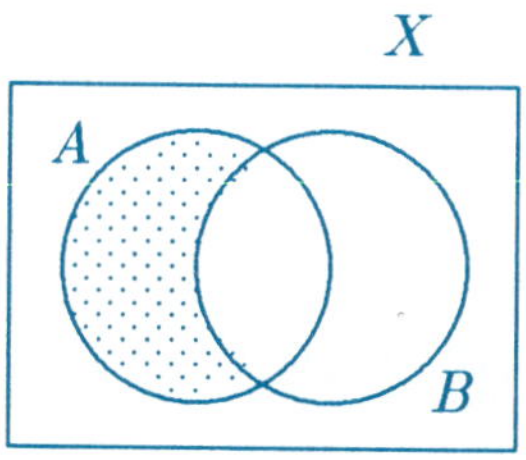

Figure 3

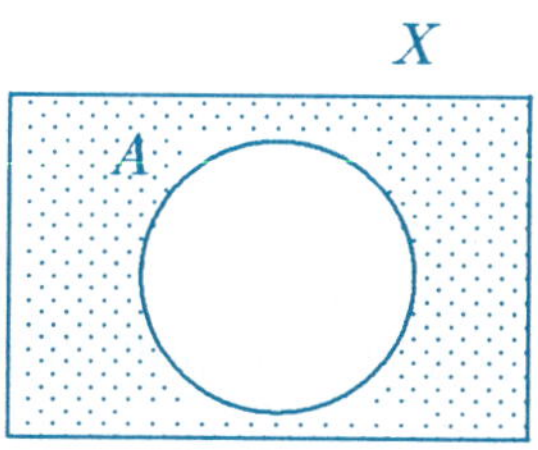

Figure 4

The shaded areas in Figures 5 and 6 indicate the sets $A \cap B \cap C$ and $A \cup (B \cap C)$, respectively.

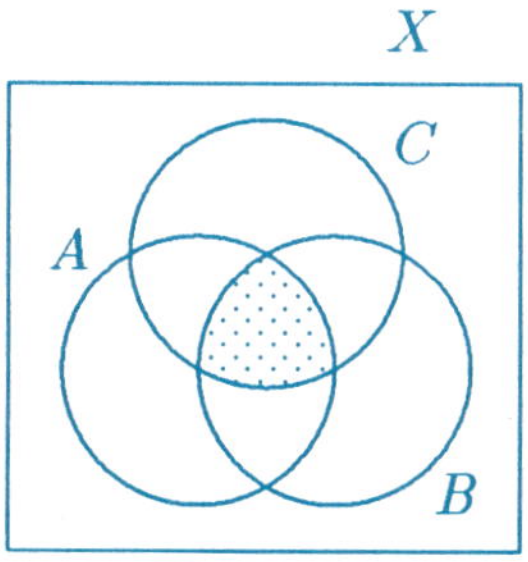

Figure 5

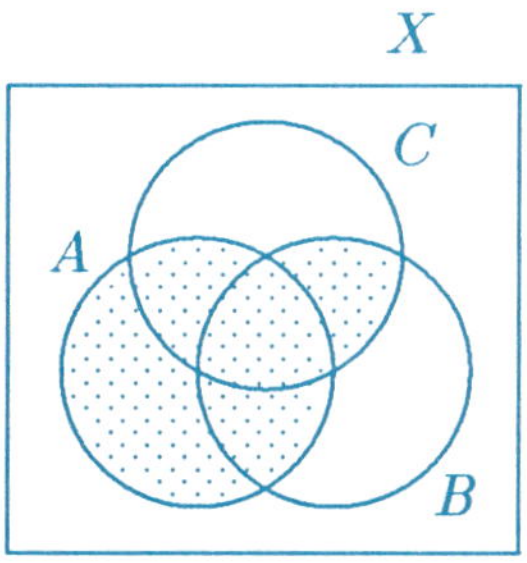

Figure 6

0.1B Exercises: Venn Diagrams

Assume that the sets A, B, C are subsets of X. For each problem draw a Venn diagram and shade the indicated region.

1. § $A \cap B$	2. $X - (A \cap C)$	3. § $(A \cap B) \cup C$
4. $X - ((A \cap B) \cup C)$	5. § $(B \cup C) \cap A$	6. $(A - B) \cup (B \cap C)$

0.2 LINEAR EQUATIONS

Recall that the *Cartesian plane* is formed by perpendicular number lines called the x and y axes. Each point in the plane corresponds to an ordered pair of numbers (a, b) and, conversely, each ordered pair of numbers (a, b) corresponds to a point in the plane. The first number a in the pair (a, b) is the x *coordinate* of the point and the second number b is the y *coordinate* of the point. In Figure 1 we describe how to plot six points. The point at the intersection of the x and y axes has coordinates (0,0) and is often called the *origin*. The point (0,3) is on the y axis and 3 units above the origin. The point (2.5,0) is on the x axis and 2.5 units to the right of the origin. The point (2,3) is 2 units to the right of the y axis and 3 units above the x axis. The point $(-2, -3)$ is 2 units to the left of the y axis and 3 units below the x axis. The point $(-2, 3)$ is 2 units to the left of the y axis and 3 units above the x axis.

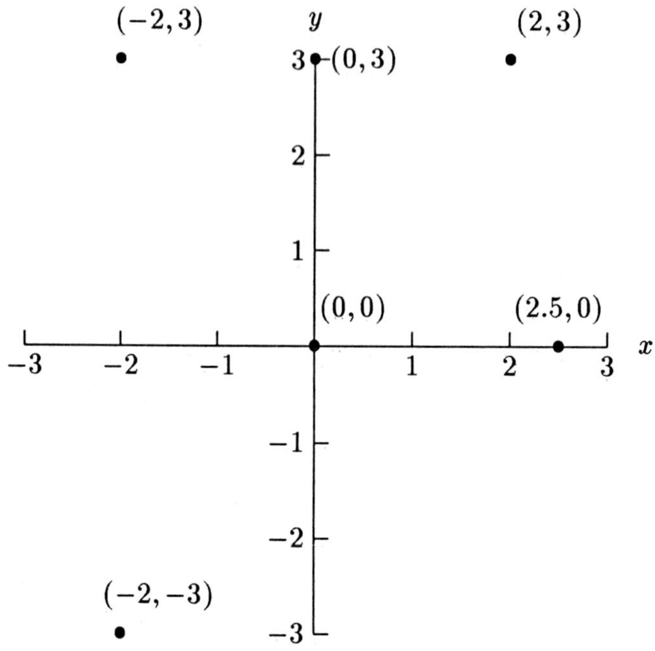

Figure 1

Equations occur in almost all areas of mathematics. Examples of equations in variables x and y include

$$\begin{array}{rcrcl} 2x & - & y & = & -4 \\ & & x & = & 6 \\ x^2 & + & 3y & = & 13 \end{array}$$

A *linear equation* in two variables x and y is an equation that can be written in the form $cx + dy = e$, where c, d, and e are fixed real numbers. The equations

$$\begin{array}{rcrcl} 2x & + & 5y & = & 12 \\ -3x & - & \frac{2}{3}y & = & 0 \\ & & 7y & = & 21 \\ & & y & = & 2x - 15 \end{array}$$

are linear equations since they can all be written in the form $cx + dy = e$. The equations

$$\begin{array}{rcrcl} x^2 & + & y^2 & = & 16 \\ 4x & + & 2xy & = & 12 \end{array}$$

are not linear equations.

The *graph* of an equation in variables x and y consists of all points (a, b) in the plane with the property that if a is substituted for x and b is substituted for y in the equation, then the two sides of the equation are equal. For instance, the point $(-2, 3)$ is a point in the graph of the equation

$$4x + 2y = -2$$

since $4(-2) + 2(3) = -2$ is true. The point $(1, 5)$ does not belong to the graph of this equation since $4(1) + 2(5) = -2$ is not true.

The graph of a linear equation is a line. Thus, to form this graph you need to find just two points that satisfy the linear equation. For instance, to graph the linear equation

$$3x - 4y = 24 \tag{1}$$

you can let $x = 0$ and solve for y. Substituting $x = 0$ in equation (1) we see that $-4y = 24$ and so $y = -6$. Therefore, the point $(0, -6)$ is on the graph of this equation. To find a second point let $y = 0$ and solve the equation for x to obtain $x = 8$. Then the point $(8, 0)$ is also point on the graph of (1). Figure 2 gives the graph of line determined by (1).

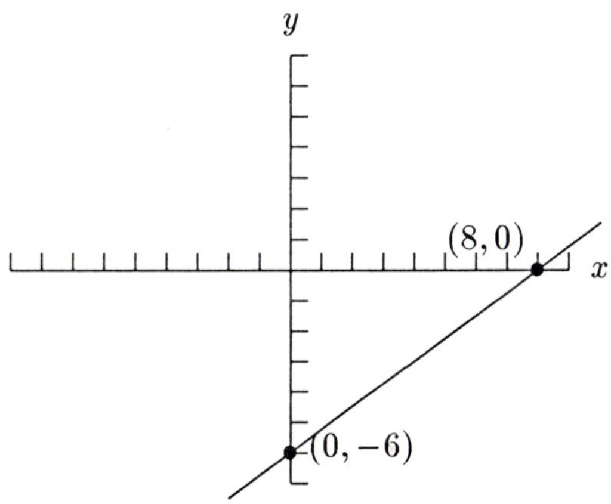

Figure 2

Example 1: Graphing Linear Equations

You can verify that the points $(0, 5)$ and $(3, 0)$ satisfy the equation $5x + 3y = 15$. Figure 3 gives the graph of this equation.

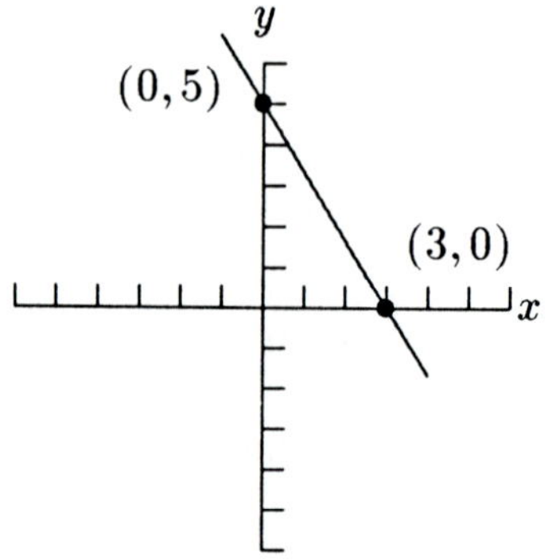

Figure 3

The graph of $x = -3$ is the vertical line indicated in Figure 4, since each point on this line has x coordinate -3.

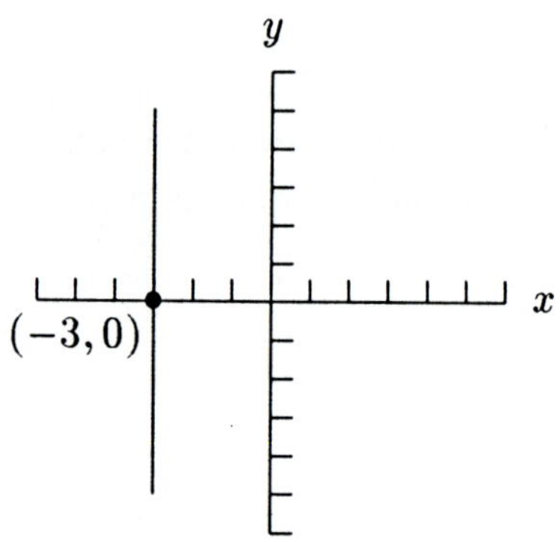

Figure 4

Frequently it is advantageous to write the equation of a nonvertical line in the *standard form*

$$y = mx + b \tag{2}$$

In this case the real number m is the *slope* of the line (which is a measure of the inclination of the line) and b is the y *intercept* of the line (the value of the y coordinate at the point the line intersects the y axis). Recall that if (x_1, y_1) and (x_2, y_2) are any two points on a line, then the slope of the line is

$$m = \frac{y_2 - y_1}{x_2 - x_1}$$

Example 2: Equations of Lines in Standard Form

The equations $y = 2x - 6$ and $y = -3x + 5$ are equations of lines written in standard form. To graph such a line find two points on the line by choosing two values for x and calculating the corresponding y values. For instance, for the line $y = 2x - 6$, you might choose $x = 0$, which gives you $y = -6$, and $x = 2$ which gives you $y = -2$. You can graph this equation by drawing the line through the points $(0, -6)$ and $(2, -2)$.

Example 3: Finding Equations of Lines in Standard Form

We find an equation of the line whose slope is 3 and which passes through the point $(4, -1)$. Since the slope is 3, we know the equation of the line is

$$y = 3x + b$$

where b is to be determined. To find b, note that since the point $(4, -1)$ lies on the line we know that for $x = 4$ and $y = -1$ we have

$$-1 = 3 \cdot 4 + b$$

and so $b = -13$. The equation of the line is

$$y = 3x - 13$$

Next we find an equation of the line that passes through the points $(7, 1)$ and $(-3, 6)$. The slope of this line is $m = \frac{6-1}{-3-7} = -\frac{1}{2}$ and so $y = -\frac{1}{2}x + b$ is the equation of this line. To find b use either of the given two points on the line, say $(-3, 6)$, to write

$$6 = -\frac{1}{2}(-3) + b$$

from which it follows that $b = \frac{9}{2}$. Therefore, the standard form of the equation of the line is $y = -\frac{1}{2}x + \frac{9}{2}$.

Example 4: Costs

The cost c (in dollars) of producing x units of Brand X is known to be

$$c = 30x + 96 \tag{3}$$

From this linear equation we see that there is a fixed cost 96 (which does not depend on the number of units produced) and a variable cost 30 (the cost for each unit manufactured). From (3) it follows that to produce 40 units of Brand X it will cost $c = 30(40) + 96 = 1296$ dollars.

Example 5: Salary Growth

Suppose you begin a job at \$200 a month and you know that your monthly salary y will increase linearly according to the equation

$$y = mx + b \tag{4}$$

where x denotes the number of months you have worked ($x = 0$ corresponds to the current month).

If your salary after six months is \$260, what will be your monthly salary after 22 months? To answer this question we must find the values of m and b in (4). When $x = 0$, your monthly salary y is 200. Substituting $x = 0$ and $y = 200$ in (4)

$$200 = m(0) + b$$

we see that $b = 200$. When $x = 6$ your monthly salary y is 260. Substituting $x = 6$, $y = 260$, and $b = 200$ in (4) we obtain

$$260 = m(6) + 200$$

and find that $m = \frac{60}{6} = 10$. Therefore, your monthly salary y is determined by the linear equation

$$y = 10x + 200$$

For $x = 22$, your monthly salary will be

$$y = 10(22) + 200 = 420$$

0.2A Exercises: Graphing and Solving Linear Equations

Graph the lines determined by the following equations.

1. $3x - 4y = 24$ 2. § $y = 6x + 3$ 3. $y = -7$ 4. § $x = 2.5$

5. $4y + 6x = 15$ 6. § $y = -2x + 6$ 7. $-x = 0.6$ 8. § $x = 2y + 3$

In Problems 9 – 13 find the slope and y intercepts of each line by converting the given equation to an equation in standard form.

9. $3x + 2y = -12$ 10. § $x = -4y + 6$ 11. $-4x - 6y = 18$

12. § $6x - 2y = 24$ 13. $-2x + 3y - 5 = 10$

14. § Find an equation of the line with slope 4 and passing through the point $(-1, 2)$.

15. Find an equation of the line with y intercept 7 and passing though the point $(3, -4)$.

16. § Find an equation of the line with slope $-1/2$ and passing through the point $(4, 0)$.

17. Find an equation of the line passing through the points $(2, -2)$ and $(4, 10)$.

18. § Find an equation of the line passing through the points $(-2, -8)$ and $(22, 52)$.

19. A large plot of ground currently has 265 trees. Each month an additional 35 trees are planted. Write an equation that will give the number of trees after x months. Use your equation to find how many trees there will be after 7 months.

20. § Assume that the relation between degrees Celsius (C) and degrees Fahrenheit (F) is linear. Suppose that 0^oC corresponds to 32^oF and 100^oC corresponds to 212^oF.

    (a) Find an equation in ^oF and ^oC that describes the relationship between Celsius and Farenheit temperatures.

    (b) Use your equation to calculate the number of degrees Celsius that corresponds to 60^oF.

    (c) Use your equation to calculate the number of degrees Farenheit that corresponds to -10^oC.

21. Suppose that you buy a car for \$12,000 and the value of the car depreciates at the rate of \$900 per year. Write a linear equation that gives the value of the car after x years, and use your equation to find the car's value after 6 years.

1 Matrices

The simple nature of matrices gives little indication of their importance. As we will see, matrices have many problem–solving applications and are easily adapted to computer use.

1.1 INTRODUCTION

A *matrix* is a rectangular array of numbers. Matrices can be described in terms of *rows* and *columns*; the rows run horizontally and the columns vertically. The *size* (or *dimension*) of a matrix is determined by the number of its rows and columns. A matrix consisting of m rows and n columns is called an $m \times n$ matrix or a matrix of *size* $m \times n$. A $1 \times n$ matrix (one row and n columns) is called a *row matrix*. An $n \times 1$ matrix is called a *column matrix*.

Example 1: The Structure and Size of a Matrix

The matrix

$$A = \begin{bmatrix} 1 & -4 & 2 & 3 \\ 2 & -\frac{1}{2} & 5 & 4 \\ 3 & 0 & 6 & 3 \end{bmatrix}$$

consists of three rows and four columns and, therefore, is a 3×4 matrix. The second row of A is $\begin{bmatrix} 2 & -\frac{1}{2} & 5 & 4 \end{bmatrix}$ and the third column is

$$\begin{bmatrix} 2 \\ 5 \\ 6 \end{bmatrix}$$

The matrix $B = \begin{bmatrix} 1 & -\frac{1}{2} & 5 & 6 & 23 \end{bmatrix}$ is a row matrix of size 1×5. The matrix

$$C = \begin{bmatrix} 1 \\ -2 \\ 5 \end{bmatrix}$$

is a column matrix of size 3×1.

The numbers found in a matrix are the *entries* of the matrix. Frequently, it will be convenient to let a_{ij} denote the entry found in the ith row and jth column of a matrix A. We use a lower-case a because the matrix was called A. We usually use the lower-case form of the letter that names the matrix; for instance, if the name of a matrix is D, we label the entries d_{ij}.

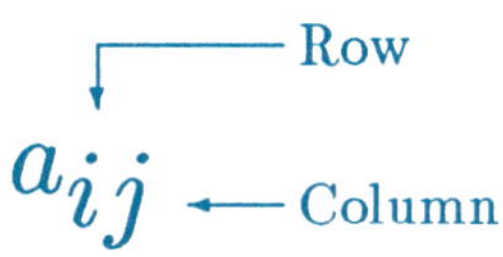

Example 2: Matrix Entries

We can refer to the entries of the matrices of Example 1 by expressions such as

$$\begin{aligned} &a_{14} = 3,\ a_{33} = 6, \text{ and } a_{24} = 4 \\ &b_{13} = 5,\ b_{15} = 23 \\ &c_{11} = 1,\ c_{21} = -2,\ c_{31} = 5 \end{aligned}$$

Two matrices A and B are *equal* if they are of the same size and their corresponding entries a_{ij} and b_{ij} are equal. In other words, $A = B$ if $a_{ij} = b_{ij}$ for each row i and each column j of the two matrices.

Example 3: Equal Matrices

If

$$A = \begin{bmatrix} \frac{1}{2} & -\frac{2}{3} \\ 1 & 0 \end{bmatrix} \quad \text{and} \quad B = \begin{bmatrix} 0.5 & -\frac{2}{3} \\ \frac{4}{4} & 0 \end{bmatrix}$$

then $A = B$.

Getting Started with CALIPSO

Here are instructions for entering matrices in CALIPSO. It is best to read these instructions while working at the computer. You can work through Exercises 1.1A below, referring to these instructions as needed.

If CALIPSO is not already loaded on your computer, you should read Appendix A.1 and A.2 and follow the instructions for loading CALIPSO on the computer you are using.

If CALIPSO is loaded on a hard disk, change to the directory that contains it. If you are using CALIPSO from a diskette, change to the drive that contains CALIPSO. Start the program by typing CALIPSO. When you are finished reading the title screen, press [*Enter*]. From the menu that appears you select an application by moving the cursor with [↑] or [↓] and pressing [*Enter*] when the cursor is on your selection. For this example, choose the *MATRIX* application.

A screen appears asking for a file name. You should always enter a file name since you may want to save your worksheet to continue your work at another time. A *file name* is any word of 8 or fewer alphabetic or numeric characters. Try to choose a name you can easily remember since you need to recall it in order to get your file back later.

After you type the file name and press [*Enter*] a screen appears asking for information about the matrix. You move between entry fields on this screen with the [↑] or [↓] keys. Here are the purposes of the entry fields.

ROWS You enter the number of rows in this field. A matrix of decimal type can have as many as 12 rows. A matrix of fraction type can have at most 8 rows.

COLUMNS You enter the number of columns in this field. A decimal matrix can have up to 12 columns and a fractional matrix up to 8 columns.

NAME Matrices are named by one of the letters from A to Z. You can use an upper- or lower-case letter for the name. Notice the list of unused letters at the top of the screen. All of the letters from A to Z appear in the list now; however, a letter will disappear from the list when it is already in use. If you do not choose a letter to name your matrix, CALIPSO will assign it the next free name from the available list.

PHRASE You can enter a phrase to describe your matrix in this field.

The phrase is optional but can be useful when you are doing a problem that involves many matrices.

FRACTION OR DECIMAL Matrices can be entered in fractional or decimal arithmetic. In a fraction type matrix you can enter numbers like 9, 1/2, and −3/5. In a decimal type matrix you can enter numbers like 9, .5, −3/5, and −0.6. Notice that you can enter fractions in a decimal matrix; such fractions will be automatically converted to their decimal equivalents. Entering fractions rather than decimals can improve accuracy. The computer will carry the decimal approximation of the fraction to 15 digits of accuracy, whereas you cannot enter so many digits yourself. You cannot use decimals in a fraction matrix. If you leave this entry field blank, the matrix will be of fraction type. To make a decimal matrix, enter any character in the field. You can also change any matrix to decimal later with the *To Decimal* option.

When you are done entering information about the matrix, press [*Enter*]. A matrix of the chosen size appears on the screen wih all of its entries set to zero. This is convenient since matrices often have a lot of zeros. You can move the cursor about the matrix with [↑], [↓],[←], or [→] keys and type a number in any field. If you make a mistake while typing an entry, you can back up with the [*Backspace*] key (on some keyboards this key is labelled [⇐]). If you want to change an entry you made after you leave a field, return to it and type the new entry.

After you type data in the matrix, press [*Esc*]. A menu appears at the bottom of the screen offering many options. We will learn to use these options later. Now we just want to discuss how to enter and view matrices. Matrices are stored in a worksheet in the order of creation. New matrices can be created with the *New Matrix* option from the menu. To use that option (or any other option) you move the cursor to it with an arrow key and press [*Enter*]. As you will see later, matrices are also added to the worksheet when you perform an operation on a matrix.

Moving About in the Worksheet

You usually move up or down in the worksheet with the [*PgUp*] or [*PgDn*] keys. You can move to the first matrix with [*Home*] or to the last matrix with [*End*]. You can also jump directly to a matrix by holding down the [*Ctrl*] key and typing the letter for the matrix you want to see.

When a matrix has more than 7 columns you cannot see all of the columns at once. You view columns that are off the screen by holding down the [Ctrl] key and pressing [→] or [←]. You can also shift the display by 1 column at a time by pressing [→] with the cursor on the rightmost column of the display or [←] with the cursor on the leftmost column of the display.

Copying Matrices

Notice the *Copy/Resize* option on the menu. You can use this to copy the *current matrix* (the one that is visible on the screen) and (optionally) change the number of rows or columns. This is useful when you need to create a second matrix that has many entries in common with the first matrix.

Saving and Printing Worksheets

You can save a worksheet by selecting the *Save Worksheet* option; however, you probably will not choose to do so this time. You may want to print your worksheet. To do this move the cursor in the menu to *Print Worksheet* and press [Enter]. You should do Exercises 1.1A to practice entering matrices. You are welcome to try any option on the menu. We will discuss each option in detail later.

1.1A CALIPSO Exercises: Entering Matrices

In these exercises we display matrix entries on line (like 1/2 rather than $\frac{1}{2}$) since that is the way you must type them for CALIPSO.

1. Create this matrix in fraction type.

$$A = \begin{bmatrix} 1 & 2/3 & -6/5 & 7 & 0 \\ -3 & 4 & 3/4 & -3 & 8 \\ 13/4 & -2 & 1 & -2/7 & 3 \end{bmatrix}$$

2. Create this matrix in decimal type.

$$M = \begin{bmatrix} 1.5 & 3.25 & -5.7 & 7 \\ -3 & 4 & 3/4 & -3 \\ 4/13 & -2 & 1 & -2/7 \\ 13.567 & -12.2 & 1.43 & 0 \end{bmatrix}$$

3. Use the *Copy/Resize* option to make a copy of matrix A in Problem 1 that has an additional row and column. You do this by getting matrix A on the screen with [PgUp] and [PgDn] keys and then moving the cursor with arrow keys to the *Copy/Resize* option and pressing [Enter]. Enter values in the new row and column to make the following matrix.

$$C = \begin{bmatrix} 1 & 2/3 & -6/5 & 7 & 0 & 1 \\ -3 & 4 & 3/4 & -3 & 8 & 2 \\ 13/4 & -2 & 1 & -2/7 & 3 & 3 \\ 1/3 & -2/3 & 4/5 & -9 & 69 & 13 \end{bmatrix}$$

4. Make a matrix with 12 rows and 12 columns. Just enter a few values, but be sure to enter some numbers in columns beyond 7 so you can experiment with display shifts.

5. Practice with the keys that allow you to view the matrices in the worksheet. You move upward (toward matrices that were created earlier) with *PgUp* and downward with *PgDn*. You can move to the top of the worksheet (to the matrix that was created first) with the *Home* key and to the bottom of the worksheet (the matrix that was created last) with the *End* key. You can move directly to any matrix in the worksheet by holding down the *Ctrl* key and pressing the letter that names the matrix.

1.2 MATRIX OPERATIONS

In this section we look at matrix arithmetic. Though there are some important differences, matrix arithmetic has many similarities to real number arithmetic.

Addition and Subtraction

Addition (or *subtraction*) of matrices A and B of the same size is done in the obvious way: you simply add (or subtract) corresponding entries. Addition (or subtraction) of different sized matrices is not defined.

Example 1: Addition and Subtraction

If

$$A = \begin{bmatrix} 2 & 0 & 4 & \frac{1}{2} \\ 3 & 4 & -2 & 5 \\ -1 & 2 & 2 & 2 \end{bmatrix} \text{ and } B = \begin{bmatrix} 4 & -3 & \frac{2}{3} & \frac{1}{3} \\ 3 & -1 & 5 & 2 \\ -4 & 2 & \frac{3}{5} & 0 \end{bmatrix}$$

then

$$A + B = \begin{bmatrix} 6 & -3 & \frac{14}{3} & \frac{5}{6} \\ 6 & 3 & 3 & 7 \\ -5 & 4 & \frac{13}{5} & 2 \end{bmatrix} \text{ and } A - B = \begin{bmatrix} -2 & 3 & \frac{10}{3} & \frac{1}{6} \\ 0 & 5 & -7 & 3 \\ 3 & 0 & \frac{7}{5} & 2 \end{bmatrix}$$

The matrices

$$A = \begin{bmatrix} 1 & -4 & 6 \\ 3 & 4 & 7 \\ 0 & 1 & 2 \end{bmatrix} \text{ and } B = \begin{bmatrix} 1 & 2 & -\frac{2}{3} & 4 \\ 0 & 0 & 4 & 7 \\ 2 & 1 & 2 & 1 \end{bmatrix}$$

can neither be added nor subtracted since they are not the same size.

Scalar Multiplication

A matrix A of any size can be multiplied by a real number r. You simply multiply each entry of A by r; rA or Ar denotes the resulting matrix. Such multiplication is called *scalar multiplication.*

Example 2: Scalar Multiplication

If

$$A = \begin{bmatrix} 2 & \frac{1}{2} & 4 & -2 \\ 4 & -1 & 0 & 6 \end{bmatrix}$$

and $r = 4$, then

$$rA = 4A = \begin{bmatrix} 8 & 2 & 16 & -8 \\ 16 & -4 & 0 & 24 \end{bmatrix}$$

If $r = -\frac{1}{2}$, then

$$Ar = A(-\frac{1}{2}) = \begin{bmatrix} -1 & -\frac{1}{4} & -2 & 1 \\ -2 & \frac{1}{2} & 0 & -3 \end{bmatrix}$$

Matrix Multiplication

This operation is a bit more complicated. At first glance, the procedure for multiplying matrices does not appear particularly reasonable. However, as we will see, it lends itself to many natural applications.

First we multiply a row matrix by a column matrix. This is done only if the number of entries in the row matrix is equal to the number of entries in the column matrix. If

$$A = [a_1 \quad a_2 \quad \cdots \quad a_n] \quad \text{and} \quad B = \begin{bmatrix} b_1 \\ b_2 \\ \vdots \\ b_n \end{bmatrix}$$

then the product $C = AB$ is a 1×1 matrix whose only entry is

$$a_1b_1 + a_2b_2 + \cdots + a_nb_n$$

In other words,

$$C = AB = [a_1b_1 + a_2b_2 + \cdots + a_nb_n]$$

Example 3: Multiplying a Row Matrix by a Column Matrix

If

$$A = [2 \quad -3 \quad 6] \quad \text{and} \quad B = \begin{bmatrix} 4 \\ 0 \\ -\frac{1}{2} \end{bmatrix}$$

then $C = AB = [(2)(4) + (-3)(0) + (6)(-\frac{1}{2})] = [5]$.

In general, to multiply a matrix A by a matrix B it must be the case that the number of columns of A is equal to the number of rows of B. If A and B satisfy this condition, then the typical entry found in the ith row and jth column of the product matrix $C = AB$ is the product of the ith row of A and the jth column of B. If A is an $m \times p$ matrix and B is a $p \times n$ matrix, then the product matrix AB is an $m \times n$ matrix.

Example 4: Multiplying Matrices

Suppose

$$A = \begin{bmatrix} 2 & 4 & -3 & 5 \\ 2 & 0 & 4 & 2 \\ 1 & 6 & 2 & 5 \end{bmatrix} \quad \text{and} \quad B = \begin{bmatrix} -2 & 5 \\ 3 & -4 \\ -2 & 0 \\ 7 & 1 \end{bmatrix}$$

The number of columns of A is equal to the number of rows of B, and therefore the product matrix $C = AB$ is defined and will be a 3×2 matrix. The entry c_{11} found in the first row and first column of $C = AB$ is obtained by multiplying the first row of A by the first column of B. Thus,

$$c_{11} = 2 \cdot (-2) + 4 \cdot 3 + (-3)(-2) + 5 \cdot 7 = 49$$

Similarly, the entry c_{32} found in the third row and second column of $C = AB$ is obtained by multiplying the third row of A by the second column of B:

$$c_{32} = 1 \cdot 5 + 6(-4) + 2 \cdot 0 + 5 \cdot 1 = -14$$

By performing more such multiplications you can verify that

$$C = AB = \begin{bmatrix} 49 & -1 \\ 2 & 12 \\ 47 & -14 \end{bmatrix}$$

1.2A Hand Exercises: Matrix Operations

Recall from the preface that the symbol § indicates that the answer to the question is given in the back of the book.

1. Let $A = \begin{bmatrix} 1 & 2 \\ -1 & 3 \end{bmatrix} \quad B = \begin{bmatrix} 5 & 4 \\ 2 & 1 \end{bmatrix}$.

   Calculate (a) § $A + B$ (b) $A - B$ (c) § $3A$ (d) AB.

2. Let $C = \begin{bmatrix} 2 & 1 \\ 3 & 2 \end{bmatrix} \quad D = \begin{bmatrix} -1 & 2 & 4 \\ 3 & -4 & 2 \end{bmatrix} \quad E = \begin{bmatrix} 1 & 3 & 2 \\ -1 & 4 & 1 \\ 2 & 1 & 0 \end{bmatrix}$.

   Calculate (a) § $3D$ (b) CD (c) § CC and (d) DE.

3. Let $R = \begin{bmatrix} 1 & -2 & 3 \\ 2 & 1 & 2 \end{bmatrix}$ $S = \begin{bmatrix} 2 & 1 & -1 \\ 3 & 1 & 2 \end{bmatrix}$ $T = \begin{bmatrix} 3 & 1 \\ 2 & -1 \end{bmatrix}$.

   Calculate (a) § $R + S$ (b) TS (c) § $S - 2R$ and (d) TR.

4. Let $M = \begin{bmatrix} 1 & 3 & 4 \\ 2 & 1 & 2 \\ -1 & 2 & 2 \end{bmatrix}$ $N = \begin{bmatrix} 1 & -1 & 2 \\ 2 & 1 & -1 \end{bmatrix}$ $P = \begin{bmatrix} 1 & -1 \\ 2 & 2 \\ 3 & 4 \end{bmatrix}$.

   Calculate (a) § NM (b) NP (c) § PN and (d) MP.

Naming a Matrix in CALIPSO

Next we use CALIPSO to carry out the basic matrix operations of addition, subtraction, scalar multiplication, and matrix multiplication. Select *MATRIX* from your CALIPSO menu. Enter the letter A to name your first matrix. We will let A be a 3×4 matrix, so enter 3 to indicate the number of rows and 4 the number of columns. For practice (but not necessary) type in "Introduction to CALIPSO" as a phrase. We will use fractional notation, so leave the option for decimal arithmetic blank. Now enter the following numbers as the entries of A.

$$A = \begin{bmatrix} 2 & -6 & 0 & 1/2 \\ 3 & 10 & 1 & 3 \\ -4 & 4 & 3/4 & -5 \end{bmatrix}$$

Next create another 3×4 matrix B by moving the cursor to *New Matrix* and repeating the procedure to create A. Let B have the following entries:

$$B = \begin{bmatrix} 2/3 & -2 & 4 & 5 \\ 6 & 0 & 0 & -2 \\ 2 & -4 & 5/6 & 1 \end{bmatrix}$$

Finally, use *New Matrix* to create the decimal matrix

$$C = \begin{bmatrix} -7.5 & 5.6 \\ 2 & 0 \\ 1 & -6 \\ -.82 & 3 \end{bmatrix}$$

To add the matrices A and B move the cursor to *Operations* and press *Enter*. A screen appears on which you can enter an operation. You may want to read some of the instructions on this screen. Eventually we will discuss all of the operations described. Type $D = A + B$ (you can also type $d = a + b$) and press [Enter]. The display will return to the worksheet with the calculated matrix D shown. To subtract B from A, choose *Operations* again and type $E = A - B$. Now try scalar multiplication by letting $F = 6B$ and $G = -3/4C$. Since A is a 3×4 matrix and C is a 4×2 matrix, the product matrix AC is defined and is a 3×2 matrix. Let $H = AC$ and use CALIPSO to find this product. Also, try to evaluate $K = AB$ to see an error message.

1.2B CALIPSO Exercises: Matrix Operations

Use CALIPSO to enter the following matrices. This same set of matrices is used in Exercises 1.2D. So to save work you should save your worksheet after you enter the matrices, but before you do these exercises. To save the worksheet, choose *Save Worksheet* from the menu. You will be asked to specify a file name for the worksheet or to verify that you want to use the same file name that you gave when starting the *MATRIX* program. You can give any name you like so long as it uses no more than 8 characters chosen from the letters A – Z and the digits 0 – 9.

$$A = \begin{bmatrix} 2 & 4 & -3 & 5 & 0 \\ 3 & 2 & 0 & -2 & 7 \\ 1 & 1 & 4 & 1 & 2 \end{bmatrix} \qquad B = \begin{bmatrix} 3 & -1 & 5 \\ 11 & 0 & -2 \\ 2 & 1 & 3 \\ 0 & 2 & -10 \\ 5 & 1 & 3 \end{bmatrix}$$

$$C = \begin{bmatrix} 4 & 4 & -4 & 3 & 2 \\ 2 & 0 & -2 & -1 & -13 \\ 1/2 & 3 & 0 & 5 & 6 \end{bmatrix} \qquad D = \begin{bmatrix} 1 & -3 & 7 \\ 0 & -4 & 6 \\ 2/3 & 5 & 1 \end{bmatrix}$$

$$E = \begin{bmatrix} 1 & 0 & 0 & -3 & 6 \\ -4 & 2 & 7 & 1/2 & 4 \\ 0 & 1 & 2 & 3 & 4 \\ 7 & 3 & -3 & 0 & 0 \end{bmatrix} \qquad F = \begin{bmatrix} 2.2 & 0 & 0 & -4.3 \\ -2.1 & 1 & 1 & 1.5 \\ -2.5 & 1.2 & 0 & 1 \end{bmatrix}$$

Move to *Save Worksheet* in the menu with the arrow keys and press [Enter]. Move to *Operations* and follow the instructions that appear on the *Operations* screen to attempt to perform the following operations. For those operations that cannot be carried out, explain why you cannot perform them.

1. $G = A + C$
2. $H = 1/2A$
3. $I = AB$
4. $J = BC$
5. $K = B + D$
6. $L = CB$
7. $M = (2 + 2/7)G$
8. $N = DD$
9. $O = F - A$
10. $P = C3$
11. $Q = BA$
12. $R = A3/5$
13. $S = -C$
14. $T = (1.5 + 2/3)D$
15. $U = 2.2 * A$
16. $V = 1/2(D + F)$

Combining Matrix Operations

You can combine the operations we have looked at thus far in various ways. For instance, if A, B, and C are matrices, then expressions such as $4A - 6B, A(B + 2C), ((1/2)A - 2B)4 + 6ABC$, etc., are meaningful provided the matrices are of the appropriate size for these operations to make sense.

1.2C Hand Exercises: Combining Operations

Using the matrices

$$A = \begin{bmatrix} 2 & -1 \\ 0 & 3 \end{bmatrix} \quad B = \begin{bmatrix} 1 & 4 \\ 2 & 6 \end{bmatrix} \quad C = \begin{bmatrix} \frac{1}{2} & -1 & 3 \\ 2 & 0 & 1 \end{bmatrix}$$

$$D = \begin{bmatrix} 3 \\ 5 \\ 1 \end{bmatrix} \quad E = \begin{bmatrix} 4 & 0 \\ -1 & \frac{2}{3} \\ 3 & 1 \end{bmatrix}$$

perform the following operations. For those operations that cannot be carried out, explain why you cannot perform them.

1.§ $2A - 3B$

2. $BA - AB$

3.§ $E(A - 3B)$

4. $EC - CE$

5.§ $(4C)(-2D)$

6. $CE - 4A$

7.§ DE

8. $CE(A + 2B)$

1.2D CALIPSO Exercises: Combining Operations

These exercises use the same matrices $A - F$ as Exercises 1.2B. Follow the instructions of *Operations* to attempt to perform the following operations. For those operations that cannot be carried out, explain why you cannot perform them.

1. $G = A - C$
2. $H = 3A + 2C$
3. $I = AB - BA$
4. $J = BC$
5. $K = (B + D)A$
6. $L = AB - 5DD$
7. $M = B(2A - (1/2)C)$
8. $N = DDD$
9. $O = F + 2/3A$
10. $P = D3$
11. $Q = 2D4 - (2D)4$
12. $R = 2A - 2C$
13. $S = -2DAB$
14. $T = FE - C$
15. $U = FEB - 4D$
16. $V = .33D6 - .45(FB)4$

Powers of a Matrix

When we consider Markov processes in Section 8.5, we will want to multiply a matrix A by itself, often many times. This is only possible if A is a *square matrix*; that is, the number of rows of A is equal to the number of columns of A. We write A^2 to denote AA; A^3 to denote AAA, etc.

Example 5: Powers of a Matrix

Using CALIPSO or by hand you can verify that if A is the square matrix

$$A = \begin{bmatrix} 4 & 1 & 2 \\ 2 & 6 & 0 \\ 3 & 2 & -4 \end{bmatrix}$$

then

$$A^2 = \begin{bmatrix} 24 & 14 & 0 \\ 20 & 38 & 4 \\ 4 & 7 & 22 \end{bmatrix} \text{ and } A^3 = \begin{bmatrix} 124 & 108 & 48 \\ 168 & 256 & 24 \\ 96 & 90 & -80 \end{bmatrix}$$

There are two ways to make these calculations with CALIPSO. You can use *Operations* to write $B = AA$ and $C = AAA$. Alternatively, you can iteratively raise a square matrix to a power while watching progress on the screen by typing the tilde symbol after the matrix name, as in $B = A\tilde{}$. The tilde symbol is entered with the key marked ˜ that is located in the upper-left corner of the keyboard. As the instructions at the bottom of the screen say: you press [Enter] (as many times as you like) to raise A to the next power. When the power you need is displayed, press [Esc] to quit.

1.2E CALIPSO Exercises: Powers of a Matrix

Use CALIPSO to enter A as a fractional matrix and B as a decimal matrix where

$$A = \begin{bmatrix} \frac{1}{2} & 3 & \frac{2}{3} & -\frac{3}{4} \\ 3 & -\frac{1}{4} & \frac{3}{4} & \frac{1}{4} \\ -\frac{1}{2} & -\frac{3}{4} & -\frac{1}{2} & \frac{1}{4} \\ -2 & \frac{3}{4} & \frac{1}{2} & -1 \end{bmatrix}$$

and

$$B = \begin{bmatrix} 2 & -1 & 3 & 4 & 5 & 6 \\ 1.2 & 3.4 & 1 & -2.2 & 2 & 3.2 \\ 1 & 3 & -4 & 2 & -5 & 1 \\ -2 & 3 & 1 & 2 & -2 & 3 \\ 1 & 1.2 & -1.2 & 3.6 & -4.8 & 0 \\ -1.2 & 3.2 & 1.4 & 2.2 & 2.1 & 3 \end{bmatrix}$$

1. Enter the operation $X = A\tilde{}$. As you will be instructed on the screen, press [Enter] to raise A to powers. After the third iteration, CALIPSO will tell you that some fractions have more than 4 digits in the numerator or denominator and advise you to start again with a decimal matrix. You can quit at this point; or, if you prefer, press [Esc] to get back to the menu, press [PgUp] to get A on the screen, change A to decimal (with the option on the menu which does that), and try again.

2. Enter the option $Y = B\tilde{}$ and raise B to powers. This time you can continue until you tire.

3. Calculate $A + A^2 + A^3$ by entering the operation $A + AA + AAA$.

Properties of Matrix Operations

Matrix addition, scalar multiplication, and matrix multiplication satisfy the following rules. If A, B, and C are appropriately sized matrices and if α

and β are real numbers, then

$$\begin{array}{rl} \text{(i)} & A+B=B+A \\ \text{(ii)} & A(BC)=(AB)C \\ \text{(iii)} & A(B+C)=AB+AC \\ \text{(iv)} & (\alpha+\beta)A=\alpha A+\beta A \\ \text{(v)} & \alpha(A+B)=\alpha A+\alpha B \\ \text{(vi)} & \text{In general, } AB \neq BA \end{array}$$

1.2F Hand Exercises: Verifying Matrix Properties

Using the matrices

$$A=\begin{bmatrix} 2 & -4 & 1 & 4 \\ 2 & 2 & 0 & 2 \\ 8 & 3 & -1 & 2 \end{bmatrix} \qquad B=\begin{bmatrix} -4 & 1 & 6 \\ 4 & -1 & 2 \\ 2 & 2 & -3 \end{bmatrix}$$

$$C=\begin{bmatrix} 4 & -4 & 3 & 3 \\ 1 & 2 & -1 & 3 \\ 2 & 0 & -3 & 6 \end{bmatrix} \qquad D=\begin{bmatrix} -1 & 3 & 0 \\ 1 & 4 & 3 \\ 2 & -5 & 1 \end{bmatrix}$$

$$E=\begin{bmatrix} 1 & 0 & 0 & -3 \\ -4 & 2 & 7 & 4 \\ 0 & 1 & 2 & 3 \end{bmatrix} \qquad F=\begin{bmatrix} 2 & 0 \\ -2 & 1 \\ -5 & 2 \end{bmatrix}$$

verify the following identities by hand calculation.

1. $A+C=C+A$
2. $B(DF)=(BD)F$
3. $D(A+C)=DA+DC$
4. $(4+7)E=4E+7E$
5. $-6(A+C)=-6A-6C$
6. $BD \neq DB$

1.3 APPLICATIONS

In this section we see how to organize data into matrices and then use matrix operations to generate information from the data. Examples 1 and 2 illustrate ways matrices are used in applications.

Example 1: Revenue from a Ball Game

A ball park has 4 seating sections, A, B, C, and D. Ticket costs are Section A, \$6.00; Section B, \$9.00; Section C, \$13.50; and Section D, \$20.00. The table

	Section A	Section B	Section C	Section D
Fri.	10,000	8,000	7,500	4,000
Sat.	13,000	10,000	9,000	5,000
Sun.	18,000	15,000	10,000	7,500

gives the estimated attendance in each of the sections for games on Friday, Saturday, and Sunday. We use scalar multiplication and matrix multiplication to answer the following questions:

(a) A recent weather forecast indicates a strong chance of rain for all of these games. It is estimated that this will result in a reduction of 25% of the estimated attendance figures. What are the new estimates?

(b) What will be the total revenue for each day if the original estimates remain valid?

(c) What is the total revenue for all of the games?

(d) What is the total attendance in each section for all of the games?

Solution:

(a) Let F be the matrix of entries in the table,

$$F = \begin{bmatrix} 10,000 & 8,000 & 7,500 & 4,000 \\ 13,000 & 10,000 & 9,000 & 5,000 \\ 18,000 & 15,000 & 10,000 & 7,500 \end{bmatrix}$$

Use CALIPSO in decimal mode to enter F. Note that we used commas to make the large numbers in array F more legible, but that you must not use commas when entering a matrix in CALIPSO. Find $G = .75F$ to obtain a table with the new attendance estimates.

(b) Use CALIPSO to create the matrix of seating costs

$$H = \begin{bmatrix} 6 \\ 9 \\ 13.5 \\ 20 \end{bmatrix}$$

Then the product matrix

$$J = FH = \begin{bmatrix} 313,250 \\ 389,500 \\ 528,000 \end{bmatrix}$$

gives the total revenue for each day. To see this, note, for instance, that by the definition of matrix multiplication the top entry in the matrix FH is equal to $10,000 \cdot 6 + 8000 \cdot 9 + 7500 \cdot 13.5 + 4000 \cdot 20 = 313,250$, which is the revenue received for Friday's game. Similarly, the other entries in FH correspond to the revenues from Saturday's and Sunday's games.

(c) Let K be the matrix $K = [1 \;\; 1 \;\; 1]$. Then $L = KJ$ is a 1×1 entry whose only entry [1,218,000] is the total revenue. To see this observe that

$$KJ = \begin{bmatrix} 1 & 1 & 1 \end{bmatrix} \begin{bmatrix} 313,250 \\ 389,500 \\ 528,000 \end{bmatrix} = [1,230,800]$$

Note that CALIPSO reports the answer as 1.2308E+06. This is the computer representation of powers of 10 (scientific) notation and is used when numbers are too long to print in the available space. The meaning of the notation is $1.2308\text{E}{+}06 = 1.2308 \cdot 10^6 = 1,230,800$.

(d) To find the total estimated attendance in each seating section calculate $M = KF$, where the matrices K and F are defined above. Again use the definition of matrix multiplication to see why the entries of M give the sums of the columns of F and, thus, the answer to this problem.

Example 2: Grades in a Finite Mathematics Course

Four instructors have been assigned to teach finite mathematics. During the semester they must give 5 one-hour exams and a two-hour final. Many instructors assign different values to tests in accord with the time spent on the material. Nevertheless, they often want each test to have 100 possible points in order to more easily compare scores. A common way of meeting these two goals is to compute the *weighted average* of the test scores. To do this you assign a fraction (a weight) to each exam in such a way that all of the fractions add to 1. You calculate the product of the weight times the student's score on the corresponding test and add the resulting numbers.

Table 1 gives the various weightings of the instructors for the finite math exams. For instance, Instructor 1 counts each exam as 15% and the final as 25% of the course grade. To calculate a student's course grade, the instructor multiplies each exam score by .15 and the final score by .25 and adds these numbers to get the total points for the course grade. Since the fractions add to 1, a student who gets a perfect score, say 100, on each exam will receive a total score of 100. A student who received scores of 75, 60, 80, 95, 90, and 85 on the 5 exams and the final gets the total points $.15 \cdot 75 + .15 \cdot 60 + .15 \cdot 80 + .15 \cdot 95 + .15 \cdot 90 + .25 \cdot 85 = 81.25$.

Table 1

	Exam 1	Exam 2	Exam 3	Exam 4	Exam 5	Final
Inst. 1	.15	.15	.15	.15	.15	.25
Inst. 2	.10	.10	.15	.15	.20	.30
Inst. 3	.05	.10	.10	.15	.15	.45
Inst. 4	.05	.05	.10	.20	.25	.35

Notice that the sum of the entries in each row is 1. You could use matrix multiplication to check this by multiplying by $\begin{bmatrix} 1 \\ 1 \\ 1 \\ 1 \\ 1 \\ 1 \end{bmatrix}$ on the right.

Suppose you received the grades of 85, 60, 80, 95, 90, and 92 on the 5 exams and the final. Find your weighted average for each of the classes. Do the same for your friend who received grades of 90, 85, 80, 80, 95, and 88.

Solution: In CALIPSO enter the matrix A corresponding to Table 1 and let B be the matrix

$$B = \begin{bmatrix} 85 & 90 \\ 60 & 85 \\ 80 & 80 \\ 95 & 80 \\ 90 & 95 \\ 92 & 88 \end{bmatrix}$$

The first column of the product matrix

$$C = AB = \begin{bmatrix} 84.50 & 86.50 \\ 86.35 & 86.90 \\ 87.40 & 86.85 \\ 88.95 & 87.30 \end{bmatrix}$$

gives your weighted averages for the various instructors; the second column of this matrix gives your friend's weighted averages. To see this, observe for instance that (by the definition of matrix multiplication) the entry in the second row and first column of AB is the product of the second row of A and the first column of B:

$$.1 \cdot 85 + .1 \cdot 60 + .15 \cdot 80 + .15 \cdot 95 + .2 \cdot 90 + .3 \cdot 92 = 86.35$$

1.3A Exercises: Applications

Do Problems 1–3 by hand and the rest with the *Operations* option of CALIPSO's *MATRIX* program. Note that many of the matrices in these exercises must be entered in decimal arithmetic mode. If you forget to set decimal mode when you create the matrix, use Esc to get to the menu, choose decimal mode from the menu, and then choose the option to change values in the matrix.

1. A small theater has three sections: A, B, and C. Tickets for Section A are \$4.00; Section B, \$5.00; and Section C, \$7.00. The table

	Section A	Section B	Sections C
Fri	200	150	100
Sat	300	200	200
Sun	200	200	150

gives the predicted attendance for a musical production. Use scalar and matrix multiplication to answer the following questions.

(a) If the predictions are valid, what will be the total revenue each day?

(b) If the predictions are valid, what will be the total revenue for all three days?

(c) If the Friday and Sunday predictions are valid but only 75% of the predicted number of people attend on Saturday, what will be the total revenue for all three days?

(d) Find the total predicted seating in each section.

2. § The table

	Exam 1	Exam 2	Exam 3	Exam 4
Instructor 1	.2	.2	.25	.35
Instructor 2	.15	.25	.30	.30

gives the weighted averages of two instructors for four exams. If you received scores 80, 75, 90, 100 on the four exams, what would be your weighted average for each of the instructors.

3. A city is divided into five voting districts. The table

	1	2	3	4	5
Democrat	.6	.4	.3	.4	.6
Republican	.3	.5	.3	.4	.3
Socialist	.1	.1	.4	.2	.1

gives the fractions of people voting for each party in each district. It is expected that in District 1, 10,000 people will vote; in District 2, 6,000; in District 3, 12,000; in District 4, 8000; in District 5, 15,000. Find the projected number of votes for each party.

4. § An auditorium has 6 seating sections A, B, C, D, E, F. Seats in section A cost \$35; in B, \$32; in C, \$27.50; in D, \$20; in E, \$15; in F, \$10. The table

	A	B	C	D	E	F
Thursday	300	500	400	600	800	200
Friday	400	350	500	200	700	400
Saturday	400	300	600	400	900	700
Sunday	200	300	300	400	250	300

gives estimated attendance for four performances. Use appropriate matrix multiplications to answer the following questions.

(a) What is the total revenue each day? (Suggestion: Multiply on the right by a column of costs.)

(b) What is the total revenue for all performances? (Suggestion: Multiply on the left by a row of 1's and on the right by a column of costs.)

(c) What is the total attendance in each section? (Suggestion: Multiply on the left by a row of 1's.)

5. The chairman of the mathematics department is concerned that the finite mathematics sections described in Example 2 may not be treating students equitably because of the different weights given each test by the instructors. He decides to examine the weighted averages that 8 students would receive from each of the 4 instructors. The 8 students (labeled A-H) received the following grades.

	Exam 1	Exam 2	Exam 3	Exam 4	Exam 5	Final
A	76	82	67	92	85	80
B	86	92	72	67	75	65
C	100	96	100	92	94	98
D	60	54	62	55	62	54
E	72	62	75	66	75	75
F	84	56	72	69	69	69
G	85	45	82	72	74	88
H	67	85	64	80	79	82

Carefully consider how you will enter these grades in a matrix, and then compute all of the weighted averages with one matrix multiplication.

6. § A clothing manufacturer has stores in New York City, Moscow, Boston, Chicago, Atlanta, and Seattle. Its sales (in thousands of dollars) are summarized in the table

	NYC	Moscow	Boston	Chicago	Atlanta	Seattle
Pants	12	24	15	18	27	22
Shirts	15	18	36	15	17	32
Sweaters	9	15	15	24	35	22
Skirts	54	43	52	36	12	63
Socks	18	22	21	15	14	17

The profit from each unit of Pants was \$12, Shirts \$9, Sweaters \$14, Skirts \$16, Socks \$5. Use one matrix multiplication to determine the profit from sales in each city. (Be careful! Consider whether to put the profits in a row or column matrix and the order in which you will multiply the matrices.)

7. A bank has investments in Brazil, Peru, Bolivia, Chile, and Mexico. The investments are divided among mortgages, stocks, government bonds, and personal loans. The amount of money (in millions of dollars) is given in the table

	Mortgages	Stocks	Bonds	Loans
Brazil	10.3	12.7	24.2	8.5
Peru	8.6	12.5	24.3	12.2
Bolivia	6.6	5.4	3.6	9.9
Chile	12.5	12.2	8.6	3.5
Mexico	10.2	9.1	3.0	4.2

The current annual return on mortgages is 12.5%, stocks 10.6%, bonds 8.8%, loans 9.5%. Use one matrix multiplication to determine the bank's earnings from each country.

8. § A flu epidemic strikes a city. Each resident is either sick, well, or a carrier. The fraction of people in each category and various age groups is given in the table

	Age						
	0-9	10-19	20-29	30-39	40-49	50-59	60-
Well	.65	.62	.45	.36	.64	.78	.82
Sick	.15	.24	.18	.12	.30	.10	.08
Carrier	.20	.14	.37	.52	.06	.12	.10

The population of the city distributed by age and sex is

		Female	Male
Age	0- 9	60,000	62,000
	10-19	82,000	85,200
	20-29	112,200	115,800
	30-39	98,000	100,500
	40-49	73,600	85,200
	50-59	64,200	54,000
	61-	58,500	46,300

Compute the product of the matrices corresponding to these tables in the appropriate order to answer the questions:

(a) How many females are sick?
(b) How many males are carriers?

9. A coffee manufacturer offers 6 coffee blends (A-F). Each blend is a mix of Columbian, Brazilian, Ecuadoran, and Costa Rican coffees. The table

	Columbian	Brazilian	Ecuadoran	Costa Rican
A	4	2.5	1.5	2
B	2	3	3	2
C	1	4	1	4
D	5	2	1.5	1.5
E	3	0	6	1
F	2	2	1	5

gives the mix for a 10-pound (10-lb) batch of each blend.

(a) If today's demand is

Blend A	500 lb	Blend B	600 lb	Blend C	900 lb
Blend D	350 lb	Blend E	450 lb	Blend F	600 lb

find the number of pounds of each kind of coffee the manufacturer needs per day.

(b) The profits on each 1-pound bag of the various blends are A, \$3.00; B, \$2.50; C, \$5.00; D, \$1.50; E, \$4.00; F, \$2.25. Find the total profit on the day described in (a) if all the coffee is sold.

10. § Suppose that you want to buy a sleeping bag, pad, tent, backpack, and camping utensils for a backpacking trip. To compare prices you called four stores and obtained the following costs for each of these items. Store 1: \$98, \$24, \$225, \$65, \$15. Store 2: \$110, \$20, \$275, \$75, \$22. Store 3: \$85, \$32, \$240, \$69, \$18. Store 4: \$124, \$27, \$200, \$62, \$15.

(a) Find the cost of purchasing these goods from each of the stores by making one matrix multiplication.

(b) Find the average cost of each of the five items by making one matrix multiplication. (Suggestion: The average cost of an item is the sum of 1/4 the cost of the item at each store.)

2 Systems of Linear Equations

2.1 INTRODUCTION

Systems of linear equations arise naturally in areas as diverse as biology, economics, and physics. An example of a system of linear equations (with 3 equations and 4 variables) is

$$\begin{array}{rcrcrcrcr} 2x & - & 3y & + & 4z & + & w & = & 18 \\ \frac{1}{2}x & & & - & 2z & - & w & = & 0 \\ x & + & 3y & - & z & + & 2w & = & -3 \end{array}$$

A *solution* to this system of equations consists of values of the variables x, y, z, w that simultaneously make all of the equations true. For instance, $x = 4, y = -2, z = 1, w = 0$ is a solution since substituting these values in the first equation yields 18; substituting these same values in the second equation yields 0; and substitution in the third equation produces -3. You can verify that $x = -2, y = -13, z = -8, w = 15$ is another solution. In fact, we will see that, for any number t, $x = 4 - \frac{2}{5}t, y = -2 - \frac{11}{15}t, z = 1 - \frac{3}{5}t, w = t$ gives a solution to this system of equations; moreover, all solutions can be written in this form.

Example 1: A System of Equations with a Unique Solution

As you will be able to determine later the system of linear equations

$$\begin{array}{rcrcrcr} 2x & + & y & - & 3z & = & -1 \\ 6x & - & 2y & + & z & = & 17 \\ -\frac{1}{2}x & - & y & - & z & = & 0 \end{array}$$

has only one solution: $x = 2, y = -2, z = 1$.

Example 2: A System of Equations with No Solution

The system of equations

$$\begin{array}{rcrcrcr} x & - & 2y & + & 4z & = & 7 \\ -3x & + & 6y & - & 12z & = & -18 \end{array}$$

has no solution. To see this notice, that the left-hand side of the second equation is -3 times the left-hand side of the first equation. However, on the right-hand side, $-18 \neq -3 \cdot 7$. This implies that no values of x, y, and z can satisfy both of these equations simultaneously.

In general a system of equations can have any number of equations and any number of variables. When a system has more than a few variables we often use $x_1, x_2, x_3, \ldots$ to name them rather than $x, y, z, \ldots$. Here is a system with 4 equations and 5 variables.

$$\begin{array}{rcrcrcrcrcr} 5x_1 & + & 3x_2 & - & 7x_3 & + & 2x_4 & - & 3x_5 & = & 12 \\ 2x_1 & + & 2x_2 & + & 6x_3 & + & 4x_4 & + & 2x_5 & = & 20 \\ -3x_1 & - & x_2 & - & 2x_3 & + & x_4 & - & 4x_5 & = & 18 \\ 4x_1 & + & 2x_2 & + & 4x_3 & + & 3x_4 & - & 5x_5 & = & 9 \end{array} \qquad (1)$$

The numbers that multiply the variables x_i are the *coefficients* of the variables, and the numbers $12, 20, 18, 9$ at the right of the equality signs form the *right-hand side* of the system. A *solution* to the system of linear equations (1) is a set of values assigned to the variables x_1, x_2, x_3, x_4, x_5 which when substituted simultaneously into all of the equations produces the desired equalities. A system of linear equations can have a unique solution, no solutions, or an infinite number of solutions; it cannot have, say, exactly five solutions. In the next few sections we will look at three commonly used matrix methods to find all solutions of a system of linear equations (or determine that no solutions exist).

In what follows it will often be convenient to write a system like (1) in matrix form as

$$AX = B$$

where

$$A = \begin{bmatrix} 5 & 3 & -7 & 2 & -3 \\ 2 & 2 & 6 & 4 & 2 \\ -3 & -1 & -2 & 1 & -4 \\ 4 & 2 & 4 & 3 & -5 \end{bmatrix}$$

$$X = \begin{bmatrix} x_1 \\ x_2 \\ x_3 \\ x_4 \\ x_5 \end{bmatrix} \qquad B = \begin{bmatrix} 12 \\ 20 \\ 18 \\ 9 \end{bmatrix}$$

Matrix A is called the *coefficient matrix*; column matrix X is called the *variable matrix*; and column matrix B is called the *right-hand side matrix.* You can verify that the matrix equation $AX = B$ is equivalent to the system of linear equations (1) by simply carrying out the multiplication AX and observing that it is equal to the column matrix B. The row entries of AX and B coincide with the equations of the given system.

We can also derive a system of equations from a matrix equation $AX = B$. For instance, the matrix equation

$$\begin{bmatrix} 2 & -4 & 3 \\ 0 & 1 & 5 \end{bmatrix} \begin{bmatrix} x_1 \\ x_2 \\ x_3 \end{bmatrix} = \begin{bmatrix} -3 \\ 2 \end{bmatrix}$$

corresponds to the system of equations

$$\begin{aligned} 2x_1 - 4x_2 + 3x_3 &= -3 \\ x_2 + 5x_3 &= 2 \end{aligned}$$

2.1A Hand Exercises: Representing a System of Equations by a Matrix Equation

1. Write the system of equations corresponding to each matrix equation.

(a) § $\begin{bmatrix} 1 & -1 \\ 3 & 2 \end{bmatrix} \begin{bmatrix} x \\ y \end{bmatrix} = \begin{bmatrix} 1 \\ 3 \end{bmatrix}$ (b) $\begin{bmatrix} 2 & 1 & -1 \\ 3 & 2 & -1 \\ 1 & 0 & 4 \end{bmatrix} \begin{bmatrix} x_1 \\ x_2 \\ x_3 \end{bmatrix} = \begin{bmatrix} 1 \\ 2 \\ 3 \end{bmatrix}$

(c) § $\begin{bmatrix} 1 & -1 & 2 & 0 & 3 \\ 2 & -1 & 2 & 4 & 3 \\ -2 & 1 & 2 & 1 & 0 \\ 3 & -3 & 2 & 1 & -2 \end{bmatrix} \begin{bmatrix} x_1 \\ x_2 \\ x_3 \\ x_4 \\ x_5 \end{bmatrix} = \begin{bmatrix} 1 \\ -2 \\ 3 \\ 3 \end{bmatrix}$

(d) $\begin{bmatrix} 2 & 0 & 5 & 1 & -3 \\ 3 & 1 & 0 & -4 & 4 \\ -3 & 2 & 1 & -1 & 0 \\ 2 & 1 & 0 & -1 & 2 \end{bmatrix} \begin{bmatrix} x_1 \\ x_2 \\ x_3 \\ x_4 \\ x_5 \end{bmatrix} = \begin{bmatrix} 2 \\ -1 \\ 0 \\ 2 \end{bmatrix}$

2. Write the matrix equation $AX = B$ corresponding to each system of equations.

(a) §

$$\begin{array}{rcrcrcr} x_1 & + & x_2 & + & 2x_3 & = & 2 \\ 2x_1 & - & 3x_2 & + & x_3 & = & -2 \\ 3x_1 & + & x_2 & - & 5x_3 & = & 4 \end{array}$$

(b)

$$\begin{array}{rcrcrcrcr} x_1 & - & 2x_2 & + & 3x_3 & + & 2x_4 & = & 3 \\ -x_1 & - & x_2 & + & 2x_3 & + & 3x_4 & = & 4 \\ 2x_1 & - & 6x_2 & - & 4x_3 & + & 3x_4 & = & 7 \end{array}$$

(c) §

$$\begin{array}{rcrcrcrcrcr} 2x_1 & + & x_2 & + & 3x_3 & - & 4x_4 & - & 2x_5 & = & 6 \\ x_1 & - & 2x_2 & & & + & 3x_4 & - & x_5 & = & 2 \\ 2x_1 & - & 2x_2 & + & 4x_3 & + & 2x_4 & - & 3x_5 & = & 8 \\ 4x_1 & - & 2x_2 & + & 4x_3 & + & x_4 & - & 2x_5 & = & 9 \end{array}$$

(d)

$$\begin{array}{rcrcrcrcrcr} 3x_1 & - & x_2 & & & - & 2x_4 & - & 5x_5 & = & 9 \\ 4x_1 & - & 3x_2 & - & 2x_3 & + & 2x_4 & - & 8x_5 & = & 5 \\ & - & 3x_2 & + & 6x_3 & - & 2x_4 & + & 2x_5 & = & 7 \\ -3x_1 & - & 3x_2 & - & 2x_3 & + & x_4 & + & 3x_5 & = & 6 \end{array}$$

3. Recall that the graph of an equation of the form $ax + by = c$ is a line.

(a) Explain why the solution of the system of equations

$$\begin{array}{rcrcr} 3x & + & 2y & = & 6 \\ -2x & + & 4y & = & 8 \end{array}$$

corresponds to the point of intersection of the line $3x + 2y = 6$ and the line $-2x + 4y = 8$.

(b) Show that the system of equations

$$\begin{aligned} 2x &- 4y = 12 \\ -6x &+ 12y = -24 \end{aligned}$$

has no solution. What can you say about the lines $2x - 4y = 12$ and $-6x + 12y = -24$?

(c) Show that the system of equations

$$\begin{aligned} 3x &+ 2y = 6 \\ -6x &- 4y = -12 \end{aligned}$$

has an infinite number of solutions. What can you say about the lines $3x + 2y = 6$ and $-6x - 4y = -12$.

(d) Explain why from a geometrical standpoint a system of two equations in two variables has a unique solution, no solution, or infinitely many solutions. Why cannot such a system have exactly four solutions?

2.2 GAUSS ELIMINATION

Gauss elimination and a related method for solving systems of linear equations, *Gauss-Jordan elimination*, both depend on what are called *elementary row operations*. Performing these operations on a system of linear equations results in a new system of equations that has precisely the same solutions as the original system, but may be easier to solve. Two systems of linear equations having the same solutions are said to be *equivalent*. The three elementary row operations are:

I. Divide or multiply any equation through by a nonzero number (for instance, divide both sides of the second equation of a system by -3).

II. Replace an equation by the sum of that equation and a multiple of any other equation (for instance, replace the third equation of a system with the sum of the third equation and $\frac{1}{2}$ times the fifth equation).

III. Interchange any two equations (for instance, put the first equation of a system fourth and the fourth equation first).

Again we emphasize that though these row operations will convert the given system of linear equations to another system, the solutions to both systems are exactly the same.

Example 1: Equivalent Systems of Equations

The systems of linear equations

$$\begin{array}{rcrcrcr} 2x_1 & - & 3x_2 & + & 4x_3 & = & 7 \\ 3x_1 & + & 5x_2 & + & x_3 & = & 9 \\ 4x_1 & - & 2x_2 & + & 7x_3 & = & -4 \end{array} \quad \text{and} \quad \begin{array}{rcrcrcr} x_1 & - & \frac{3}{2}x_2 & + & 2x_3 & = & \frac{7}{2} \\ 3x_1 & + & 5x_2 & + & x_3 & = & 9 \\ -5x_1 & - & 17x_2 & + & 4x_3 & = & -31 \end{array}$$

System 1 System 2

are equivalent systems of linear equations since System 2 is obtained from System 1 by dividing the first equation of System 1 by 2, and replacing the third equation of System 1 with the sum of this equation and −3 times the second equation.

To facilitate calculations in what follows it is useful to work with *augmented matrices.* The augmented matrix of a system of equations is obtained by introducing a column containing the right-hand sides of the equations to the right of the coefficient matrix. We can apply elementary row operations to the augmented matrix just as we would to the system of equations. Two matrices are *equivalent* if one is obtained from the other through a series of matrix row operations, where a *matrix row operation* consists of an elementary row operation I, II, or III applied to a row of a matrix rather than to an equation.

In examples we put a vertical bar before the column of the augmented matrix that corresponds to the right-hand side of the system of equations. This helps you see that the matrix is augmented. CALIPSO does not display matrices with the vertical bar.

Example 2: An Augmented Matrix

The augmented matrix corresponding to the system of equations found in Example 1 of Section 2.1 is

$$A = \left[\begin{array}{rrr|r} 2 & 1 & -3 & -1 \\ 6 & -2 & 1 & 17 \\ -\frac{1}{2} & -1 & -1 & 0 \end{array}\right]$$

Gauss elimination involves using elementary row operations to attempt to convert an augmented matrix of the form

$$A = \left[\begin{array}{ccc|c} a & b & c & d \\ e & f & g & h \\ i & j & k & l \end{array}\right]$$

to an equivalent matrix of the form

$$\hat{A} = \left[\begin{array}{ccc|c} \hat{a} & \hat{b} & \hat{c} & \hat{d} \\ 0 & \hat{f} & \hat{g} & \hat{h} \\ 0 & 0 & \hat{k} & \hat{l} \end{array}\right] \tag{1}$$

Here the letters with hats indicate numbers that we calculated in the course of performing elementary row operations I, II, and III. The coefficients $\hat{a}, \hat{f}$, and $\hat{k}$ are called *leading entries* of their rows. In general, the leading entry of a row is the leftmost nonzero entry in the row. The system of equations corresponding to $\hat{A}$ is

$$\begin{array}{rcrcrcl} \hat{a}x_1 & + & \hat{b}x_2 & + & \hat{c}x_3 & = & \hat{d} \\ & & \hat{f}x_2 & + & \hat{g}x_3 & = & \hat{h} \\ & & & & \hat{k}x_3 & = & \hat{l} \end{array} \tag{2}$$

As we see next, systems of equations having the form (2) are particularly easy to solve with the use of a technique called *back substitution.* To illustrate these ideas consider the system of equations

$$\begin{array}{rcrcrcr} 2x_1 & + & x_2 & - & x_3 & = & -4 \\ 4x_1 & + & 3x_2 & + & x_3 & = & 2 \\ 2x_1 & - & x_2 & + & 2x_3 & = & 12 \end{array} \tag{3}$$

The corresponding augmented matrix is

$$\left[\begin{array}{rrr|r} 2 & 1 & -1 & -4 \\ 4 & 3 & 1 & 2 \\ 2 & -1 & 2 & 12 \end{array}\right] \tag{4}$$

You can verify that applying Gauss elimination to (4) gives the augmented matrix

$$\left[\begin{array}{rrr|r} 2 & 1 & -1 & -4 \\ 0 & 1 & 3 & 10 \\ 0 & 0 & 9 & 36 \end{array}\right]$$

which corresponds to the system of equations

$$\begin{array}{rcrcrcr} 2x_1 & + & x_2 & - & x_3 & = & -4 \\ & & x_2 & + & 3x_3 & = & 10 \\ & & & & 9x_3 & = & 36 \end{array} \tag{5}$$

We now apply back substitution to (5). This is done by working backwards from the last equation to the first equation. The last equation in (5) gives $x_3 = 4$. To find x_2 we substitute $x_3 = 4$ in the second equation to obtain $x_2 = -2$. We continue working backwards to find x_1 from the first equation of (5). Substituting $x_3 = 4$ and $x_2 = -2$ in this equation we find that $x_1 = 1$. Therefore, $x_1 = 1, x_2 = -2, x_3 = 4$ is the solution of (5), and since (5) and (3) are equivalent we have found the solution to (3) as well.

In Sections 2.4 and 2.5 we will view Gauss elimination from a more general perspective, but for the present we focus on matrices of the form $\hat{A}$ in (1). Examples 3 and 4 illustrate these ideas.

Example 3: The Backward Pass

Suppose we obtained the following matrix by applying Gauss elimination to an augmented matrix of a system of equations.

$$\left[\begin{array}{rrr|r} 2 & -2 & 3 & 1 \\ 0 & -3 & -2 & 3 \\ 0 & 0 & 4 & -2 \end{array}\right]$$

The system of equations corresponding to this matrix is

$$\begin{array}{rcrcrl} 2x & - & 2y & + & 3z & = 1 \\ & & -3y & - & 2z & = 3 \\ & & & & 4z & = -2 \end{array}$$

We begin the backward pass with the last equation, which gives $z = -\frac{2}{4} = -\frac{1}{2}$. Next we move up to the next equation, solve for the leading variable

y, and use the previously determined value of z to get

$$y = \frac{2z+3}{-3} = \frac{2(-\frac{1}{2})+3}{-3} = -\frac{2}{3}$$

Finally we solve the first equation for x and then substitute the known values of y and z to get

$$x = \frac{2y-3z+1}{2} = \frac{2(-\frac{2}{3})-3(-\frac{1}{2})+1}{2} = \frac{7}{12}$$

Example 4: Applying Gauss Elimination

We apply a series of elementary row operations to the matrix

$$A = \left[\begin{array}{rrr|r} 2 & 1 & -3 & -1 \\ 6 & -2 & 1 & 17 \\ -\frac{1}{2} & -1 & -1 & 0 \end{array}\right]$$

given in Example 2 to convert this matrix to an equivalent matrix having the form of matrix $\hat{A}$ above.

First we create 0's below the entry 2 in the first column. To do this, replace the second row of the matrix A by a row obtained by adding the second row of A to -3 times the first row. This gives the equivalent matrix

$$B = \left[\begin{array}{rrr|r} 2 & 1 & -3 & -1 \\ 0 & -5 & 10 & 20 \\ -\frac{1}{2} & -1 & -1 & 0 \end{array}\right]$$

Now replace the third row of B with the row obtained by adding the third row of B to $\frac{1}{4}$ times the first row. We get the equivalent matrix

$$C = \left[\begin{array}{rrr|r} 2 & 1 & -3 & -1 \\ 0 & -5 & 10 & 20 \\ 0 & -\frac{3}{4} & -\frac{7}{4} & -\frac{1}{4} \end{array}\right]$$

Now obtain a 0 below the entry -5 in the second column by replacing the third row of C with the sum of this row and $-\frac{3}{20}$ times the second row. This results in the following equivalent matrix of the desired form.

$$D = \left[\begin{array}{rrr|r} 2 & 1 & -3 & -1 \\ 0 & -5 & 10 & 20 \\ 0 & 0 & -\frac{13}{4} & -\frac{13}{4} \end{array}\right]$$

The matrix D corresponds to the system of linear equations

$$\begin{array}{rcrcrcr} 2x & + & y & - & z & = & -1 \\ & & -5y & - & 10z & = & 20 \\ & & & & -\frac{13}{4}z & = & -\frac{13}{4} \end{array}$$

This system is equivalent to the original system. Moreover, solving this system with backward substitution is easy. From the third equation we see that $z = 1$. Substituting $z = 1$ into the second equation we find that $y = -2$. Finally substituting $z = 1, y = -2$ into the first equation shows that $x = 2$, and we have found a solution $x = 2, y = -2, z = 1$. In fact it should be clear that this is the only solution. Since to obtain this solution we used only elementary row operations, we can conclude that the (unique) solution to the system of equations found in Example 1 of Section 2.1 is $x = 2, y = -2, z = 1$.

Note: On occasion it is advantageous to use Gauss elimination to obtain a matrix of the form (2) where all of the leading entries are equal to 1. As you can check, doing this in Example 4 will give the matrix

$$\left[\begin{array}{rrr|r} 1 & \frac{1}{2} & -\frac{3}{2} & -\frac{1}{2} \\ 0 & 1 & -2 & -4 \\ 0 & 0 & 1 & 1 \end{array}\right] \tag{6}$$

from which it is easy to get the previously obtained solution $x = 2, y = -2, z = 1$.

Before generalizing the procedure used in Example 4 let us see how you can easily carry out these steps with CALIPSO.

Doing Row Operations With CALIPSO

Here is how to use CALIPSO to do the calculations in Example 4. Move the cursor to *By Command*. To get 0's below the entry 2 found in the first column and first row, use elementary row operation II twice. Begin by replacing the second row of the original matrix by itself plus -3 times the first row; then replace the third row of the matrix just obtained by adding to the third row $1/4$ times the first row. Finally replace the third row with the sum of it and $-3/20$ times the second row. You now have obtained the matrix we used to solve the system of equations.

CALIPSO Demonstrations

The instructor of your course should have an "Instructor's Answer Disk." This disk contains a program to demonstrate row reduction of matrices and other applications we will see later. If a computer is not available in your classroom, your instructor may make this program available to you. Even though this program was written for use as a classroom demonstration, it is a useful tutorial for learning Gauss elimination. The program is named GAUSS and you start it by typing its name.

2.2A Exercises: Solving Systems of Equations

The systems of equations corresponding to the matrices in Exercises 1–7 all have unique solutions. Use the elementary row operations to convert each of the matrices to the form

$$\begin{bmatrix} * & * & * & * & * & * \\ 0 & * & * & * & * & * \\ 0 & 0 & * & * & * & * \\ 0 & 0 & 0 & * & * & * \\ 0 & 0 & 0 & 0 & * & * \end{bmatrix}$$

(where an $*$ stands for any number) and then use back substitution to solve (by hand) the equivalent system of equations corresponding to your final matrix. This procedure gives the solution to the system of equations corresponding to the original matrix. Perform the row operations for Problems 1–5 and 10–12 by hand and use CALIPSO's *By Command* operation to do the rest.

1. $\begin{bmatrix} 1 & 2 & 8 \\ -3 & 4 & 6 \end{bmatrix}$

2.§ $\begin{bmatrix} 2 & 6 & -4 & 22 \\ 1 & 3 & 5 & 10 \\ 4 & 0 & 2 & 10 \end{bmatrix}$

3. $\begin{bmatrix} 2 & 3 & -2 & 4 \\ -1 & 4 & 12 & 15 \\ -3 & 12 & 16 & 8 \end{bmatrix}$

4.§ $\begin{bmatrix} 2 & 2 & 1 & 4 \\ 3 & 1 & 2 & 6 \\ 1 & 3 & 1 & 0 \end{bmatrix}$

5. $\begin{bmatrix} 3 & 4 & 1 & 2 \\ 1 & 2 & 3 & 6 \\ 2 & 5 & 1 & 2 \end{bmatrix}$

6.§ $\begin{bmatrix} \frac{1}{2} & \frac{2}{3} & \frac{3}{4} & -\frac{1}{2} \\ \frac{2}{3} & \frac{1}{2} & -\frac{1}{4} & \frac{2}{3} \\ \frac{3}{4} & \frac{1}{2} & -\frac{1}{2} & \frac{2}{3} \end{bmatrix}$

7. $$\begin{bmatrix} 1 & 2 & -\frac{2}{3} & \frac{1}{4} & 2 \\ -\frac{1}{2} & \frac{1}{4} & 8 & \frac{1}{2} & 6 \\ \frac{1}{2} & -\frac{1}{4} & \frac{2}{3} & -\frac{1}{4} & \frac{2}{3} \\ -\frac{1}{4} & \frac{2}{3} & \frac{1}{4} & \frac{1}{2} & \frac{3}{4} \end{bmatrix}$$

8. § $$\begin{bmatrix} 2 & 0 & -2 & 4 & 2 & 3 \\ -1 & 4 & -2 & 3 & 4 & 8 \\ 2 & 8 & 0 & 0 & 2 & 4 \\ -3 & 1 & 2 & 6 & -4 & 2 \\ 4 & 2 & 4 & -2 & 3 & 4 \end{bmatrix}$$

9. $$\begin{bmatrix} -\frac{1}{2} & 2 & -2 & 1 & -4 & \frac{3}{4} \\ -1 & \frac{1}{4} & -\frac{1}{2} & 2 & 2 & -1 \\ 2 & -1 & \frac{1}{2} & -\frac{1}{2} & \frac{1}{2} & \frac{3}{4} \\ -\frac{3}{2} & \frac{1}{2} & 2 & 1 & -4 & 2 \\ \frac{1}{2} & -\frac{1}{4} & \frac{1}{4} & -2 & \frac{1}{3} & \frac{1}{4} \end{bmatrix}$$

Form the augmented matrix corresponding to the following systems of linear equations and find the solution by Gauss elimination.

10. §

$$\begin{aligned} 2x - 3y &= 6 \\ 3x + 2y &= 7 \end{aligned}$$

11.

$$\begin{aligned} x - 3y + 4z &= 9 \\ 2x + y - 3z &= 8 \\ -x - 2y + z &= 4 \end{aligned}$$

12. §

$$\begin{aligned} x_1 + x_2 + 2x_3 &= 4 \\ 2x_1 - 2x_3 &= 0 \\ -x_1 + 3x_2 - 4x_3 &= 4 \end{aligned}$$

13.

$$\begin{aligned} -x_1 + 2x_2 - 2x_3 + x_4 &= 4 \\ 2x_1 + x_2 - 2x_3 - 3x_4 &= 9 \\ x_2 - 6x_3 - 2x_4 &= 4 \\ 3x_1 - x_2 + 2x_4 &= 5 \end{aligned}$$

14. §

$$\begin{array}{rcrcrcrcrl} x_1 & + & x_2 & + & x_3 & + & x_4 & + & x_5 & = 4 \\ 2x_1 & + & x_2 & - & 2x_3 & - & 3x_4 & + & 2x_5 & = 9 \\ -x_1 & + & x_2 & + & 4x_3 & - & 2x_4 & - & 3x_5 & = 2 \\ -3x_1 & + & x_2 & & & + & 3x_4 & - & 2x_5 & = 5 \\ x_1 & & & + & 2x_3 & - & x_4 & + & 3x_5 & = 8 \end{array}$$

15.

$$\begin{array}{rcrcrcrcrl} 3x_1 & - & 2x_2 & + & 4x_3 & + & 3x_4 & - & x_5 & = 8 \\ 2x_1 & - & 3x_2 & + & x_3 & - & 2x_4 & - & x_5 & = 12 \\ 2x_1 & - & 3x_2 & + & 2x_3 & - & x_4 & + & 2x_5 & = 3 \\ 3x_1 & - & 3x_2 & & & + & 2x_4 & + & x_5 & = 6 \\ -x_1 & & & - & x_3 & + & 2x_4 & + & 4x_5 & = 6 \end{array}$$

Using Row Interchanges

Sometimes you cannot obtain a matrix in the form you would like because a 0 occurs in the place you want to create a nonzero leading entry. When this happens you try to find a row *below* the one you are working on that has a nonzero entry in the desired column. You then interchange the two rows. We defer until Sections 2.4 and 2.5 consideration of the difficulty that arises when there are no nonzero entries to pick. Example 5 requires two row interchanges.

Example 5: Row Interchanges

$$\begin{bmatrix} 0 & 2 & 1 & 1 \\ -2 & -1 & 3 & 2 \\ 2 & 1 & \frac{1}{3} & -2 \end{bmatrix} \rightarrow \begin{bmatrix} 2 & 1 & \frac{1}{3} & -2 \\ -2 & -1 & 3 & 2 \\ 0 & 2 & 1 & 1 \end{bmatrix}$$

STEP 1. Interchange rows 1 and 3.

$$\rightarrow \begin{bmatrix} 2 & 1 & \frac{1}{3} & -2 \\ 0 & 0 & \frac{10}{3} & 0 \\ 0 & 2 & 1 & 1 \end{bmatrix}$$

STEP 2. Row 2 plus 1 times Row 1.

$$\rightarrow \begin{bmatrix} 2 & 1 & \frac{1}{3} & -2 \\ 0 & 2 & 1 & 1 \\ 0 & 0 & \frac{10}{3} & 0 \end{bmatrix}$$

STEP 3. Interchange row 2 and row 3.

2.2B Exercises: Problems Requiring Row Interchanges

Follow the same instructions as in Exercises 2.2A for the following matrices. Do Problems 1–3 by hand and the rest with CALIPSO.

1. $\begin{bmatrix} 0 & 2 & 3 & 2 \\ 2 & 3 & 4 & 3 \\ 1 & 2 & 3 & 1 \end{bmatrix}$ 2. $\begin{bmatrix} 2 & 4 & -6 & -10 \\ -4 & -8 & 3 & -11 \\ 1 & 1 & 1 & 6 \end{bmatrix}$ 3.§ $\begin{bmatrix} 1 & 2 & 4 & 3 \\ 3 & 6 & 13 & 11 \\ 2 & 2 & 1 & 5 \end{bmatrix}$

4. $\begin{bmatrix} 0 & 1 & 2 & -1 & 3 \\ 0 & 0 & 0 & 2 & 0 \\ 0 & 3 & 2 & 3 & 1 \\ 1 & 2 & 1 & -2 & 3 \end{bmatrix}$ 5.§ $\begin{bmatrix} 2 & 1 & 3 & 4 & 5 \\ 6 & 3 & 9 & 13 & 19 \\ -6 & -4 & -4 & -12 & -16 \\ 0 & 0 & 4 & 1 & 6 \end{bmatrix}$

2.3 GAUSS-JORDAN ELIMINATION

Gauss-Jordan elimination also uses elementary row operations to convert a given matrix to one for which solutions are easy to find. Its advantage is that back substitution is not required to find the final solution. Let

$$\left[\begin{array}{ccc|c} a & b & c & d \\ e & f & g & h \\ i & j & k & l \end{array}\right]$$

be an augmented matrix corresponding to a system of linear equations having a unique solution. The Gauss-Jordan method will terminate with a matrix of the form

$$\left[\begin{array}{ccc|c} 1 & 0 & 0 & d' \\ 0 & 1 & 0 & h' \\ 0 & 0 & 1 & l' \end{array}\right]$$

From this matrix we easily solve for $x_1 = d'$, $x_2 = h'$, $x_3 = l'$.

Example 1: Applying Gauss-Jordan Elimination

To apply Gauss-Jordan elimination to the matrix

$$A = \left[\begin{array}{rrr|r} 2 & 1 & -3 & -1 \\ 6 & -2 & 1 & 17 \\ -\frac{1}{2} & -1 & -1 & 0 \end{array}\right]$$

first carry out Gauss elimination to convert this matrix to the equivalent matrix

$$D = \left[\begin{array}{rrr|r} 1 & \frac{1}{2} & -\frac{3}{2} & -\frac{1}{2} \\ 0 & 1 & -2 & -4 \\ 0 & 0 & 1 & 1 \end{array}\right]$$

as indicated in (6) of Section 2.2.

Next create 0's above the leading entry 1 in the third row (by adding appropriate multiples of the third row to the second and first rows) to obtain the equivalent matrix

$$E = \left[\begin{array}{rrr|r} 1 & \frac{1}{2} & 0 & 1 \\ 0 & 1 & 0 & -2 \\ 0 & 0 & 1 & 1 \end{array}\right]$$

Now create a 0 above the leading entry 1 in the second row by adding $-\frac{1}{2}$ times the second row to the first row. This results in the equivalent matrix

$$F = \left[\begin{array}{rrr|r} 1 & 0 & 0 & 2 \\ 0 & 1 & 0 & -2 \\ 0 & 0 & 1 & 1 \end{array}\right]$$

which corresponds to the system of equations

$$\begin{array}{rcr} x & = & 2 \\ y & = & -2 \\ z & = & 1 \end{array}$$

whose obvious solution is $x = 2, y = -2, z = 1$.

In general, to carry out the Gauss-Jordan method for a system of linear equations having a unique solution

- Use Gauss elimination to create leading entries equal to 1 and with 0's below these leading entries. The leading entries are located further to the right as you move down the matrix.

- Create 0's above the leading entry in the rightmost column containing a leading entry.

- Move to the left to create 0's above the leading entry in the next column containing a leading entry 1.

- Continue this procedure of creating 0's above the leading entries until all other entries in a column containing a leading entry 1 are 0.

We show how this method proceeds for a matrix corresponding to a system of 4 equations and 4 variables having a unique solution. We add a prime to the letters indicating numbers each time we perform an arithmetic operation on that number. The first step is the result of Gauss elimination.

$$\left[\begin{array}{cccc|c} a & b & c & d & e \\ f & g & h & i & j \\ k & l & m & n & o \\ p & q & r & s & t \end{array}\right] \rightarrow \left[\begin{array}{cccc|c} 1 & b' & c' & d' & e' \\ 0 & 1 & h' & i' & j' \\ 0 & 0 & 1 & n' & o' \\ 0 & 0 & 0 & 1 & t' \end{array}\right] \rightarrow \left[\begin{array}{cccc|c} 1 & b' & c' & 0 & e'' \\ 0 & 1 & h' & 0 & j'' \\ 0 & 0 & 1 & 0 & o'' \\ 0 & 0 & 0 & 1 & t' \end{array}\right]$$

$$\rightarrow \left[\begin{array}{cccc|c} 1 & b'' & 0 & 0 & e''' \\ 0 & 1 & 0 & 0 & j''' \\ 0 & 0 & 1 & 0 & o'' \\ 0 & 0 & 0 & 1 & t' \end{array}\right] \rightarrow \left[\begin{array}{cccc|c} 1 & 0 & 0 & 0 & e'''' \\ 0 & 1 & 0 & 0 & j''' \\ 0 & 0 & 1 & 0 & o'' \\ 0 & 0 & 0 & 1 & t' \end{array}\right]$$

It is tempting to carry out the steps of the Gauss-Jordan method by obtaining zeros above the leading 1 in the same way that you obtain 0's below the leading 1. This is inferior to the method we describe because you must make many more calculations. The GAUSS program on the Instructor's Answer Disk contains a demonstration of Gauss-Jordan elimination. This demonstration is worth using as a tutorial because it illustrates the importance of following the steps of the reduction in order.

2.3A Exercises: Gauss-Jordan Elimination

The systems corresponding to the following matrices all have unique solutions. Use the Gauss-Jordan method to reduce the matrices to the form

$$\left[\begin{array}{cccc|c} 1 & 0 & 0 & 0 & * \\ 0 & 1 & 0 & 0 & * \\ 0 & 0 & 1 & 0 & * \\ 0 & 0 & 0 & 1 & * \end{array}\right]$$

Give the unique solution to the original problem. Do Problems 1, 2, and 3 by hand and use the *By Command* of CALIPSO for the others.

1. $\left[\begin{array}{rrr|r} 3 & -1 & 2 & -1 \\ -2 & 4 & 1 & -32 \\ 3 & 2 & 2 & -2 \end{array}\right]$

2.§ $\left[\begin{array}{rrrr|r} 1 & 2 & -3 & 4 & 6 \\ 3 & -6 & 1 & 0 & 4 \\ 1 & 2 & 5 & 7 & -2 \\ 4 & 4 & 2 & 1 & 3 \end{array}\right]$

3. $\left[\begin{array}{rrr|r} 1 & 2 & 4 & 6 \\ 3 & 6 & 13 & 8 \\ 2 & 2 & 1 & 4 \end{array}\right]$

4.§ $\left[\begin{array}{rrrr|r} 1 & -2 & 3 & -1 & 5 \\ -2 & \frac{3}{2} & \frac{2}{3} & -2 & 4 \\ 0 & 1 & -1 & 4 & 6 \\ 4 & 2 & 1 & -\frac{1}{2} & 3 \end{array}\right]$

5. $\left[\begin{array}{rrrr|r} 0 & 1 & 2 & -1 & 5 \\ 0 & 0 & 0 & 2 & 6 \\ 0 & 3 & 2 & 3 & 1 \\ 1 & 2 & 1 & -2 & 6 \end{array}\right]$

6.§ $\left[\begin{array}{rrrrr|r} 1.5 & 2.3 & 4 & 5.5 & 7.2 & 3.3 \\ 2.4 & 3.65 & 2.2 & -3.65 & 2.8 & 10.5 \\ 0.6 & 2.8 & -3.4 & 3.4 & -7.2 & 2.8 \\ -1.5 & 3.2 & 4.2 & 5.65 & -2.3 & -2 \\ 1.02 & 3.01 & -4.21 & 6.05 & 9.2 & -4.3 \end{array}\right]$

Example 2: Gambling with the Proceeds

We use Gauss-Jordan elimination to solve the following problem. Having won the \$500,000 lottery Fred wants to invest his hard-earned money in precious metals, stocks, bonds, and a bank account. He estimates that he will earn 10% on his investment in precious metals, 14% on his stock investments, 8% on the money placed in bonds, and 6% on the money left in his bank account. He wants to earn \$50,000 on his total investment. To be safe, however, he decides to put one-half as much in the bank as in the other three investments. He also will invest in bonds three-fifths the amount that he does in stocks and precious metals together. How much money should he assign to each of the alternatives?

To set up this problem let x_1 be the amount invested in precious metals, x_2 the amount invested in stocks, x_3 the amount invested in bonds, and x_4 the amount put in the bank. We have the following equations:

$$\begin{aligned} x_1 + x_2 + x_3 + x_4 &= 500,000 \\ .10x_1 + .14x_2 + .08x_3 + .06x_4 &= 50,000 \\ x_4 &= \tfrac{1}{2}(x_1 + x_2 + x_3) \\ x_3 &= \tfrac{3}{5}(x_1 + x_2) \end{aligned}$$

Rewriting, we get the system of equations

$$\begin{array}{rcrcrcrcr} x_1 & + & x_2 & + & x_3 & + & x_4 & = & 500,000 \\ .10x_1 & + & .14x_2 & + & .08x_3 & + & .06x_4 & = & 50,000 \\ .5x_1 & + & .5x_2 & + & .5x_3 & - & x_4 & = & 0 \\ .6x_1 & + & .6x_2 & - & x_3 & & & = & 0 \end{array}$$

Applying Gauss-Jordan elimination to the augmented matrix

$$\begin{bmatrix} 1 & 1 & 1 & 1 & 500,000 \\ .10 & .14 & .08 & .06 & 50,000 \\ .5 & .5 & .5 & -1 & 0 \\ .6 & .6 & -1 & 0 & 0 \end{bmatrix}$$

we obtain the equivalent matrix

$$\begin{bmatrix} 1 & 0 & 0 & 0 & 0 \\ 0 & 1 & 0 & 0 & 222,220 \\ 0 & 0 & 1 & 0 & 111,110 \\ 0 & 0 & 0 & 1 & 166,670 \end{bmatrix}$$

from which it follows that $x_1 = 0, x_2 = 222,220, x_3 = 111,110, x_4 = 166,670$. If you run this example on CALIPSO, you will find the answer for x_1 to be $x_1 =$ 8.7311E-11. This is an example of the way computers display numbers in scientific notation. The "E" indicates you are to multiply 8.7311 by 10 to the power that follows E. That is, 8.7311E-11 = $8.7311 \times 10^{-11} = .000000000087311$. This number should be 0; it is slightly in error due to round-off in intermediate calculations. You will often encounter tiny numbers in place of zeros when working with computers.

There is often a subtle difficulty in modeling a problem like this one with systems of equations. The real-world problem implies that the amounts invested will be nonnegative, but nothing in the model (the system of equations) implies that. As it happens, the solution to our problem has all nonnegative values. See Problem 10 in Exercises 2.3B for a different outcome.

We will learn how to model problems with nonnegative variables when we take up linear programming in Chapter 4.

2.3B Exercises: Applied Problems

Model each of these problems with a system of equations. Use Gauss-Jordan elimination to solve the system by hand or by using the *By Command* operation of CALIPSO. You need to use decimal data type for most of the matrices.

1. Suppose that each ounce of Food A contains 2 units of fat and 3 units of protein, and that each ounce of Food B comtains 3 units of fat and 5 units of protein. If a meal is to consist of 41 units of fat and 66 units of protein, how many ounces of each food should you eat?

2. § The sum of two numbers is 35. The difference between 3 times the larger number and 2 times the smaller number is 40. Find the numbers.

3. A law firm wants to invest a $160,000 trust fund in municipal bonds paying 9% interest, certificates of deposit paying 7.5%, and mortgages paying 12%. The trust requires that 3 times more be invested in bonds and certificates of deposit than in mortgages. The firm wants to realize $15,000 annually from the investments. How much should it invest in each of the options?

4. § A small shipping firm has three fleets of vans: Fleet A, Fleet B, and Fleet C at its disposal for shipping three sizes of cargo: letters, packages, and crates. The following table gives the carrying capacity of a van in each fleet.

	Letter	Package	Crate
Fleet A van	800	60	20
Fleet B van	1000	30	10
Fleet C van	400	100	40

   If the firm must carry 14,000 letters, 1,060 packages, and 380 crates, how many vans from each fleet should it use?

5. Prince Cruise Lines (PCL) can carry 1000 passengers on its liner. PCL charges a fare of $300 per man, $100 per woman, and $500 per child.

Food charges are \$120 per man, \$240 per woman, and \$300 per child. If PCL takes in \$230,000 for fares and \$192,000 for food, how many men, women, and children went sailing?

6. § Adeline, the cook, wants to mix a cheap granola with two more expensive varieties to offer a nutritious breakfast to her dormitory residents. Each pound of Chintz contains 4 units of protein, 4 units of carbohydrate, and 6 units of fat. Each pound of Scrumch contains 12 units of protein, 4 units of carbohydrate, and 8 units of fat. Each pound of Tallow contains 6 units of protein, 2 units of carbohydrate, and 24 units of fat. Adeline feels that healthful granola should contain 6 units of protein, 4 units of carbohydrate, and 12 units of fat. In what proportions should she mix the three types of granola to realize that healthful mixture? (Hint: Let x_1, x_2, x_3 be the respective amounts of Chintz, Scrumch, Tallow required to obtain a bowlful containing exactly the required units of each of the three nutrients. You calculate the proportion of Chintz to use to be $\frac{x_1}{x_1 + x_2 + x_3}$.)

7. At the end of the first year of his investment program, Fred, of whom we learned in Example 2, has netted a capital of \$542,500. Because he is somewhat disappointed with his profits, Fred decides to alter his investment strategy. He still plans to invest in precious metals, stocks, and bonds; he estimates that his return on these will be the same as he estimated last year. However, he is dissatisfied with the 6% return on his bank account and decides to move this account to a savings and loan, where he will get a 7% return and all the safety of a bank. This year he wants to invest all his capital and earn \$58,000. Instead of the restrictions he imposed last year, he will now invest the same amount in stocks as in precious metals and bonds together. Also, he will only put 1/4 as much in the bank as in the other 3 investments. How much money should he assign to each alternative?

8. § A family plans to develop 15 acres near their home into a truck garden in order to sell the vegetables at the local farmers market. They plan to grow tomatoes, beans, peas, corn, and carrots. The table gives their estimates of the various resources the project will require; the last row gives the amount of each resource they can expend.

Crop	Labor (h/acre)	Water (gal/acre)	Fertilizer (lb/acre)	Pesticides (lb/acre)
Tomatoes	90	350	12	6.2
Beans	18	300	8	4.5
Peas	24	250	10	3.8
Corn	30	220	12	5.6
Carrots	20	150	9	3.5
Resources	478	3480	151	70

What acreage should they sow in each crop?

9. Bill and Jim engage in a race over a 30-mile course in which they ride a horse for the first segment, a llama for the second segment, and in a donkey cart for the third segment. Bill's horse runs at 12 miles per hour (mph), his llama runs at 6 mph, and his donkey walks at 3 mph. Jim's horse runs at 10 mph, his llama runs at 5 mph, and his donkey trots at 6 mph. If Bill completes the course in $6\frac{1}{2}$ hours and Jim completes it in 5 hours, how long was each segment of the course? (Hint: If the first segment is x miles long, Bill's horse covers it in $\frac{1}{12}x$ hours.)

10. § Solve Example 2 with the assumption that the amount invested in bonds be $\frac{4}{5}$ the amount put in stocks and precious metals together (rather than $\frac{3}{5}$ as in the example). Explain why the solution you find is a sensible solution to the system of equations but not to the real-world problem.

2.4 ECHELON AND REDUCED ECHELON FORMS

In Section 2.5 we will use echelon and reduced echelon forms of matrices to find all solutions of a system of linear equations (or determine that no solutions exist).

Definition. If all the entries in a row of a matrix are 0, we call the row a *zero row*. The *leading entry* in a nonzero row of a matrix is the first nonzero entry in the row. A matrix is in *echelon form* **A** if it satisfies conditions 1 and 2:

1. For any two rows that have nonzero entries, the leading entry in the lower row occurs farther to the right than does the leading entry in the higher row.

2. All zero rows are found grouped at the bottom of the matrix.

The matrix is in *echelon form* **B** if, in addition to conditions 1 and 2, it satisfies:

3. The leading entries are all 1's.

We make this distinction between echelon forms because it takes slightly more operations to solve a system when the third condition is imposed. On the other hand, it is sometimes easier to use echelon form **B** when you are working by hand.

Example 1: Echelon Form

These matrices are in echelon form **A**:

$$\begin{bmatrix} -1 & \frac{1}{2} & -\frac{3}{2} & -\frac{1}{2} \\ 0 & 2 & -2 & -4 \\ 0 & 0 & 3 & 1 \end{bmatrix} \quad \begin{bmatrix} 0 & 2 & 6 & 2 & 3 & 4 \\ 0 & 0 & 0 & 1 & 5 & 2 \\ 0 & 0 & 0 & 0 & 4 & 3 \end{bmatrix} \quad \begin{bmatrix} 0 & 1 & 3 & 6 & 4 \\ 0 & 0 & 0 & 2 & 5 \\ 0 & 0 & 0 & 0 & 0 \\ 0 & 0 & 0 & 0 & 0 \end{bmatrix}$$

and these are in echelon form **B**

$$\begin{bmatrix} 1 & 2 & 4 & 0 & 4 \\ 0 & 1 & -3 & 2 & 2 \\ 0 & 0 & 0 & 1 & 0 \\ 0 & 0 & 0 & 0 & 0 \\ 0 & 0 & 0 & 0 & 0 \end{bmatrix} \quad \begin{bmatrix} 1 & 0 & 4 & 5 & 3 \\ 0 & 0 & 1 & 0 & 6 \\ 0 & 0 & 0 & 1 & 2 \\ 0 & 0 & 0 & 0 & 0 \end{bmatrix}$$

Note that the matrices you obtained in Exercise 2.2A were all in echelon form **A**. Generally speaking, Gauss elimination is the process that converts (using elementary row operations) a given matrix to an equivalent matrix in echelon form. Back substitution can then be used to solve the corresponding system of equations.

Definition. A matrix is in *reduced echelon form* if it is in echelon form **B** and each column that contains a leading entry 1 has 0's everywhere else.

Example 2: Reduced Echelon Form

The following matrices are in reduced echelon form.

$$\begin{bmatrix} 1 & 0 & 0 & 2 \\ 0 & 1 & 0 & -4 \\ 0 & 0 & 1 & 1 \end{bmatrix} \quad \begin{bmatrix} 1 & 0 & 0 & 0 & 4 \\ 0 & 0 & 1 & 0 & 2 \\ 0 & 0 & 0 & 1 & 3 \\ 0 & 0 & 0 & 0 & 0 \\ 0 & 0 & 0 & 0 & 0 \end{bmatrix}$$

$$\begin{bmatrix} 1 & 0 & 0 & 0 & 3 \\ 0 & 0 & 1 & 0 & 6 \\ 0 & 0 & 0 & 1 & 2 \\ 0 & 0 & 0 & 0 & 0 \end{bmatrix} \quad \begin{bmatrix} 0 & 1 & 9 & 0 & 0 & 4 \\ 0 & 0 & 0 & 1 & 0 & 2 \\ 0 & 0 & 0 & 0 & 1 & 3 \end{bmatrix}$$

Note that the matrices you obtained in Exercises 2.3A are all in reduced echelon form. Gauss-Jordan elimination is the process that converts (using elementary row operations) a given matrix to an equivalent matrix in reduced echelon form.

2.4A Exercises: Echelon Form

Reduce the following matrices to echelon form **A** or **B**. Note that the matrices in these exercises are also used in Exercises 2.4C. If you use CALIPSO, you may want to enter all of the matrices below and save them in a worksheet *before* you do the exercises. You save a worksheet by selecting *Save Worksheet* from the menu.

1. $\begin{bmatrix} 2 & 1 & 1 & 3 & 4 \\ -1 & 2 & 4 & 1 & 2 \\ 3 & 4 & 6 & 7 & 10 \end{bmatrix}$ 2.§ $\begin{bmatrix} 1 & 6 & 0 \\ 0 & 0 & -1 \\ -3 & 4 & -3 \\ 4 & 2 & 3 \end{bmatrix}$

3. $\begin{bmatrix} -1 & 2 & -3 & 4 & 3 \\ 2 & 1 & -4 & 6 & 12 \\ -3 & 4 & -1 & 4 & 8 \\ 4 & 12 & 4 & -4 & 2 \\ 6 & 2 & -1 & -3 & 6 \\ -2 & 3 & 9 & -2 & 4 \end{bmatrix}$ 4.§ $\begin{bmatrix} -2 & 1 & 3 \\ 4 & 0 & 2 \\ 6 & -2 & 0 \\ 1 & 0 & 2 \\ 1 & 0 & 5 \end{bmatrix}$

5. $\begin{bmatrix} -3 & 2 & -3 & 4 & 3 & 2 & -3 \\ -12 & 1 & -4 & 8 & 12 & 10 & -4 \\ -4 & 4 & -2 & 5 & 9 & 2 & 6 \end{bmatrix}$ 6.§ $\begin{bmatrix} 0 & 6 & 2 & 1 \\ 1 & 4 & 3 & 2 \\ -1 & 2 & -1 & -1 \\ 3 & 18 & 11 & 7 \end{bmatrix}$

7. $\begin{bmatrix} -1 & 2 & -3 & 4 & 3 \\ 2 & 1 & -4 & 6 & 12 \\ -3 & 4 & -1 & 5 & 9 \\ 1 & 3 & -7 & 10 & 15 \\ -5 & -2 & 13 & -15 & -21 \\ 0 & 0 & 0 & 6 & -2 \end{bmatrix}$ 8.§ $\begin{bmatrix} 2 & 1 & -3 \\ 3 & 1 & 0 \\ -2 & 1 & 3 \\ 4 & -1 & 3 \end{bmatrix}$

2.4B Exercises: Reduced Echelon Form

Reduce the following matrices to reduced echelon form. Note that the matrices in these exercises are also used in Exercises 2.4C. If you use CALIPSO, you may want to enter all of the matrices below and save them in a worksheet before you do the exercises.

1. $\begin{bmatrix} 2 & 4 & -1 & 3 \\ -1 & 4 & -13 & 3 \\ 1 & 0 & 4 & 2 \end{bmatrix}$ 2.§ $\begin{bmatrix} 0 & 2 & 1 \\ 1 & 4 & 2 \\ 5 & 0 & 6 \\ 2 & -2 & 2 \end{bmatrix}$

3. $\begin{bmatrix} 0 & 2 & -3 & 4 & 3 \\ 2 & 2 & -5 & 6 & 15 \\ -4 & 4 & -1 & 4 & -9 \\ 1 & 4 & -3 & 12 & 15 \\ 0 & -2 & 14 & -15 & -21 \\ 0 & 0 & 0 & 6 & -2 \end{bmatrix}$ 4.§ $\begin{bmatrix} 3 & 2 & 1 \\ -2 & 4 & 6 \\ 1 & 2 & 5 \\ 0 & 2 & 0 \\ -1 & 3 & 4 \end{bmatrix}$

5. $$\begin{bmatrix} 2 & 1 & -3 & 3 & 3 & 1 & -3 \\ -1 & 1 & 2 & 8 & 9 & 8 & -4 \\ -3 & 2 & -1 & 3 & 9 & 2 & 5 \end{bmatrix}$$

6.§ $$\begin{bmatrix} 0 & 2 & -3 & 4 \\ -3 & 1 & -1 & 3 \\ 3 & 1 & -2 & 1 \\ 3 & 5 & -8 & 9 \end{bmatrix}$$

7. $$\begin{bmatrix} -1.5 & 2.8 & -3.6 & 4 & 3 \\ 2.3 & 1 & -4.8 & 6.6 & 12.2 \\ -3.4 & 4.5 & -2.5 & 3.6 & 9.2 \\ 4.3 & 12.9 & 6.9 & -3.9 & 2.5 \\ 6.4 & 2.5 & -1 & -3.3 & 6.2 \\ -2.34 & 3.56 & 9.82 & -2.45 & 4.69 \end{bmatrix}$$

8.§ $$\begin{bmatrix} 0 & 1 & 1 \\ 2 & 3 & 6 \\ 2 & 1 & 5 \\ -1 & 3 & 0 \\ 1 & 2 & 4 \end{bmatrix}$$

Automatic Reduction to Echelon Forms

CALIPSO provides you the opportunity to convert automatically a matrix to either echelon and/or reduced echelon form. If you move the cursor in the menu to *Echelon Form A* and press [*Enter*], the matrix on the screen is reduced to the echelon form in which leading entries are not necessarily 1's. The reduced matrix appears under the next available matrix name. To reduce a matrix to the echelon form with 1's for leading entries, use *Echelon Form B*. To obtain reduced echelon form, choose *Reduced Echelon*.

2.4C CALIPSO Exercises: Automatic Row Reduction

Use *Echelon Form A or B* and *Reduced Echelon* to check your answers in Exercises 2.4A and 2.4B.

2.5 SYSTEMS WITH NONUNIQUE SOLUTIONS

You can use Gauss-Jordan elimination (and reduced echelon forms) to discover if a system of linear equations has no solutions. You can also use this procedure to determine if a system has an infinite number of solutions.

If in applying the Gauss-Jordan method to the augmented matrix corresponding to a system of equations a row occurs of the form

$$\begin{bmatrix} 0 & 0 & \cdots & 0 & * \end{bmatrix}$$

where $*$ represents a nonzero number, then the system of equations has no solutions (substituting the 0's in for the variables results in the equation $0 = *$, which is impossible; therefore, no solutions can exist).

Example 1: No Solutions

Applying *Reduced Echelon* of *MATRIX* to the system

$$\begin{array}{rcrcrcrcr} 2x_1 & - & 3x_2 & + & 4x_3 & + & 5x_4 & = & 4 \\ x_1 & + & 2x_2 & - & 2x_3 & + & x_4 & = & 3 \\ -x_1 & - & 9x_2 & + & 10x_3 & + & 2x_4 & = & -14 \end{array}$$

results in the matrix

$$\left[\begin{array}{cccc|c} 1 & 0 & \frac{2}{7} & \frac{13}{7} & 0 \\ 0 & 1 & -\frac{8}{7} & -\frac{3}{7} & 0 \\ 0 & 0 & 0 & 0 & 1 \end{array}\right]$$

The bottom row shows that the system of equations has no solutions.

Determining solutions of systems having an infinite number of solutions is a bit more complicated. To illustrate the basic ideas we consider the system

$$\begin{array}{rcrcrcr} 2x & - & 3y & - & 13z & = & 14 \\ 3x & - & 8y & - & 30z & = & 28 \\ x & + & 2y & + & 4z & = & 0 \end{array}$$

Applying *Reduced Echelon* we obtain the matrix

$$\left[\begin{array}{ccc|c} 1 & 0 & -2 & 4 \\ 0 & 1 & 3 & -2 \\ 0 & 0 & 0 & 0 \end{array}\right]$$

For matrices in reduced echelon form, variables corresponding to a leading entry 1 (x and y in the above matrix) are called *basic variables.* The other variables (z in this example) are *nonbasic* or *free variables.* If a system of linear equations leads to a matrix in reduced echelon form where every leading entry corresponds to a basic variable, then the system has a unique solution. If, however, there are also nonbasic variables, then there will be an infinite number of solutions (unless, as we saw, there are no solutions). To find all of the solutions, you can assign any values you want to the nonbasic variables. Thus, for instance, in the above example we can let $z = r$, where r is any number (r is called a *parameter*). Then from the second row of the matrix we have that

$$y + 3r = -2, \text{ or, equivalently,}$$
$$y = -3r - 2$$

Similarly, the first row of this matrix tells us that

$$x - 2r = 4, \text{ or, equivalently,}$$
$$x = 2r + 4$$

Thus, all solutions have the form

$$z = r, y = -3r - 2, x = 2r + 4$$

where the parameter r can be any number. You can check that if, for example, $r = \frac{1}{2}$, then $z = \frac{1}{2}, y = -\frac{7}{2}, x = 5$ is a solution.

Example 2: Infinitely Many Solutions

Reduced Echelon shows that the reduced echelon form of the matrix associated with the system of linear equations

$$\begin{array}{rcrcrcrcrcr} 2x_1 & - & 3x_2 & + & 4x_3 & + & x_4 & - & x_5 & = & 2 \\ 3x_1 & - & 2x_2 & + & 6x_3 & & & + & x_5 & = & 4 \\ x_1 & + & x_2 & + & 2x_3 & - & x_4 & + & 2x_5 & = & 2 \\ 8x_1 & - & 7x_2 & + & 16x_3 & + & x_4 & + & x_5 & = & 10 \end{array}$$

is

$$\left[\begin{array}{ccccc|c} 1 & 0 & 2 & -\frac{2}{5} & 1 & \frac{8}{5} \\ 0 & 1 & 0 & -\frac{3}{5} & 1 & \frac{2}{5} \\ 0 & 0 & 0 & 0 & 0 & 0 \\ 0 & 0 & 0 & 0 & 0 & 0 \end{array}\right]$$

From this matrix we see that the basic variables are x_1 and x_2; the non-basic or free variables are x_3, x_4, x_5. This means that the variables x_3, x_4, x_5

can take on any values. Let $x_3 = r, x_4 = s, x_5 = t$. Then from the second row we have

$$x_2 - \frac{3}{5}s + t = \frac{2}{5}, \text{ or, equivalently, } x_2 = \frac{2}{5} + \frac{3}{5}s - t$$

and from the first row

$$x_1 + 2r - \frac{2}{5}s + t = \frac{8}{5}, \text{ or, equivalently, } x_1 = \frac{8}{5} - 2r + \frac{2}{5}s - t$$

Therefore any solution of the original system of equations can be written in the form

$$\begin{aligned} x_1 &= \tfrac{8}{5} - 2r + \tfrac{2}{5}s - t \\ x_2 &= \tfrac{2}{5} + \tfrac{3}{5}s - t \\ x_3 &= r \\ x_4 &= s \\ x_5 &= t \end{aligned}$$

where the parameters r, s, and t can be any numbers whatsoever.

2.5A Hand Exercises: Systems With No Solution or Infinitely Many Solutions

Show that the systems of equations in Problems 1 and 2 have no solutions by reducing them by hand to reduced echelon form.

1. §

$$\begin{array}{rcrcrcr} x_1 & - & x_2 & + & 4x_3 & = & 6 \\ -2x_1 & & & + & 2x_3 & = & 4 \\ 2x_1 & - & x_2 & + & 3x_3 & = & 2 \end{array}$$

2.

$$\begin{array}{rcrcrcrcr} 2x_1 & - & 2x_2 & + & 4x_3 & + & 6x_4 & = & 8 \\ 2x_1 & - & 2x_2 & + & 6x_3 & & & = & 4 \\ 2x_1 & - & 2x_2 & + & 5x_3 & + & 3x_4 & = & 10 \end{array}$$

Determine all solutions to Problems 3–7 by reducing them by hand to reduced echelon form.

3. §

$$\begin{array}{rcrcrcr} x_1 & + & 2x_2 & + & 4x_3 & = & 5 \\ -2x_1 & - & x_2 & + & x_3 & = & 3 \end{array}$$

4.

$$\begin{array}{rcrcrcrcr} 2x_1 & - & 2x_2 & + & 4x_3 & + & 6x_4 & = & 8 \\ 2x_1 & - & 2x_2 & + & 6x_3 & - & 3x_4 & = & 4 \\ 2x_1 & - & 2x_2 & + & 5x_3 & + & 3x_4 & = & 10 \end{array}$$

5.

$$\begin{array}{rcrcrcrcr} x_1 & - & 3x_2 & + & 2x_3 & + & x_4 & = & 4 \\ 2x_1 & - & x_2 & + & 2x_3 & + & x_4 & = & 4 \end{array}$$

6. §

$$\begin{array}{rcrcrcr} x_1 & + & x_2 & - & 3x_3 & = & 3 \\ 5x_1 & + & x_2 & - & 4x_3 & = & 18 \\ 3x_1 & - & x_2 & + & 2x_3 & = & 12 \end{array}$$

7.

$$\begin{array}{rcrcrcr} 4x_1 & + & 2x_2 & + & x_3 & = & 4 \\ 6x_1 & & & + & 2x_3 & = & 14 \\ x_1 & - & x_2 & + & x_3 & = & 5 \\ -3x_1 & - & 3x_2 & & & = & 1 \end{array}$$

2.5B CALIPSO Exercises: Finding All Solutions of a System

Use the *Reduced Echelon* option to determine all solutions of the following problems. Introduce parameters and write out the solutions as we did in Example 2.

1. §

$$\begin{array}{rcrcrcrcrcr} -2x_1 & + & 3x_2 & + & 3x_3 & + & 2x_4 & + & x_5 & = & 4 \\ 3x_1 & - & 4x_2 & + & 2x_3 & + & 3x_4 & + & x_5 & = & 6 \\ x_1 & + & 5x_2 & + & 3x_3 & - & 2x_4 & & & = & 3 \\ 4x_1 & + & 3x_2 & + & 6x_3 & - & x_4 & - & 2x_5 & = & 12 \end{array}$$

2. §

$$\begin{array}{rcrcrcrcrcr} -2x_1 & + & 3x_2 & + & 3x_3 & + & 2x_4 & + & x_5 & = & 5 \\ 3x_1 & - & 4x_2 & + & 2x_3 & + & 3x_4 & + & x_5 & = & 3 \\ x_1 & - & x_2 & + & 5x_3 & + & 5x_4 & + & 2x_5 & = & 8 \\ x_1 & + & 4x_2 & - & 6x_3 & + & 2x_4 & - & 3x_5 & = & 4 \end{array}$$

3.

$$\begin{array}{rcrcrcrcrcrcr}
2.5x_1 & + & 2.3x_2 & + & 3x_3 & + & 2.4x_4 & + & x_5 & - & 3.4x_6 & = & 4.2 \\
3.6x_1 & - & 4.3x_2 & + & 2.4x_3 & + & 3.6x_4 & + & 1.2x_5 & - & 2.2x_6 & = & 6.5 \\
x_1 & + & 5.1x_2 & + & 3.4x_3 & - & 2.6x_4 & & & + & x_6 & = & 3.8 \\
-4.2x_1 & + & 3.5x_2 & + & 6.9x_3 & - & x_4 & - & 2.3x_5 & & & = & 12.6 \\
 & + & 2.3x_2 & & & + & 2.4x_4 & + & x_5 & - & 6.4x_6 & = & 4.2
\end{array}$$

4. Verify the solution for the system of linear equations given at the beginning of this chapter.

How Computers Report Solutions

Computers are not readily adaptable to giving the solution of a system in terms of parameters. We describe a way of presenting the solution which is easier to implement on a computer and which also has significant advantages over the parametric representation. We begin by rewriting the solution given in Example 2 as a matrix equation.

$$\begin{bmatrix} x_1 \\ x_2 \\ x_3 \\ x_4 \\ x_5 \end{bmatrix} = \begin{bmatrix} \frac{8}{5} - 2r + \frac{2}{5}s - t \\ \frac{2}{5} + \frac{3}{5}s - t \\ r \\ s \\ t \end{bmatrix} = t \begin{bmatrix} -1 \\ -1 \\ 0 \\ 0 \\ 1 \end{bmatrix} + s \begin{bmatrix} \frac{2}{5} \\ \frac{3}{5} \\ 0 \\ 1 \\ 0 \end{bmatrix} + r \begin{bmatrix} -2 \\ 0 \\ 1 \\ 0 \\ 0 \end{bmatrix} + \begin{bmatrix} \frac{8}{5} \\ \frac{2}{5} \\ 0 \\ 0 \\ 0 \end{bmatrix} \quad (1)$$

Since the parameters r, s, t are arbitrary numbers, the solution is completely specified by the four column matrices. Note that the fourth column is not to be multiplied by a constant. It is interesting to examine the solution more closely to see why one column matrix is treated differently than the others. To this end, introduce the matrices

$$A = \begin{bmatrix} 2 & -3 & 4 & 1 & -1 \\ 3 & -2 & 6 & 0 & 1 \\ 1 & 1 & 2 & -1 & 2 \\ 8 & -7 & 16 & 1 & 1 \end{bmatrix} \quad \text{and } B = \begin{bmatrix} 2 \\ 4 \\ 2 \\ 10 \end{bmatrix}$$

and write the problem in Example 2 as $AX = B$. You should check that the matrix $\begin{bmatrix} \frac{8}{5} \\ \frac{2}{5} \\ 0 \\ 0 \\ 0 \end{bmatrix}$ is a solution to $AX = B$ but that the three matrices

$$\begin{bmatrix} -1 \\ -1 \\ 0 \\ 0 \\ 1 \end{bmatrix} \quad \begin{bmatrix} \frac{2}{5} \\ \frac{3}{5} \\ 0 \\ 1 \\ 0 \end{bmatrix} \quad \begin{bmatrix} -2 \\ 0 \\ 1 \\ 0 \\ 0 \end{bmatrix}$$

are not solutions to $AX = B$; instead they are solutions to $AX = \mathbf{0}$ where $\mathbf{0}$ is a column matrix of 0's. The system of linear equations $AX = \mathbf{0}$ is called the *homogeneous* system corresponding to $AX = B$.

In general, for any system of equations $AX = B$, the most general solution can be written as the sum of *any* particular solution to $AX = B$ plus a sum of multiples of the set of *basic* solutions to $AX = \mathbf{0}$. The *basic* solutions to $AX = \mathbf{0}$ are those we obtain from the reduced echelon form matrix as in Example 2. For that example, the general solution is

$$t\begin{bmatrix} -1 \\ -1 \\ 0 \\ 0 \\ 1 \end{bmatrix} + s\begin{bmatrix} \frac{2}{5} \\ \frac{3}{5} \\ 0 \\ 1 \\ 0 \end{bmatrix} + r\begin{bmatrix} -2 \\ 0 \\ 1 \\ 0 \\ 0 \end{bmatrix} + \begin{bmatrix} \frac{8}{5} \\ \frac{2}{5} \\ 0 \\ 0 \\ 0 \end{bmatrix}$$

Since $x_1 = 0, x_2 = 1, x_3 = 1, x_4 = 1, x_5 = 0$ is also a solution of our example problem (as you can check), we can also write the general solution as

$$c_1\begin{bmatrix} -1 \\ -1 \\ 0 \\ 0 \\ 1 \end{bmatrix} + c_2\begin{bmatrix} \frac{2}{5} \\ \frac{3}{5} \\ 0 \\ 1 \\ 0 \end{bmatrix} + c_3\begin{bmatrix} -2 \\ 0 \\ 1 \\ 0 \\ 0 \end{bmatrix} + \begin{bmatrix} 0 \\ 1 \\ 1 \\ 1 \\ 0 \end{bmatrix}$$

The values of r, s, t and c_1, c_2, c_3 needed to express a given solution will be different; however, both representations work equally well.

2.5C Exercises: Representing Solutions of $AX = B$

1. Suppose

$$c_1 \begin{bmatrix} 2 \\ 3 \\ \frac{1}{3} \\ 2 \\ 0 \end{bmatrix} + c_2 \begin{bmatrix} \frac{2}{3} \\ -\frac{1}{3} \\ -1 \\ 1 \\ 0 \end{bmatrix} + c_3 \begin{bmatrix} 4 \\ 3 \\ -2 \\ 1 \\ 2 \end{bmatrix} + \begin{bmatrix} 0 \\ -2 \\ -1 \\ 2 \\ 1 \end{bmatrix}$$

is the general solution of a system $AX = B$. Give the values of $x_1, \ldots, x_5$ when $c_1 = 2, c_2 = -1, c_3 = 4$.

2. § Suppose

$$c_1 \begin{bmatrix} 0 \\ 0 \\ \frac{1}{2} \\ 0 \\ 0 \end{bmatrix} + c_2 \begin{bmatrix} \frac{2}{5} \\ -\frac{2}{5} \\ 0 \\ 1 \\ 0 \end{bmatrix} + c_3 \begin{bmatrix} -1 \\ 1 \\ 2 \\ -1 \\ 1 \end{bmatrix} + \begin{bmatrix} 2 \\ -1 \\ 1 \\ 2 \\ 0 \end{bmatrix}$$

is the general solution of a system $AX = B$. Give the values of $x_1, \ldots, x_5$ when $c_1 = 2, c_2 = -1, c_3 = 4$.

3. Suppose

$$c_1 \begin{bmatrix} 1 \\ -2 \\ \frac{1}{4} \\ 2 \\ 5 \end{bmatrix} + c_2 \begin{bmatrix} \frac{1}{3} \\ -\frac{2}{3} \\ 1 \\ 1 \\ 0 \end{bmatrix} + c_3 \begin{bmatrix} 0 \\ -2 \\ 2 \\ 1 \\ 4 \end{bmatrix} + c_4 \begin{bmatrix} -2 \\ 0 \\ 2 \\ -1 \\ 3 \end{bmatrix} + \begin{bmatrix} 1 \\ 0 \\ 1 \\ -3 \\ -1 \end{bmatrix}$$

is the general solution of a system $AX = B$. Give the values of $x_1, \ldots, x_5$ when $c_1 = 2, c_2 = -1, c_3 = 4, c_4 = 2$.

4. § Suppose

$$c_1 \begin{bmatrix} 2 \\ -1 \\ \frac{2}{3} \\ -1 \\ 0 \end{bmatrix} + c_2 \begin{bmatrix} 2 \\ \frac{1}{4} \\ -\frac{1}{4} \\ 1 \\ 0 \end{bmatrix} + c_3 \begin{bmatrix} 1 \\ -2 \\ 3 \\ -1 \\ 1 \end{bmatrix} + c_4 \begin{bmatrix} -2 \\ 1 \\ -1 \\ 1 \\ 1 \end{bmatrix} + \begin{bmatrix} 1 \\ 2 \\ -1 \\ 3 \\ 4 \end{bmatrix}$$

is the general solution of a system $AX = B$. Give the values of $x_1, \ldots, x_5$ when $c_1 = 2, c_2 = -1, c_3 = 4, c_4 = -1$.

5. Suppose $x_1 = 5 + 2r - 6s, x_2 = -3 - 4r + s, x_3 = 2 + 3r$ is the general solution of $AX = B$. Write this solution as a particular solution to $AX = B$ plus a sum of multiples of two basic solutions to $AX = \mathbf{0}$.

6. § Suppose $x_1 = 3 - 2r - 3s + 4t, x_2 = -4r + s, x_3 = 2 + 3r - t$ is the general solution of $AX = B$. Write this solution as a particular solution to $AX = B$ plus a sum of multiples of three basic solutions to $AX = \mathbf{0}$.

7. Suppose $x_1 = -5 + 6r - 2s - 4t, x_2 = 3 + 4r - 2s, x_3 = 9 + 3r + t$ is the general solution of $AX = B$. Write this solution as a particular solution to $AX = B$ plus a sum of multiples of three basic solutions to $AX = \mathbf{0}$.

8. § Suppose $x_1 = -1 + 7r - 4s + t + 2u, x_2 = 3r - 4s + 6u, x_3 = -2 - 2r + t - u, x_4 = 2r + 3s + 4u$ is the general solution of $AX = B$. Write this solution as a particular solution to $AX = B$ plus a sum of multiples of four basic solutions to $AX = \mathbf{0}$.

9. Suppose $x_1 = 5 - r - 4t + 2u, x_2 = 3 - 2r - s, x_3 = 9 + 3r + t + 2u, x_4 = -2 + 3r - 2s + u$ is the general solution of $AX = B$. Write this solution as a particular solution to $AX = B$ plus a sum of multiples of four basic solutions to $AX = \mathbf{0}$.

Using CALIPSO To Solve Systems Automatically

One of the options of *Operations* allows you to solve a system $AX = B$ by writing the formula $X = B/A$. B must be a column matrix with the same number of rows as A. The solution is given as a matrix whose last column is a particular solution of $AX = B$ and whose other columns are a basic set of solutions to $AX = \mathbf{0}$. For instance, the solution to Example 2 given in (1) as

$$t\begin{bmatrix} -1 \\ -1 \\ 0 \\ 0 \\ 1 \end{bmatrix} + s\begin{bmatrix} \frac{2}{5} \\ \frac{3}{5} \\ 0 \\ 1 \\ 0 \end{bmatrix} + r\begin{bmatrix} -2 \\ 0 \\ 1 \\ 0 \\ 0 \end{bmatrix} + \begin{bmatrix} \frac{8}{5} \\ \frac{2}{5} \\ 0 \\ 0 \\ 0 \end{bmatrix}$$

is reported by CALIPSO as

$$X = \begin{bmatrix} -1 & \frac{2}{5} & -2 & \frac{8}{5} \\ -1 & \frac{3}{5} & 0 & \frac{2}{5} \\ 0 & 0 & 1 & 0 \\ 0 & 1 & 0 & 0 \\ 1 & 0 & 0 & 0 \end{bmatrix}$$

2.5D CALIPSO Exercises: Automatic Solution

Use the "Division" command $X = B/A$ of *Operations* to solve the problems in Exercises 2.5B.

---

3 Matrix Inverses and LU Factors

3.1 MATRIX INVERSES

Definition. An *identity matrix* is a square matrix of the form

$$I = \begin{bmatrix} 1 & 0 & 0 & \cdots & 0 \\ 0 & 1 & 0 & \cdots & 0 \\ 0 & 0 & 1 & \cdots & 0 \\ \vdots & & & & \vdots \\ 0 & 0 & 0 & \cdots & 1 \end{bmatrix}$$

An identity matrix is any square matrix that has 1's along the diagonal and 0's elsewhere. Identity matrices play essentially the same role in multiplying matrices that the number 1 plays in multiplying real numbers; that is, if I is an $n \times n$ identity matrix and A is a $p \times n$ matrix, then $AI = A$ (and if A is an $n \times q$ matrix, then $IA = A$).

Definition. If A and B are $n \times n$ matrices and $AB = I$ and $BA = I$, then we say A is the *inverse* of B and write $B = A^{-1}$. In this case we can also say that B is the inverse of A and write $A = B^{-1}$.

Note: It is shown in more advanced books that $AB = I$ implies $BA = I$. Thus, if A and B are square matrices, to conclude that B is the inverse of A you only need to check either that $AB = I$ or that $BA = I$.

Example 1: An Inverse

If

$$A = \begin{bmatrix} -2 & 2 & -1 \\ 3 & -5 & 4 \\ 5 & -6 & 4 \end{bmatrix}$$

then

$$B = \begin{bmatrix} 4 & -2 & 3 \\ 8 & -3 & 5 \\ 7 & -2 & 4 \end{bmatrix}$$

is the inverse of A since (as you can check) $AB = I$ and $BA = I$.

Not all square matrices have inverses (consider, for instance a square matrix consisting of only 0's). If a square matrix does have an inverse, we say the matrix is *invertible.*

Before looking at the usefulness of matrix inverses we show how you can find the inverse of an invertible matrix. We apply this procedure to a 4×4 matrix; the procedure is the same for any sized square matrix. Letting $*$ stand for any entry in a 4×4 matrix, we augment it with a 4×4 identity matrix to create the 4×8 matrix

$$\left[\begin{array}{cccc|cccc} * & * & * & * & 1 & 0 & 0 & 0 \\ * & * & * & * & 0 & 1 & 0 & 0 \\ * & * & * & * & 0 & 0 & 1 & 0 \\ * & * & * & * & 0 & 0 & 0 & 1 \end{array}\right]$$

Original Identity

Apply the Gauss-Jordan procedure to this matrix to convert the original matrix to the identity matrix. As you do this the identity matrix is transformed to some other matrix (whose entries we also denote with $*$); this new matrix turns out to be the inverse of the original matrix. Here is a diagram of the result

$$\left[\begin{array}{cccc|cccc} 1 & 0 & 0 & 0 & * & * & * & * \\ 0 & 1 & 0 & 0 & * & * & * & * \\ 0 & 0 & 1 & 0 & * & * & * & * \\ 0 & 0 & 0 & 1 & * & * & * & * \end{array}\right]$$

Identity Inverse

Example 2: Finding an Inverse

To find the inverse of

$$A = \begin{bmatrix} 5 & -2 & -2 \\ -1 & 1 & 0 \\ -1 & 0 & 1 \end{bmatrix}$$

form

$$\left[\begin{array}{rrr|rrr} 5 & -2 & -2 & 1 & 0 & 0 \\ -1 & 1 & 0 & 0 & 1 & 0 \\ -1 & 0 & 1 & 0 & 0 & 1 \end{array}\right]$$

Apply the Gauss-Jordan elimination method to obtain

$$\left[\begin{array}{rrr|rrr} 1 & 0 & 0 & 1 & 2 & 2 \\ 0 & 1 & 0 & 1 & 3 & 2 \\ 0 & 0 & 1 & 1 & 2 & 3 \end{array}\right]$$

It follows that

$$A^{-1} = \begin{bmatrix} 1 & 2 & 2 \\ 1 & 3 & 2 \\ 1 & 2 & 3 \end{bmatrix}$$

Sometimes it is impossible to reduce the original matrix to an identity matrix as depicted above because you encounter a row of zeros. This happens if the original matrix is not invertible.

Example 3: A Noninvertible Matrix

To find the inverse of

$$A = \begin{bmatrix} 1 & 2 \\ -2 & -4 \end{bmatrix}$$

form

$$\left[\begin{array}{rr|rr} 1 & 2 & 1 & 0 \\ -2 & -4 & 0 & 1 \end{array}\right]$$

To begin reducing this matrix to echelon form add 2 times the first row to the second row to get

$$\left[\begin{array}{rr|rr} 1 & 2 & 1 & 0 \\ 0 & 0 & 2 & 1 \end{array}\right]$$

Since both entries in the first two columns of the second row are 0, you cannot obtain an identity matrix in the first two columns. Therefore, this matrix has no inverse.

3.1A Exercises: Finding Inverses

Augment the matrices in Problems 1–8 with an identity matrix and find the inverse by performing row operations. Do Problems 1–5 by hand and do problems 6–8 with the *By Command* function of CALIPSO.

1.§ $\begin{bmatrix} 2 & -4 \\ 4 & 6 \end{bmatrix}$

2. $\begin{bmatrix} 1 & 4 \\ 0 & 2 \end{bmatrix}$

3.§ $\begin{bmatrix} 0 & 4 \\ -2 & 2 \end{bmatrix}$

4. $\begin{bmatrix} 2 & -2 & 3 \\ -2 & 3 & -6 \\ 3 & 2 & -1 \end{bmatrix}$

5.§ $\begin{bmatrix} 0 & 0 & 1 \\ 1 & 2 & 0 \\ -2 & 1 & 1 \end{bmatrix}$

6. $\begin{bmatrix} -1 & -4 & 2 \\ 1 & 3 & -1 \\ 0 & 1 & 0 \end{bmatrix}$

7.§ $\begin{bmatrix} -1.24 & 2.86 & 1.2 & -2.4 \\ 1.6 & 2.4 & -24.24 & 12.25 \\ 2.56 & -2.56 & 12.3 & 2.8 \\ 1.2 & 3.4 & -1.2 & 0 \end{bmatrix}$

8. $\begin{bmatrix} -\frac{1}{2} & 0 & 1 & \frac{1}{4} \\ 6 & -1 & 0 & -\frac{5}{2} \\ \frac{3}{2} & 0 & 1 & -\frac{1}{4} \\ -9 & 1 & 0 & 4 \end{bmatrix}$

Use *Operations* to find the inverse of the following matrices. As indicated in the instructions there you enter your matrix, say A, and write $X = A\hat{}$, where X is any unused matrix name. Check your answer by multiplying each matrix by the computed inverse.

9. $\begin{bmatrix} 0 & -1.24 & 2.86 & 1.2 \\ -2.4 & 2.2 & -1.4 & 1.6 \\ 2.4 & -24.24 & 12.25 & 0 \\ -1 & 2.2 & -1.2 & 2.56 \end{bmatrix}$

10. $\begin{bmatrix} 1 & 2 & -2 & 2 & -2 \\ -1 & 0 & 3 & 4 & -2 \\ -2 & 1 & 0 & 4 & -1 \\ 0 & -2 & 1 & 6 & -4 \\ 0 & 1 & -2 & 3 & 5 \end{bmatrix}$

11. Enter this array as fractions in decimal mode.

$$\begin{bmatrix} \frac{1}{3} & -\frac{1}{2} & \frac{2}{3} & \frac{1}{2} & \frac{2}{3} & -\frac{1}{4} & 3 & \frac{1}{3} \\ 1 & 2 & -2 & 4 & -3 & 0 & -\frac{1}{2} & -\frac{1}{3} \\ -\frac{1}{3} & \frac{1}{4} & -\frac{2}{3} & \frac{1}{4} & \frac{4}{5} & \frac{2}{3} & \frac{1}{2} & \frac{1}{4} \\ \frac{1}{4} & -\frac{1}{2} & \frac{2}{5} & \frac{1}{3} & -\frac{1}{5} & \frac{1}{2} & \frac{2}{3} & \frac{1}{2} \\ \frac{2}{3} & -\frac{1}{2} & \frac{2}{3} & \frac{1}{4} & -\frac{1}{4} & \frac{2}{3} & -\frac{1}{4} & -\frac{1}{2} \\ 0 & 1 & 3 & -4 & 2 & -1 & 1 & 2 \\ 1 & 2 & -1 & 3 & 4 & 0 & \frac{1}{3} & -\frac{1}{4} \\ -\frac{1}{4} & -\frac{1}{2} & -\frac{2}{5} & \frac{1}{3} & -\frac{1}{5} & \frac{1}{2} & -\frac{1}{2} & \frac{2}{3} \end{bmatrix}$$

Show that the following matrices do not have an inverse by the method used in Example 3.

12. $\begin{bmatrix} 2 & 4 & 6 \\ 4 & 5 & 6 \\ 14 & 16 & 18 \end{bmatrix}$

13. $\begin{bmatrix} 1 & 2 & -1 & 3 \\ 2 & 1 & 1 & 2 \\ -1 & 5 & 1 & 2 \\ 2 & 4 & -2 & 6 \end{bmatrix}$

Encoding Messages

Matrix inversion can be used to encode and decode sensitive messages for transmission. We assign a number to each letter of the alphabet and the punctuation marks. To make messages easier to read, we code a space (a blank character) as 0. The simplest correspondence is to let each letter of the alphabet be associated with its position in the alphabet; that is,

0	5 E	10 J	15 O	20 T	25 Y
1 A	6 F	11 K	16 P	21 U	26 Z
2 B	7 G	12 L	17 Q	22 V	27 ,
3 C	8 H	13 M	18 R	23 W	28 .
4 D	9 I	14 N	19 S	24 X	29 ;

For instance "I AM HIDING" is encoded as [9 0 1 13 0 8 9 4 9 14 7]. Now this simple-minded scheme would not fool anyone for long. We will use a matrix and its inverse to scramble the messages.

We group the characters of the message in strings of 4 letters (as you will see, we could use any number of letters). Now make up a 4×4 matrix.

This matrix will be the cipher used to encode the message. Let us use the cipher

$$C = \begin{bmatrix} 1 & 0 & 2 & 5 \\ 5 & 12 & 3 & 9 \\ 6 & 8 & 4 & 10 \\ 8 & 4 & 3 & 2 \end{bmatrix} \tag{1}$$

To send our message we extend it with spaces until its length is a multiple of 4, say to "I AM HIDING" and encode the message by making a matrix whose columns are the numerical equivalents of the letters. For our message, let

$$M = \begin{bmatrix} 9 & 0 & 9 \\ 0 & 8 & 14 \\ 1 & 9 & 7 \\ 13 & 4 & 0 \end{bmatrix}$$

We encode the message in matrix S computed as

$$S = CM = \begin{bmatrix} 1 & 0 & 2 & 5 \\ 5 & 12 & 3 & 9 \\ 6 & 8 & 4 & 10 \\ 8 & 4 & 3 & 2 \end{bmatrix} \begin{bmatrix} 9 & 0 & 9 \\ 0 & 8 & 14 \\ 1 & 9 & 7 \\ 13 & 4 & 0 \end{bmatrix} = \begin{bmatrix} 76 & 38 & 23 \\ 165 & 159 & 234 \\ 188 & 140 & 194 \\ 101 & 67 & 149 \end{bmatrix}$$

Since we want to be as deceptive as possible, we hide the fact that we are using groups of 4 in our coding scheme by sending the message as a string of numbers formed from the columns of S. We send

$$\begin{bmatrix} 76 & 165 & 188 & 101 & 38 & 159 & 140 & 67 & 23 & 234 & 194 & 149 \end{bmatrix}$$

The receiver needs to know

$$C^{-1} = \begin{bmatrix} -1.074 & -0.78571 & 1.2857 & -0.21429 \\ 0.28571 & 0.39286 & -0.51786 & 0.10714 \\ 3 & 2 & -3.5 & 1 \\ -0.78571 & -0.64286 & 1.1429 & -0.35714 \end{bmatrix}$$

because then he or she can decode the message by calculating

$$M = C^{-1}S = C^{-1} \begin{bmatrix} 76 & 38 & 23 \\ 165 & 159 & 234 \\ 188 & 140 & 194 \\ 101 & 67 & 149 \end{bmatrix} = \begin{bmatrix} 9 & 0 & 9 \\ 0 & 8 & 14 \\ 1 & 9 & 7 \\ 13 & 4 & 0 \end{bmatrix}$$

3.1B Exercises: Hand Exercises on Secret Messages

1. § Encode the message "She is no spy." with the cipher

$$C = \begin{bmatrix} 1 & 4 \\ 4 & 6 \end{bmatrix}$$

2. Decode the message in Problem 1.

3. Encode the mesaage "Math is fun." with the cipher

$$C = \begin{bmatrix} -1 & 3 \\ 2 & -1 \end{bmatrix}$$

4. Decode the message in Problem 3.

3.1C Exercises: CALIPSO Exercises on Secret Messages

Use decimal mode to enter these messages in CALIPSO since most of the calculations will overflow fraction mode.

1. § This secret message was coded with the cipher C given in (1) of the example above. Decode it.
[79 261 254 144 85 277 290 206 42 314 248 122 98 218 256 186
48 324 260 118 136 302 348 212 14 406 308 224]

2. Encode the message "The poison is behind the sofa." with the cipher

$$C = \begin{bmatrix} \frac{1}{3} & -\frac{1}{3} & 6 \\ 8 & \frac{1}{4} & -\frac{3}{5} \\ \frac{2}{5} & \frac{5}{6} & -\frac{2}{3} \end{bmatrix}$$

3. Decode the message in Problem 2.

4. § Encode the message "Sell all holdings in IBM." with the cipher

$$C = \begin{bmatrix} -5 & 2 & 6 & 7 \\ 4 & 2 & -3 & -2 \\ 7 & -1 & -3 & -8 \\ 1 & -1 & 0 & 1 \end{bmatrix}$$

5. Decode the message in Problem 4.

6. Extend our 29 symbol code for letters and punctuation marks to a 39 symbol code that includes the digits 0–9. Encode the message "Buy 1,524 shares of Cantun Holidays" with the cipher

$$\begin{bmatrix} 1 & 2 & -2 & 2 & -2 \\ -1 & 0 & 3 & 4 & -2 \\ -2 & 1 & 0 & 4 & -1 \\ 0 & -2 & 1 & 6 & -4 \\ 0 & 1 & -2 & 3 & 5 \end{bmatrix}$$

7. Decode the message in Problem 6.

Solving Systems of Equations with an Inverse

Suppose A is an invertible matrix and $AX = B$ is a system of linear equations. To find X we multiply both sides of $AX = B$ by A^{-1} to get

$$A^{-1}(AX) = A^{-1}B$$

Since $A^{-1}(AX) = (A^{-1}A)X = IX = X$, we have found the solution

$$X = A^{-1}B$$

to the system of equations.

Example 4: Using the Inverse Matrix to Solve a System

We use the matrix equation $X = A^{-1}B$ to solve the system of linear equations

$$\begin{array}{rcrcrcrcr} 2x_1 & - & x_2 & + & \frac{1}{2}x_3 & + & x_4 & = & 14 \\ 3x_1 & + & 2x_2 & - & 4x_3 & + & 2x_4 & = & -10 \\ x_1 & & & + & 2x_3 & - & x_4 & = & 14 \\ & & 3x_2 & + & 3x_3 & + & 4x_4 & = & 23 \end{array}$$

The coefficient matrix, variable column, and right-and side column for this system are

$$A = \begin{bmatrix} 2 & -1 & \frac{1}{2} & 1 \\ 3 & 2 & -4 & 2 \\ 1 & 0 & 2 & -1 \\ 0 & 3 & 3 & 4 \end{bmatrix} \quad X = \begin{bmatrix} x_1 \\ x_2 \\ x_3 \\ x_4 \end{bmatrix} \quad B = \begin{bmatrix} 14 \\ -10 \\ 14 \\ 23 \end{bmatrix}$$

Using *Reduced Echelon*, or, more directly, the inverse operation of CALIPSO you can verify that the inverse of A is

$$A^{-1} = \begin{bmatrix} .1588 & .1336 & .2816 & -.0361 \\ -.3971 & .1661 & .2960 & .09025 \\ .05054 & -.09386 & .1805 & .07942 \\ .2599 & -.05415 & -.3574 & .1227 \end{bmatrix}$$

Therefore, the solution of the system of linear equations is given by

$$X = A^{-1}B = \begin{bmatrix} .1588 & .1336 & .2816 & -.0361 \\ -.3971 & .1661 & .2960 & .09025 \\ .05054 & -.09386 & .1805 & .07942 \\ .2599 & -.05415 & -.3574 & .1227 \end{bmatrix} \begin{bmatrix} 14 \\ -10 \\ 14 \\ 23 \end{bmatrix} = \begin{bmatrix} 4 \\ -1 \\ 6 \\ 2 \end{bmatrix}$$

In general, using Gauss-Jordan elimination to solve systems of linear equations is preferred to using matrix inverses, since it involves fewer calculations. However, it may be the case that you need to solve $AX = B$ many times with the same A matrix and different B matrices. In this situation, it is advantageous to compute A^{-1} once and use it repeatedly.

Example 5: Weekly Scheduling of Refineries

An oil company has four refineries, each of which produces gasoline, kerosene, diesel, and heating oil. In thousand of gallons per hour, the production capacities of refineries A, B, C, D are given by the table

	Gasoline	Kerosene	Diesel	Heating
A	5.2	7.3	3.8	4.2
B	2.4	3.2	4.1	2.2
C	3.2	3.3	2.5	3.6
D	1.8	0.9	2.1	1.5

The company receives weekly orders for a certain quantity of each product and must determine how many hours each refinery must run to meet this demand. Since the weekly orders vary but the production capacity does not, it is advantageous to make a one-time calculation of the inverse of the production matrix and use the inverse to calculate each week's schedule. We use CALIPSO to calculate the inverse of

$$P = \begin{bmatrix} 5.2 & 7.3 & 3.8 & 4.2 \\ 2.4 & 3.2 & 4.1 & 2.2 \\ 3.2 & 3.3 & 2.5 & 3.6 \\ 1.8 & 0.9 & 2.1 & 1.5 \end{bmatrix}$$

to be

$$P^{-1} = \begin{bmatrix} 0.45207 & -0.7438 & -0.66889 & 1.4305 \\ 0.13338 & 0.2083 & -0.02558 & -0.61758 \\ -0.13947 & 0.42146 & -0.11024 & 0.036949 \\ -0.42725 & 0.17754 & 0.97235 & -0.73106 \end{bmatrix}$$

Suppose this week's demand for gasoline, kerosene, diesel, and heating oil is 615, 355, 368, and 179 thousand gallons, respectively. We make a *demand* matrix R of these values by

$$R = \begin{bmatrix} 615 \\ 355 \\ 368 \\ 179 \end{bmatrix}$$

Then the schedule of hours of operation, X, is the solution to the problem $PX = R$, or, equivalently, to $X = P^{-1}R$. The schedule for this week is

$$P^{-1}R = \begin{bmatrix} 0.45207 & -0.7438 & -0.66889 & 1.4305 \\ 0.13338 & 0.2083 & -0.02558 & -0.61758 \\ -0.13947 & 0.42146 & -0.11024 & 0.036949 \\ -0.42725 & 0.17754 & 0.97235 & -0.73106 \end{bmatrix} \begin{bmatrix} 615 \\ 355 \\ 368 \\ 179 \end{bmatrix} = \begin{bmatrix} 23.9 \\ 36.0 \\ 29.9 \\ 27.2 \end{bmatrix}$$

Thus, for this demand they should operate plant A for 23.9 hours, B for 36 hours, C for 29.9 hours, and D for 27.2 hours.

3.1D Exercises: Hand Exercises Using the Inverse Matrix to Solve Systems

1. Solve using an inverse matrix

$$\begin{aligned} 2x - 4y &= 6 \\ x + 3y &= -1 \end{aligned}$$

2. § Solve the system in Problem 1 with the right-hand side replaced by 4, 0.

3. Solve using an inverse matrix

$$\begin{array}{rcrcl} -x & + & 3y & = & 8 \\ 2x & - & y & = & 1 \end{array}$$

4. § Solve the system in Problem 3 with the right-hand side replaced by 6, 2.

3.1E Exercises: CALIPSO Exercises Using the Inverse Matrix to Solve Systems

1. Solve using an inverse matrix

$$\begin{array}{rcrcrcr} 3x & + & 2y & - & 3z & = & 2 \\ x & + & 3y & + & 2z & = & 10 \\ -2x & + & y & + & 9z & = & 9 \end{array}$$

2. § Solve the system in Problem 1 with the right-hand side replaced by 5, 2, 4.

3. Solve using an inverse matrix.

$$\begin{array}{rcrcrcrcr} x_1 & - & 2x_2 & + & -\frac{1}{2}x_3 & - & 3x_4 & = & 8 \\ 3x_1 & - & 2x_2 & + & 2x_3 & + & 3x_4 & = & -1 \\ -x_1 & & & + & x_3 & - & 3x_4 & = & 4 \\ & & 2x_2 & - & 4x_3 & - & x_4 & = & 2 \end{array}$$

4. § Solve the system in Problem 3 with the right-hand side replaced by 3, −3, 1, 0.

5. Solve using an inverse matrix.

$$\begin{array}{rcrcrcrcr} x_1 & - & 2.2x_2 & + & 0.5x_3 & - & 3x_4 & = & 8.12 \\ 3.6x_1 & - & 8.2x_2 & - & 3.1x_3 & + & 3.2x_4 & = & -1.6 \\ -x_1 & + & 2.6x_2 & - & 1.2x_3 & - & 2.3x_4 & = & 4 \\ 2.2x_1 & + & 2.2x_2 & - & 3.4x_3 & - & 1.25x_4 & = & 2.6 \end{array}$$

6. § Solve the system in Problem 5 with the right-hand side replaced by $3.2, -5.3, 1.6, 0$.

7. A coffee manufacturer offers 4 coffee blends (A-D). Each blend is a mix of Columbian, Brazilian, Ecuadoran, and Costa Rican coffees. The table

	Columbian	Brazilian	Ecuadoran	Costa Rican
A	4	2.5	1.5	2
B	2	3	3	2
C	1	4	1	4
D	5	2	1.5	1.5

gives the mix found in each 10-pound can of each blend. Suppose the daily demand for 10-pound cans of blends A, B, C, D is given by a column matrix R. Carefully consider how to form a matrix from the data and use the inverse of this matrix to find the number of pounds of each kind of coffee the manufacturer needs to meet the demands when

$$R = \begin{bmatrix} 131 \\ 138 \\ 152 \\ 125 \end{bmatrix} \text{ and when } R = \begin{bmatrix} 1345 \\ 1380 \\ 1331 \\ 1345 \end{bmatrix}$$

8. § A town of 75,000 people has experienced a flu epidemic for the past month. If s_1 and w_1 represent the number of people sick and well in a given week, it is found that the number of people sick (s_2) and well (w_2) in the following week is given by

$$\begin{aligned} \tfrac{1}{4}s_1 + \tfrac{1}{3}w_1 &= s_2 \\ \tfrac{3}{4}s_1 + \tfrac{2}{3}w_1 &= w_2 \end{aligned}$$

(a) Write this system in matrix form and calculate the inverse of the coefficient matrix.

(b) If 24,167 people are sick this week, how many were sick last week?

CALIPSO

In Section 2.5, you learned to solve $AX = B$ for X by selecting *Operations* and entering the formula $X = B/A$. If matrix A is invertible, CALIPSO will give the unique solution $X = A^{-1}B$ as a column matrix X.

Example 6 uses this operation in a powerful way to solve an instance of an important class of economic problems.

Example 6: Leontieff Input-Output Analysis

Input-output analysis begins by dividing an enterprise into two (often overlapping) groups — the input group and the output group. Commonly the input group is a set of consumers who use products produced by the output group (the producers). For a simple instance, we consider a community of farmers and ironworkers. The farmers are producers of food and also consumers of some of the food they produce and of iron for their tools. Thus, the farmers are both an input and output group. The ironworkers are also both an input and output group since they produce iron and consume food and some of the iron to make their own tools. We can model this activity handily with matrices.

Suppose for every \$1.00 worth of food they produce, the farmers consume \$0.16 dollars worth of food and use \$0.21 worth of the food to buy tools. And suppose that for each \$1.00 worth of iron they produce, the ironworkers use \$0.24 worth of the iron to buy food and consume \$0.12 worth of the iron for their tools. We record these data in a table called the *input-output table.*

	Food	Iron
Food	.16	.24
Iron	.21	.12

The matrix of numbers with the row and column labels deleted is called the *input-output matrix* and is often denoted by the letter Q. Here

$$Q = \begin{bmatrix} .16 & .24 \\ .21 & .12 \end{bmatrix}$$

Suppose the community of farmers and ironworkers can market \$12,000 worth of food and \$30,000 worth of iron (by *market* we mean the amount they can sell beyond what they consume). Suppose they produce x_1 dollars worth of food and x_2 dollars worth of iron. Then the equations that balance production and consumption are of the form

$$\begin{aligned} \text{Production} = \; & (\text{Market Demand}) + (\text{Product for Farmers}) \\ & +(\text{Product for Ironworkers}) \end{aligned}$$

For our example the equations are

$$\begin{aligned} x_1 &= 12{,}000 + .16x_1 + .24x_2 \\ x_2 &= 30{,}000 + .21x_1 + .12x_2 \end{aligned}$$

We can use matrices to write these equations by introducing two column matrices

$$X = \begin{bmatrix} x_1 \\ x_2 \end{bmatrix} \quad \text{and} \quad D = \begin{bmatrix} 12{,}000 \\ 30{,}000 \end{bmatrix}$$

Then the equations are equivalent to the single matrix equation

$$X = D + QX$$

To solve this equation for X, we first write it as $X - QX = D$ and solve for X by letting I be an identity matrix and calculating

$$\begin{aligned} &X - QX = D \\ &IX - QX = D \\ &(I - Q)X = D \\ &X = (I - Q)^{-1}D \end{aligned}$$

where we obtained the last step by multiplying both sides on the left by $(I - Q)^{-1}$. We can use the *Operations* of CALIPSO to calculate X by creating a 2×2 identity matrix I, the matrices Q and D, and entering the formula $X = (I - Q)\hat{}\,D$ [alternatively, you can use $X = D/(I - Q)$]. When we do so, we find that

$$X = \begin{bmatrix} x_1 \\ x_2 \end{bmatrix} = \begin{bmatrix} 25{,}784 \\ 40{,}244 \end{bmatrix}$$

3.1F Exercises: Hand Exercises on Input–Output Problems

1. Let

$$Q = \begin{bmatrix} .2 & .2 \\ .1 & .4 \end{bmatrix}, \quad X = \begin{bmatrix} x_1 \\ x_2 \end{bmatrix}, \quad D = \begin{bmatrix} 50 \\ 100 \end{bmatrix}$$

Solve the matrix equation $(I - Q)X = D$.

2. § Let

$$Q = \begin{bmatrix} \frac{1}{4} & \frac{1}{2} \\ \frac{1}{2} & 0 \end{bmatrix}, \quad X = \begin{bmatrix} x_1 \\ x_2 \end{bmatrix}, \quad D = \begin{bmatrix} 40 \\ 20 \end{bmatrix}$$

Solve the matrix equation $(I - Q)X = D$.

3. A community of two groups produces iron and butter. For every \$1.00 worth of iron they produce, the iron workers consume \$0.20 worth of iron and \$0.10 worth of butter. For every \$1.00 worth of butter they produce, the butter makers consume \$0.25 worth of iron and \$0.15 worth of butter. The community can market \$15,000 worth of iron and \$10,000 worth of butter.

   (a) Form the input–output table.

   (b) Solve the equation $X = D + QX$ for X to find the production schedule that meets the given conditions.

4. § Suppose a community produces lumber and steel. Suppose that it takes 0.3 unit of lumber and 0.2 unit of steel to produce 1 unit of lumber, and that it takes 0.25 unit of lumber and 0.4 unit of steel to produce 1 unit of steel. Find the production schedule that will meet an external demand of 10 units of lumber and 15 units of steel.

5. A community produces three forms of energy: oil, coal, and electricity according to the input–output table

	Oil	Coal	Electricity
Oil	.2	.1	.4
Coal	.1	.3	.3
Electricity	.3	.2	.2

The external demands for oil, coal, and electricity are 150, 100, and 250 units, respectively. Find the required total output for each energy form that meets the given conditions.

3.1G Exercises: CALIPSO Exercises on Input–Output Problems

1. Let

$$Q = \begin{bmatrix} 0.1 & 0 & 0.1 \\ 0.1 & 0.2 & 0 \\ 0.2 & 0.1 & 0.1 \end{bmatrix}, \quad X = \begin{bmatrix} x_1 \\ x_2 \\ x_3 \end{bmatrix}, \quad D = \begin{bmatrix} 100 \\ 150 \\ 200 \end{bmatrix}$$

Solve the matrix equation $(I - Q)X = D$.

2. § A community of five groups consumes and produces five commodities according to the input-output table

	Group 1	Group 2	Group 3	Group 4	Group 5
Group 1	.20	.30	.12	.15	.20
Group 2	.15	.20	.40	.15	.25
Group 3	.20	0	.12	.30	0
Group 4	.18	.14	.09	0	.16
Group 5	0	.12	.13	.15	.22

Suppose the market demands for the commodities are \$10,000, \$12,000, \$8,000, \$24,000, \$15,000, respectively. Calculate the column matrix X giving the production schedule.

3. Suppose in a rather uncomplicated society there are three major industries A, B, and C. In addition to producing goods for the members of this society, each of the industries requires inputs from the others to operate. To produce one unit of output, industry A requires .3 unit of its own output, .2 unit of the output of industry B, and .3 unit of the output of industry C. Similarly, to produce one unit of output, industry B requires .3 unit of its own output, .4 unit of the output of industry A, and .1 unit of the output of industry C. Industry C needs .2 unit of its own output, .3 unit of the output of A, and .25 unit of the output of B to produce one unit of output. Finally suppose that the society needs 50 units of the output of industry A, 75 units of the output of industry B, and 40 units of the output of industry C. Set up an input-output table and determine the total output of each of these industries to meet these conditions.

4. § A commune is divided into eight groups by activity. Each group produces a commodity that can be used by any of the eight groups or marketed for a profit. The groups and the estimated annual dollar demand for their production in excess of the commune's consumption are

   - Farmers who produce vegetables with market value \$8,500.
   - Herders who produce dairy products with market value \$9,000, meat with market value \$10,500, and wool with market value

$6,500. Since this group produces three distinct products that are consumed in different quantities by groups in the commune, we regard the herders as three groups: Dairymen, Cattlemen, and Sheepherders.

- Weavers who produce wool cloth for making clothes. The cloth has an annual market value of $13,500.
- Teachers who staff a school for children of the commune or from a neighboring community. The tuition paid by children from outside the commune is $7,200 per year.
- Foresters who manage woodlands. Since they produce both firewood and lumber that are consumed in different quantities by groups in the commune, we regard foresters as two groups: loggers and millworkers. Loggers produce firewood with a market value of $12,200, and millworkers produce lumber with market value $14,000.

The following table gives the estimated fraction of each dollar's worth of goods produced that is consumed by each group. The columns and rows are headed by the first letter of each group's name. Thus, the column headed W gives the fraction of a dollar's worth of wool cloth consumed by each group (each row).

	F	D	C	S	W	T	L	M
F	.08	.03	0	.05	.15	.04	.02	.05
D	.11	.12	.13	.09	.12	.07	.09	.12
C	.03	.05	.03	.03	.07	0	.05	.07
S	.08	.04	.05	.11	.06	.05	.06	.01
W	.04	.05	0	.06	.10	.12	.09	.10
T	.03	.03	.06	.02	.06	0	.02	.03
L	.10	.12	.09	.11	.06	0	.10	.14
M	.06	.06	.07	.08	.04	.02	.05	.06

Compute the dollar value of each group's production that can be marketed outside the commune.

3.2 LU FACTORS *

Matrices of very large size frequently occur in applications. Some applications in economics, statistics, and engineering require hundreds or even thousands of rows and columns. Calculating the inverse of a large matrix is expensive in computer time and often inaccurate because of accumulated round-off error. Luckily, it is possible to perform almost any calculation involving matrices in ways that are less expensive. The LU factorization of a matrix substitutes for the inverse in solving systems of equations.

The best way to solve a single system of equations $AX = B$ is to use the Gauss or Gauss-Jordan method. As we have said before, it can be advantageous to calculate A^{-1} and use the equation $X = A^{-1}B$ to solve the system when you need to solve many systems with the same A matrix and several B column matrices (which is a very common thing to do in practice). We will show how a factorization of A can be used instead of A^{-1}, and that doing so is more efficient. We begin with a definition and an example.

Definition. A matrix that has only 0's above its diagonal is *lower triangular.* A matrix that has only 0's below its diagonal is *upper triangular.*

Example 1: Lower and Upper Triangular Matrices.

With any numbers replacing the $*$, L is lower triangular and U is upper triangular.

$$L = \begin{bmatrix} * & 0 & 0 & 0 \\ * & * & 0 & 0 \\ * & * & * & 0 \\ * & * & * & * \end{bmatrix} \qquad U = \begin{bmatrix} * & * & * & * \\ 0 & * & * & * \\ 0 & 0 & * & * \\ 0 & 0 & 0 & * \end{bmatrix}$$

Definition. A square matrix A has an *LU factorization* if $A = LU$ where L is lower triangular and U is upper triangular.

Example 2: An LU Factorization

The matrix

$$A = \begin{bmatrix} 2 & 4 & -4 \\ 1 & -4 & 3 \\ -6 & -9 & 10 \end{bmatrix}$$

has an LU factorization since

$$A = \begin{bmatrix} 2 & 4 & -4 \\ 1 & -4 & 3 \\ -6 & -9 & 10 \end{bmatrix} = \begin{bmatrix} 1 & 0 & 0 \\ \frac{1}{2} & 1 & 0 \\ -3 & -\frac{1}{2} & 1 \end{bmatrix} \begin{bmatrix} 2 & 4 & -4 \\ 0 & -6 & 5 \\ 0 & 0 & \frac{1}{2} \end{bmatrix}$$

Next we show how you can use an LU factorization to solve a system of equations. Let L denote the lower triangular and U the upper triangular matrix in the factorization of A in Example 2. We have $A = LU$. Now suppose $B = \begin{bmatrix} 8 \\ 6 \\ 5 \end{bmatrix}$ and suppose we are interested in solving

$$AX = LUX = B \tag{2}$$

Let Y be the solution of $LY = B$ and X be the solution of $UX = Y$. By substitution, we obtain $AX = (LU)X = L(UX) = LY = B$. Thus, we can solve $AX = B$ by solving two systems $LY = B$ and $UX = Y$. However, this would only be a good plan if it were relatively easy to solve the two new problems, and it is! We solve $LY = B$ by a *forward substitution.* (A forward substitution is made like a back substitution, except we work from top to bottom.)

$$LY = \begin{bmatrix} 1 & 0 & 0 \\ \frac{1}{2} & 1 & 0 \\ -3 & -\frac{1}{2} & 1 \end{bmatrix} \begin{bmatrix} y_1 \\ y_2 \\ y_3 \end{bmatrix} = \begin{bmatrix} 8 \\ 6 \\ 5 \end{bmatrix}$$

or, written as equations,

$$\begin{aligned} y_1 \qquad\qquad\qquad\qquad &= 8 \\ \tfrac{1}{2}y_1 + \quad y_2 \qquad\qquad &= 6 \\ -3y_1 - \tfrac{1}{2}y_2 + y_3 &= 5 \end{aligned}$$

The forward substitution yields the solution

$$\begin{aligned} y_1 &= 8 \\ y_2 &= 6 - \tfrac{1}{2}y_1 = 2 \\ y_3 &= 5 + 3y_1 + \tfrac{1}{2}y_2 = 30 \end{aligned}$$

Next we make a back substitution to solve $UX = Y$.

$$UX = \begin{bmatrix} 2 & 4 & -4 \\ 0 & -6 & 5 \\ 0 & 0 & \frac{1}{2} \end{bmatrix} \begin{bmatrix} x_1 \\ x_2 \\ x_3 \end{bmatrix} = \begin{bmatrix} 8 \\ 2 \\ 30 \end{bmatrix}$$

or, written as equations,

$$\begin{aligned} 2x_1 + 4x_2 - 4x_3 &= 8 \\ -6x_2 + 5x_3 &= 2 \\ \tfrac{1}{2}x_3 &= 30 \end{aligned}$$

The back substitution yields

$$\begin{aligned} x_3 &= 60 \\ x_2 &= \frac{2 - 5x_3}{-6} = \frac{149}{3} \\ x_1 &= \frac{8 - 4x_2 + 4x_3}{2} = \frac{74}{3} \end{aligned}$$

which is the solution to the original problem (1).

3.2A Exercises: Using LU Factorizations to Solve Systems of Equations

Solve the systems of equations $AX = B$ using the given LU factorization of A. Do Problems 1 and 2 by hand and the others with CALIPSO.

1. §

$$A = \begin{bmatrix} 2 & -3 \\ 6 & -5 \end{bmatrix} = \begin{bmatrix} 1 & 0 \\ 3 & 1 \end{bmatrix} \begin{bmatrix} 2 & -3 \\ 0 & 4 \end{bmatrix}; \quad B = \begin{bmatrix} 18 \\ 24 \end{bmatrix}$$

2.

$$A = \begin{bmatrix} 2 & 2 \\ -1 & 2 \end{bmatrix} = \begin{bmatrix} 1 & 0 \\ -\frac{1}{2} & 1 \end{bmatrix} \begin{bmatrix} 2 & 2 \\ 0 & 3 \end{bmatrix}; \quad B = \begin{bmatrix} 12 \\ 18 \end{bmatrix}$$

3. §

$$A = \begin{bmatrix} 1 & 2 & 1 \\ 2 & 2 & 3 \\ -1 & -4 & 2 \end{bmatrix} = \begin{bmatrix} 1 & 0 & 0 \\ 2 & 1 & 0 \\ -1 & 1 & 1 \end{bmatrix} \begin{bmatrix} 1 & 2 & 1 \\ 0 & -2 & 1 \\ 0 & 0 & 2 \end{bmatrix}$$

$$B = \begin{bmatrix} 3 \\ -1 \\ 2 \end{bmatrix}$$

4.

$$A = \begin{bmatrix} 6 & -4 & 2 & -2 \\ 6 & 0 & 4 & -1 \\ 9 & -14 & 0 & -3 \\ 3 & 6 & 6 & 5 \end{bmatrix} = \begin{bmatrix} 1 & 0 & 0 & 0 \\ 1 & 1 & 0 & 0 \\ \frac{3}{2} & -2 & 1 & 0 \\ \frac{1}{2} & 2 & 1 & 1 \end{bmatrix} \begin{bmatrix} 6 & -4 & 2 & -2 \\ 0 & 4 & 2 & 1 \\ 0 & 0 & 1 & 2 \\ 0 & 0 & 0 & 2 \end{bmatrix}$$

$$B = \begin{bmatrix} 4 \\ 0 \\ 2 \\ 3 \end{bmatrix}$$

Some matrices do not have LU factorizations; however, as we shall see, we can rearrange the rows of such a matrix in such a way that the rearrangement does have an LU factorization.

How to Form the LU Factorization

It is very easy to form the LU factorization of a square matrix A. We obtain the U matrix by performing row operations on A to reduce it to the echelon form **A** (the one that does not require leading entries to be 1's). As we perform the row operations we record the multipliers used in adding a multiple of one row to another in a way that allows us to form the L matrix. Example 3 shows you how to do it.

Example 3: Finding an LU Decomposition

We obtain the LU decomposition of the matrix A of Example 2 by forming a sequence $A \to A_1 \to A_2 \to A_3 = U$. We form L as we form the sequence starting with a lower triangular matrix L with 1's on its main diagonal and unknown entries below its diagonal. We denote the unknown entries by $*$. Here are the initial matrices.

$$A = \begin{bmatrix} 2 & 4 & -4 \\ 1 & -4 & 3 \\ -6 & -9 & 10 \end{bmatrix} \qquad L = \begin{bmatrix} 1 & 0 & 0 \\ * & 1 & 0 \\ * & * & 1 \end{bmatrix}$$

We begin by adding $-\frac{1}{2}$ times row 1 to row 2. We place the *negative* of this multiplier in the same position in the L matrix that the just created 0 occupies in the A matrix to obtain

$$A_1 = \begin{bmatrix} 2 & 4 & -4 \\ 0 & -6 & 5 \\ -6 & -9 & 10 \end{bmatrix} \qquad L = \begin{bmatrix} 1 & 0 & 0 \\ \frac{1}{2} & 1 & 0 \\ * & * & 1 \end{bmatrix}$$

Next we add 3 times row 1 to row 3 and put -3 in the position of the L matrix corresponding to the 0 we just created.

$$A_2 = \begin{bmatrix} 2 & 4 & -4 \\ 0 & -6 & 5 \\ 0 & 3 & -2 \end{bmatrix} \qquad L = \begin{bmatrix} 1 & 0 & 0 \\ \frac{1}{2} & 1 & 0 \\ -3 & * & 1 \end{bmatrix}$$

Finally, we add $\frac{1}{2}$ times row 2 to row 3 to get a zero below the -6 in A_2. We put $-\frac{1}{2}$ in the corresponding position in L.

$$A_3 = \begin{bmatrix} 2 & 4 & -4 \\ 0 & -6 & 5 \\ 0 & 0 & \frac{1}{2} \end{bmatrix} \qquad L = \begin{bmatrix} 1 & 0 & 0 \\ \frac{1}{2} & 1 & 0 \\ -3 & -\frac{1}{2} & 1 \end{bmatrix}$$

Now let $U = A_3$. We have obtained the LU factorization we saw in Example 2.

In reducing a matrix to echelon form you sometimes encounter a 0 on the diagonal of some A_i. In this case the Gauss elimination procedure attempts to get a nonzero entry on the diagonal by interchanging the current row with a row below it. When this happens the procedure given in Example 3 fails, and, in fact, the matrix does not have an LU decomposition. There is a related factorization called the *PLU factorization* that deals with this circumstance. We will not consider it in this book; however, since CALIPSO can perform the PLU factorization, we give an example of it in Exercises 3.2C, Problems 7–10.

3.2B Exercises: Finding the LU Factorization

In Problems 1–6 find an LU factorization of the given matrix by hand.

1.§ $\begin{bmatrix} 2 & 1 \\ 2 & 4 \end{bmatrix}$ 2. $\begin{bmatrix} 3 & -5 \\ 4 & 1 \end{bmatrix}$

3.§ $\begin{bmatrix} 2 & 3 & -1 \\ -2 & 2 & 4 \\ 1 & -1 & 2 \end{bmatrix}$ 4. $\begin{bmatrix} 3 & 2 & 4 & -1 \\ 1 & -2 & 4 & 4 \\ -2 & 1 & -2 & 3 \\ 1 & -1 & 2 & 3 \end{bmatrix}$

In each step of the sequence below we performed several row operations to get 0's below the leading entry. Determine which multipliers we used and fill in the entries in the L matrix.

5. §

$$\begin{bmatrix} 2 & 8 & 1 & 1 \\ 4 & 13 & 3 & -1 \\ -2 & -5 & -3 & 3 \\ -6 & -18 & -1 & 1 \end{bmatrix} \to \begin{bmatrix} 2 & 8 & 1 & 1 \\ 0 & -3 & 1 & -3 \\ 0 & 3 & -2 & 4 \\ 0 & 6 & 2 & 4 \end{bmatrix} \to$$

$$\begin{bmatrix} 2 & 8 & 1 & 1 \\ 0 & -3 & 1 & -3 \\ 0 & 0 & -1 & 1 \\ 0 & 0 & 4 & -2 \end{bmatrix} \to \begin{bmatrix} 2 & 8 & 1 & 1 \\ 0 & -3 & 1 & -3 \\ 0 & 0 & -1 & 1 \\ 0 & 0 & 0 & 2 \end{bmatrix}$$

6.

$$\begin{bmatrix} 1 & -1 & 0 & 0 & 0 \\ -1 & 3 & -1 & 0 & 0 \\ 0 & -1 & 1 & -1 & 0 \\ 0 & 0 & -1 & 1 & -1 \\ 0 & 0 & 0 & -1 & 1 \end{bmatrix} \to \begin{bmatrix} 1 & -1 & 0 & 0 & 0 \\ 0 & 2 & -1 & 0 & 0 \\ 0 & -1 & 1 & -1 & 0 \\ 0 & 0 & -1 & 1 & -1 \\ 0 & 0 & 0 & -1 & 1 \end{bmatrix} \to$$

$$\begin{bmatrix} 1 & -1 & 0 & 0 & 0 \\ 0 & 2 & -1 & 0 & 0 \\ 0 & 0 & \frac{1}{2} & -1 & 0 \\ 0 & 0 & -1 & 1 & -1 \\ 0 & 0 & 0 & -1 & 1 \end{bmatrix} \rightarrow \begin{bmatrix} 1 & -1 & 0 & 0 & 0 \\ 0 & 2 & -1 & 0 & 0 \\ 0 & 0 & \frac{1}{2} & -1 & 0 \\ 0 & 0 & 0 & -1 & -1 \\ 0 & 0 & 0 & -1 & 1 \end{bmatrix} \rightarrow$$

$$\begin{bmatrix} 1 & -1 & 0 & 0 & 0 \\ 0 & 2 & -1 & 0 & 0 \\ 0 & 0 & \frac{1}{2} & -1 & 0 \\ 0 & 0 & 0 & -1 & -1 \\ 0 & 0 & 0 & 0 & 2 \end{bmatrix}$$

How to Find LU Factorizations with CALIPSO

You can find the LU factorization of any square matrix (or discover that none exists) by entering the matrix and selecting *LU FACTORS* from the menu. L and U will be reported as separate matrices (whose names are unlikely to be L and U).

The classroom demonstration program GAUSS (found on the instructor's solution disk) can demonstrate LU factorization far better than examples in a book. To start the program type *Gauss*, enter a matrix, and select *LU Factors* from the menu.

3.2C Exercises: Using CALIPSO to Find LU Factors

For each of the systems $AX = B$ in Problems 1–4

(a) Use *LU Factors* to write $A = LU$.

(b) Use *Operations* to solve $LY = B$ and then $UX = Y$.

(c) Use *Operations* to solve $AX = B$.

(d) Compare your answers in (b) and (c).

1.

$$A = \begin{bmatrix} 4 & 1 & 3 & 0 \\ -1 & 2 & -2 & 1 \\ 0 & 2 & 3 & 3 \\ 1 & 2 & -1 & 4 \end{bmatrix} \qquad B = \begin{bmatrix} 8 \\ 0 \\ 8 \\ 6 \end{bmatrix}$$

2. §

$$A = \begin{bmatrix} 4.2 & 1.3 & 3.1 & 0 \\ -1.3 & 3.2 & -2.2 & 4.1 \\ 1.2 & 2.4 & 3.1 & 3.2 \\ 1.5 & 3.2 & -5.1 & 4.5 \end{bmatrix} \qquad B = \begin{bmatrix} 18.2 \\ 10.6 \\ 4.9 \\ 15.3 \end{bmatrix}$$

3.

$$A = \begin{bmatrix} 2 & 1 & 2 & 3 \\ 1 & -2 & 4 & -1 \\ 0 & 2 & -4 & 3 \\ 1 & 0 & 0 & 0 \end{bmatrix} \qquad B = \begin{bmatrix} 121 \\ 210 \\ 99 \\ 110 \end{bmatrix}$$

4. §

$$A = \begin{bmatrix} 6.5 & -2.1 & 3.6 & 2.1 \\ -1 & 0 & 2.4 & -3.2 \\ 1 & -3.6 & 4.1 & 2.2 \\ -1.6 & 2.2 & 4.1 & -3.2 \end{bmatrix} \qquad B = \begin{bmatrix} 8.2 \\ -0.6 \\ 3.9 \\ 2.3 \end{bmatrix}$$

5.

$$A = \begin{bmatrix} 1 & -1 & 0 & 2 \\ 5 & 2 & -3 & 1 \\ -1 & 3 & 4 & -3 \\ -1 & 1 & -2 & 1 \end{bmatrix} \qquad B = \begin{bmatrix} 21 \\ 20 \\ 19 \\ 10 \end{bmatrix}$$

6.

$$A = \begin{bmatrix} 2.34 & -1.56 & 2.82 & 3.45 & 9.43 & 12.62 \\ 1.56 & -2.32 & 4.82 & -1.26 & 0 & 3.86 \\ 0 & 2.82 & -4.52 & 3.86 & 2.22 & 0 \\ 1 & 2.69 & 0 & -1.56 & -12.76 & -3.45 \\ 2 & 21.36 & 22.45 & 21.63 & 19.76 & 18.69 \\ 12.24 & -23.65 & 12.45 & 15.62 & 13.85 & 0 \end{bmatrix}$$

$$B = \begin{bmatrix} 121.45 \\ 210.34 \\ 99.63 \\ 110.45 \\ 115.72 \\ 98.69 \end{bmatrix}$$

When it is necessary to interchange rows in finding an LU factorization, CALIPSO forms a third matrix P in the factorization. P is called a *permutation* matrix and its purpose is to rearrange rows in the A matrix to account for the row interchanges that were made. Specifically, CALIPSO factors $A = PLU$ where P is a rearrangement of the rows of an identity matrix. The following exercises show how to use this factorization.

7. If $A = PLU$, explain why you can solve the system $AX = B$ by solving the three systems $PZ = B$, $LY = Z$, $UX = Y$. [Hint: The solution is similar to the one we gave in solving (1). Consider $AX = P(L(UX)) = B$.]

8. § Solve the system

$$AX = PLUX = \begin{bmatrix} 0 & 0 & 1 \\ 1 & 0 & 0 \\ 0 & 1 & 0 \end{bmatrix} \begin{bmatrix} 1 & 0 & 0 \\ -2 & 1 & 0 \\ 3 & 2 & 1 \end{bmatrix} \begin{bmatrix} -2 & 3 & 1 \\ 0 & 4 & -1 \\ 0 & 0 & 2 \end{bmatrix} X = \begin{bmatrix} 2 \\ -1 \\ 3 \end{bmatrix}$$

9. (a) Use *LU Factors* of CALIPSO to find the PLU factorization of

$$A = \begin{bmatrix} -2 & 3 & -1 & 2 \\ 4 & -6 & 2 & 3 \\ 6 & -1 & -2 & 3 \\ -8 & -4 & -3 & -1 \end{bmatrix}$$

(b) Use the PLU factorization to solve $AX = B$ when

$$B = \begin{bmatrix} 27 \\ 32 \\ 42 \\ 22 \end{bmatrix}$$

(c) Check your solution in (c) by solving $AX = B$ directly.

10. (a) Use *LU Factors* of CALIPSO to find the PLU factorization of

$$A = \begin{bmatrix} 0 & -1 & -2 & 1 \\ 3 & 6 & -3 & 2 \\ -1 & 0 & 2 & -4 \\ 6 & 4 & -2 & 3 \end{bmatrix}$$

(b) Use the PLU factorization to solve $AX = B$ when

$$B = \begin{bmatrix} 12 \\ 10 \\ 8 \\ 9 \end{bmatrix}$$

(c) Check your solution in (c) by solving $AX = B$ directly.

---

4 Introductory Linear Programming

4.1 SYSTEMS OF LINEAR INEQUALITIES

Recall that the equation of a *line* can be written in the form $ax + by = c$ where a, b, and c represent constants and x and y represent variables. A *linear inequality* in two variables x and y is an expression of the form $ax + by \leq c$ or of the form $ax + by \geq c$. The *graph* of an inequality is the set of all points (x, y) that satisfy the inequality.

Example 1: Graph of an Inequality

The graph of the inequality $x + 2y \leq 6$ consists of all points on or below the line $x + 2y = 6$.

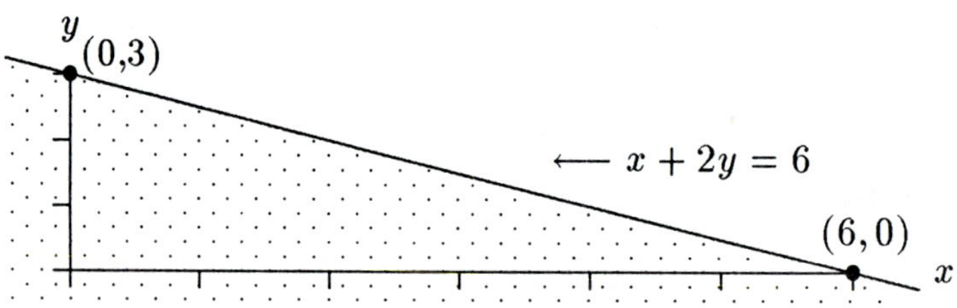

You can tell that the graph lies below the line by substituting the coordinates of a particular point (x, y) in the inequality. If x and y satisfy the inequality, then the graph consists of all points lying on the same side of the line as the point (x, y). If x and y do not satisfy the inequality, then the graph consists of all points on the side of the line not containing (x, y). For instance, you can substitute $(x, y) = (0, 0)$ in $x + 2y \leq 6$ to get $0 + 2 \cdot 0 \leq 6$. Since this inequality is true, all points on the same side of the line as (0,0) satisfy $x + 2y \leq 6$.

Definitions. The *graph* of a system of inequalities is the set of all points (x, y) that simultaneously satisfy all of the inequalities. We often call the inequalities *constraints* and the graph of the constraints the *feasible region.* The *boundary line* of a constraint $ax + by \leq c$ or $ax + by \geq c$ is the line $ax + by = c$. A *corner point* of a feasible region is a point that is in the feasible region and is the intersection of two or more boundary lines. An *edge* of the feasible region is a line segment joining two adjacent corner points.

Example 2: Graph of a System of Inequalities

The feasible region for the constraints

$$\begin{array}{rcrcr} 5x & + & 2y & \leq & 15 \\ 2x & + & 3y & \leq & 12 \\ x & + & y & \geq & 2 \\ x & & & \geq & 0 \\ & & y & \geq & 0 \end{array}$$

is the shaded region

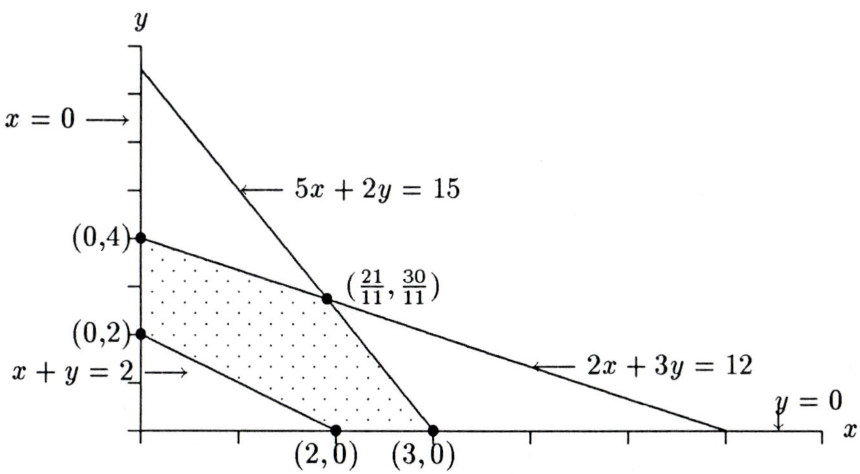

The corner points of the feasible region are $(2, 0), (3, 0), (\frac{21}{11}, \frac{30}{11}), (0, 4)$, and $(0, 2)$. Note that each corner point is the intersection of two lines; you find these points by solving systems of two equations (corresponding to the two lines) in two variables. Note too that the inequalities $x \geq 0$, $y \geq 0$ ensure that the graph will be in the first quadrant. The edges are the segments joining adjacent corner points.

4.1A Exercises: Graphing Feasible Regions

Graph the inequalities or systems of inequalities in the following problems.

1.§ $4x + 2y \geq 3$

2. $2x - y \leq 6$

3.§ $3x - 4y \leq 8$

4. $2x - 4y \leq 0$

5.§ $y \leq 6$

6. $x \geq 9$

7.§ $$\begin{array}{rcrcr} 6x & - & 2y & \leq & 14 \\ 3x & + & y & \leq & 11 \\ \multicolumn{5}{c}{x \geq 0, y \geq 0} \end{array}$$

8. $$\begin{array}{rcrcr} x & + & 2y & \leq & 4 \\ -x & + & y & \geq & -1 \\ \multicolumn{5}{c}{x \geq 0, y \geq 0} \end{array}$$

9.§ $$\begin{array}{rcrcr} \frac{1}{2}x & - & y & \geq & 2 \\ x & + & y & \geq & 7 \\ \multicolumn{5}{c}{x \geq 0, y \geq 0} \end{array}$$

10. $$\begin{array}{rcrcr} x & - & 3y & \leq & 8 \\ 2x & - & y & \leq & 9 \\ -2x & + & 3y & \leq & 3 \\ \multicolumn{5}{c}{x \geq 0, y \geq 0} \end{array}$$

11.§ $$\begin{array}{rcrcr} -2x & + & 5y & \leq & 12 \\ 2x & + & 3y & \geq & 4 \\ 3x & - & 2y & \leq & 2 \\ \multicolumn{5}{c}{x \geq 0, y \geq 0} \end{array}$$

12. $$\begin{array}{rcrcr} -\frac{1}{2}x & - & \frac{1}{3}y & \leq & 3 \\ 2x & + & 3y & \leq & 21 \\ -2x & + & 15y & \geq & 3 \\ \multicolumn{5}{c}{x \geq 0, y \geq 0} \end{array}$$

In Section 4.2 you will find that systems of linear inequalities play an important role in linear programming. In this section you will learn to use CALIPSO to graph systems of linear inequalities in two variables.

The *GRAPH* Program of CALIPSO

You start the *GRAPH* program by moving the cursor in the CALIPSO menu to *GRAPH* and pressing [Enter]. If you want to print your graph, you must run the DOS program GRAPHICS before starting CALIPSO. You will find more information on this subject in the appendix on using CALIPSO.

The next screen is the familiar one asking for a file name. If you are making a new graph, you type a file name that you have not used before. To work with a graph you saved earlier, type the name you saved it under. In the example that follows we suppose that you are making a new graph named POTATO. After you type POTATO and press [Enter] a screen appears that contains 5 boxes positioned approximately as in Figure 1 below. All of the words inside the boxes in Figure 1 do not appear on the screen; we have put them in to describe the purpose of the boxes.

- **Message Area** is where messages appear requesting you to respond. Some of the messages are instructive, e.g., "You can type a constraint" and others are commands, e.g., "Press a key to erase the level lines." You should always check the message box to see what you can do next.

- **Typing Area** is where letters appear as you type them. When you press [Enter] the information you type will be used in one of several ways depending on the message currently in the Message Area. If the message says you can type a constraint, the contents of the Typing Area will be transferred to the box that holds the constraints. If the message says you should respond to a query, your response will be taken from the typing area.

- **Obj. Function** designates the area containing the objective function. We do not use objective functions in this section nor discuss them until Section 4.2. Since the *GRAPH* program expects an objective function, just type x in this entry box.

- **Constraint Area** contains the inequalities (constraints) that you have entered in the Typing Area. To the left of one of the constraints (or possibly to the left of the objective function) you will see the symbol >. This symbol, which we call the *cursor*, serves as a pointer to a constraint or to the objective function. You can move the cursor among constraints or between the constraints and the objective function with the [↑] and [↓] keys. As you will see in the example below, the boundary lines of the feasible set will be highlighted in a color.

- [*F1*] = **Display Menu** appears in the box at the lower right of the screen. This menu tells you which function keys perform operations. Here are the available operations.

  [*F2*] **Change Scale of Viewport.** When you enter a constraint, CALIPSO chooses a scale for the coordinate system that will display the *intercepts* (the points where lines cross the coordinate axes) of the boundary lines. You can change this scale in order to magnify parts of the feasible set. When you press [*F2*], the message in the Message Area asks you to type a number. This number will be used as the largest x and y value that will be displayed in the viewport. For instance, if the graph indicates that the largest x or y value is currently 10 and you change the scale to 5, the feasible set will appear larger and, perhaps, part of it will not appear on the screen. Alternatively, you can shrink the size of the feasible set by increasing the scale.

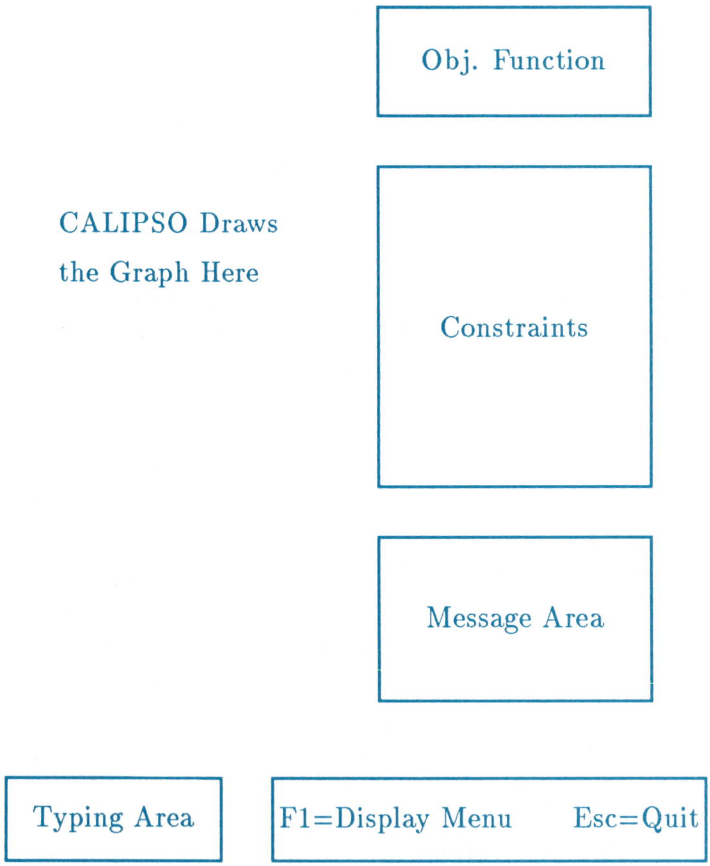

Figure 1

- $\boxed{F3}$ **Zoom on the Feasible Region.** This function changes the scale to make the feasible region appear as large as possible. It only works when the inequalities $x >= 0$ and $y >= 0$ are included as constraints.

[F4] **Press once to select an edge, again to deselect.** You can use this function to find the corner points of the feasible region. As you will see in Section 4.2, these corner points play an important role in linear programming.

When you move the cursor to a constraint, the edge of the feasible region that corresponds to that constraint is highlighted in a color. When you press [F4], the color of the edge is fixed (it does not change back to white when you move the cursor again). We say that the edge has been *selected.* When you move the cursor after selecting an edge, a different color is used to highlight edges. When the selected edge and the edge marked by the cursor intersect, the coordinates of the intersection appear in the message area. You will see how this works in Example 3.

[F5] **Draw level lines of the objective function.** We will not discuss this function until Section 4.2.

[F6] **Remove the edge pointed at by the cursor.** You use this function to delete an inequality from the list of constraints.

[F7] **Print the Screen.** You can print the screen anytime. As we discussed earlier, you can only print the screen if you have a printer with graphics capability and have activated GRAPHICS.COM when you started the *GRAPH* program.

[F8] **Save the problem to use again.** The constraints are saved under the file name you specified when you started the *GRAPH* program.

[F9] **Help.** This key allows you to read an explanation of each function.

Example 3: Graphing with CALIPSO

We walk through the steps of graphing a feasible region with CALIPSO. Note that computer keyboards do not have keys for $\leq$ and $\geq$. We use <= for $\leq$ and >= for $\geq$.

- Select *GRAPH* from the main CALIPSO menu by moving the cursor to it and pressing [Enter]. On the next screen specify the file name POTATO.

- The cursor initially points to the objective function. We do not discuss the objective function until the next section. To make CALIPSO function, enter $x + y$ as an objective function. Since we are only graphing feasible sets now, the objective function will be of no concern.

- Type $5x + 3y <= 15$ and press [Enter]. Notice that the constraint appears in the constraint box and the line $5x + 3y = 15$ is drawn on the graph. The side of the line corresponding to points (x, y) that satisfy the constraint is colored.
- Type $2x + 2y <= 8$ and press [Enter]. Notice that the line $2x + 2y = 8$ is added to the graph. The colored region now indicates the feasible region, which is the set of points (x, y) that simultaneously satisfy the constraints $5x + 3y <= 15$ and $2x + 2y <= 8$.
- Type $x \geq 0$ and press [Enter]. Then type $y \geq 0$ and press [Enter]. The feasible region of all 4 constraints is now colored.
- Press [F3] to zoom in on the feasible region.
- Experiment with moving the cursor up and down the list of constraints. You will see the edge corresponding to the constraint change color.
- Select the constraint $5x + 3y <= 15$ by moving the cursor to it and pressing [F4]. Then move the cursor to the constraint $2x + 2y <= 8$. Notice that the edges corresponding to these constraints appear in different colors. Also notice that the intersection of these edges $(1.5, 2.5)$ appears in the message box. The message also says $z = 4$; z gives the value of the objective function (which we are ignoring until Section 4.2).
- If you move the cursor to another constraint and select it, the constraint you selected previously ($5x + 3y <= 15$) is no longer selected.
- You change a constraint by moving the cursor to the one you want to change and typing a new constraint.

4.1B Exercises: Graphing Feasible Regions

Graph the feasible region of each system of constraints with the *GRAPH* program of CALIPSO. Note that *GRAPH* expects you to type an expression called an "objective function" in its first entry box. We do not use the objective function in this application; however, since you must enter something, type $x + y$ in this box.

1. The constraints of Example 2.

2. $$\begin{aligned} x + 3y &\le 9 \\ 2x + 3y &\le 12 \\ 5x + 2y &\le 15 \\ x &\ge 0 \\ y &\ge 0 \end{aligned}$$

3. $$\begin{aligned} -3x + y &\le 6 \\ 3x + 5y &\ge 15 \\ x &\ge 0 \\ y &\ge 0 \end{aligned}$$

4. $$\begin{aligned} x + y &\le 3 \\ 3x + y &\le 6 \\ x + 2y &\le 5 \\ x &\ge 0 \\ y &\ge 0 \end{aligned}$$

5. $$\begin{aligned} x - y &\le 2 \\ x + y &\ge 3 \\ 2x + 3y &\le 12 \\ 2x - y &\ge 1 \\ x &\ge 0 \\ y &\ge 0 \end{aligned}$$

6. $$\begin{aligned} 2x - y &\ge 2 \\ x + 3y &\ge 5 \\ x + 3y &\le 9 \\ x + y &\le 6 \\ x &\ge 0 \\ y &\ge 0 \end{aligned}$$

4.2 LINEAR PROGRAMMING PROBLEMS WITH TWO VARIABLES

In this section we consider some examples of linear programming (LP) problems and see how to formulate them mathematically. For the moment we restrict our attention to problems involving just two variables. In Section 4.3 you will see how to use CALIPSO to solve these problems.

Generally speaking, LP problems involve finding the maximum or minimum value of a cost function, a profit function, a size function, etc., subject to some sort of constraints. The function to be maximized or minimized is called the *objective function.*

Example 1: Planting Two Crops

The following situation leads to a typical LP problem. A farmer has 640 acres available for planting wheat and corn. He estimates that each acre planted in wheat will yield a profit of \$60 and each acre planted in corn will yield a profit of \$80. Each acre of wheat requires 45 hours of labor and capital costs of \$100. Each acre of corn requires 60 hours of labor and capital costs of \$80. The farmer estimates that he has 36,000 hours of labor and \$60,000 capital at his disposal. Under these conditions how many acres of wheat and corn should the farmer plant to maximize his profits? For clarity, we summarize the data in Table 1.

Table 1

	Labor h/acre	Capital \$/acre	Profit \$/acre
Wheat	45	100	60
Corn	60	80	80
Available	36,000	60,000	

To formulate this problem as an LP problem we first select variables to represent the unknown quantities we want to find. In this example we are trying to determine the amount of wheat and corn that should be planted, so we let x be the number of acres of wheat to be planted and y the number of acres of corn.

Since the profit per acre for wheat is \$60 and for corn is \$80, it follows that the profit obtained by planting x acres of wheat and y acres of corn is given by the function

$$z = 60x + 80y$$

It is this profit function (the objective function) that we want to maximize.

Next consider the various constraints. Since the farmer has 36,000 labor-hours available, and since each acre planted in wheat requires 45 hours of labor and each acre planted in corn requires 60 hours, we have the constraint

$$45x + 60y \leq 36,000$$

Here we use the inequality $\leq$ rather than $=$ since other constraints may prevent the use of all of the available labor hours. Similar reasoning gives the capital constraint

$$100x + 80y \leq 60,000$$

The farmer has 640 acres available for planting and so

$$x + y \leq 640$$

Finally, it is obvious that x and y cannot be negative, so we have

$$x \geq 0 \text{ and } y \geq 0$$

Combining the above information we have an LP problem in the form

$$\begin{aligned} \text{Maximize: } z = \; & 60x + 80y \\ \text{Subject to: } \; & 45x + 60y \leq 36,000 \\ & 100x + 80y \leq 60,000 \\ & x + y \leq 640 \\ & x, y \geq 0 \end{aligned}$$

The feasible region of this LP problem is the set of all points that satisfy simultaneously all of the constraints.

In the next few sections we will see how to find the values of x and y that correspond to points in the feasible region and make z as large as possible. In the meantime you can use CALIPSO to picture this feasible region and find the corner points.

Modeling LP Problems

Formulating word problems as LP problems requires considerable practice. Although each word problem has its own characteristics, the following step-by-step procedure is often useful.

Step 1. Read the problem carefully. If appropriate, organize the data into a table as in Example 1 and Example 2 below.

Step 2. Determine the variables. These represent the unknown quantities whose values you want to find.

Step 3. Formulate the objective function. This is what you want to maximize or minimize.

Step 4. Formulate the constraint inequalities. These are the expressions that limit the amount of resources you can use.

Step 5. Solve the problem. (In the next few sections you are given a variety of methods for doing this.)

Example 2: Manufacturing Furniture

A woodworker produces beds and tables. The manufacture of these products involves three basic operations: cutting, assembly, and finishing.

Table 2 gives the hours required for each of these activities to produce one bed or one table.

Table 2

Item	Cutting time h/unit	Assembly time h/unit	Finishing time h/unit	Profit $/unit
Bed	$\frac{6}{5}$	1	$\frac{3}{2}$	75
Table	1	$\frac{1}{2}$	2	50
Available hrs	72	50	120	

The woodworker profits $75 for each bed and $50 for each table. Since the woodworker's products are in great demand, there are no limits on the number of products she can sell. Suppose a maximum of 72 work-hours per week is available for cutting, 50 work-hours per week for assembly, and 120 work-hours per week for finishing. How many of each product should she manufacture in order to maximize profit?

Let x be the number of beds and y be the number of tables manufactured in a week. Then the profit z is given by the objective function

$$z = 75x + 50y$$

From Table 2 we see that the cutting time in terms of x and y is $\frac{6}{5}x + 1 \cdot y$. Since the cutting time cannot exceed 72, we obtain the constraint

$$\tfrac{6}{5}x + y \le 72$$

In a similar way we obtain the constraints for assembly and finishing as

$$\begin{aligned} x + \tfrac{1}{2}y &\le 50 \\ \tfrac{3}{2}x + 2y &\le 120 \end{aligned}$$

Finally, since it is impossible to manufacture negative units of these products, we include the constraints $x \ge 0$ and $y \ge 0$. Combining this information, we obtain the LP problem

$$\begin{array}{lrcrcl} \text{Maximize: } z = & 75x & + & 50y & & \\ \text{Subject to:} & \tfrac{6}{5}x & + & y & \le & 72 \\ & x & + & \tfrac{1}{2}y & \le & 50 \\ & \tfrac{3}{2}x & + & 2y & \le & 120 \\ & x, y \ge 0 & & & & \end{array}$$

Example 3: Feeding Garfield

Jon has decided to give Garfield a mixture of two cat foods. Cat food A costs \$.60/can and cat food B costs \$1.20/can. Each can of cat food A contains 4 units of protein, 3 units of carbohydrate, and 2 units of fat. Each can of cat food B contains 3 units of protein, 5 units of carbohydrate, and 3 units of fat. Jon, to Garfield's dismay, has determined that Garfield should have at least 40 units of protein and 32 units of carbohydrate each day, but no more than 24 grams of fat per day.

How many ounces of each cat food should Garfield devour each day to meet these requirements but minimize Jon's cost?

As the first step in setting up this problem we make a table of the data.

Cat Food	Protein	Carbohydrate	Fat	Cost per Pound
A	4	3	2	\$0.60
B	3	5	3	\$1.20
Units needed	≥ 40	≥ 32	≤ 24	

The first step in setting up this problem is to identify the variables. Let x be the number of ounces of cat food A and let y be the number of ounces of cat food B to be consumed by Garfield. The objective function

$$z = 0.60x + 1.20y$$

is to be minimized in this example. The nutritional constraints are

$$\begin{aligned} 4x + 3y &\geq 40 && \text{(Protein constraint)} \\ 3x + 5y &\geq 32 && \text{(Carbohydrate constraint)} \\ 2x + 3y &\leq 24 && \text{(Fat constraint)} \\ x, y &\geq 0 \end{aligned}$$

4.2A Exercises: Modeling LP Problems

You can solve these problems by drawing graphs of the feasible region by hand or by using the *GRAPH* program.

1. Graph the feasible region and determine the corner points for the LP problem in Example 1.

2. Graph the feasible region and determine the corner points for the LP problem in Example 2. If you use CALIPSO, note that *GRAPH*, unlike other CALIPSO procedures, does not accept fractions. You must convert the fractions in this example to decimals. (Alternatively, you can multiply the first constraint in Example 2 by 5 and the others by 2 to obtain an equivalent system of constraints that involves no fractions.)

3. Grunwilda, the cook, wants to mix two types of granolas to make a nutritious breakfast for her fraternity residents. Each pound of Greaseball contains 4 units of protein, 4 units of carbohydrate, and 24 units of fat. Each pound of Sugarpop contains 12 units of protein, 4 units of carbohydrate, and 8 units of fat. A 100-pound bag of Greaseball costs \$50 and a 60-pound bag of Sugarpop costs \$48. Grunwilda feels that each student should have for breakfast at least 6 units of protein, 4 units of carbohydrate, and 12 units of fat.

   (a) § Fill in the table with the conditions of this problem.

Granola	Protein	Carbohydrate	Fat	Cost per Pound
Greaseball				
Sugarpop				
Units needed				

   (b) § Set up an LP problem to answer the question: How many pounds of Greaseball and how many pounds of Sugarpop should Grunwilda serve to provide the minimum dietary needs at the least cost?

   (c) Graph the feasible region and determine the corner points for the LP problem in (b).

4. A manufacturer has two plants, P_1 and P_2. Each plant manufactures four products: A, B, C, and D. The manufacturer has a contract to supply a distributor with 1000 units of A, 800 units of B, 900 units of C, and 780 units of D each day. Suppose each day plant P_1 can manufacture 200, 600, 90, 130 units of product A, B, C, D, respectively; and plant P_2 can manufacture 100, 200, 150, 80 units of A, B, C, D, respectively. Finally, suppose it costs \$8,000 per day to operate plant P_1 and \$11,000 per day to operate plant P_2.

   (a) Fill in the table with the conditions of this problem.

Product	Plant P_1 (units/day)	Plant P_2 (units/day)	Required Output (units/day)
A			
B			
C			
D			
Cost/day			

(b) Set up an LP problem to determine how many days each plant should operate to minimize cost but still meet the demands of the distributor.

(c) Graph the feasible region and determine the corner points for the LP problem in (b).

5. A firm produces two products A and B. Each product consists of ingredients R, S, and T. Each unit of A contains 5 g of ingredient R, 4 g of ingredient S, and 1 g of ingredient T. Each unit of B contains 3 g of ingredient A, 4 g of ingredient S, and 3 g of ingredient T. On a daily basis, the firm has 9 kg of R, 9 kg of S, and 4 kg of T available. The profit for each unit of A sold is \$6, and the profit for each unit of B sold is \$6.

   (a) § Set up a table summarizing the conditions of the problem.

   (b) § Set up an LP problem to answer the question: How many units of products A and B should the firm produce in order to maximize profit?

   (c) Graph the feasible region and determine the corner points.

6. A hospital meal consists of two foods, I and II. Food I provides 6 units of protein, 6 units of carbohydrate, and 4 units of fat. Food II provides 8 units of protein, 4 units of carbohydrate, and 2 units of fat. Each ounce of food I costs 50 cents, and each ounce of food II costs 40 cents. A meal should provide at least 32 units of protein, 30 units of carbohydrate, and no more than 18 units of fat.

   (a) Set up a table summarizing the information.

(b) Set up an LP problem to answer the question: How much of each food should be served to meet the nutritional constraints at the least cost?

(c) Graph the feasible region and determine the corner points.

4.3 GEOMETRIC SOLUTIONS TO TWO-VARIABLE LP PROBLEMS

Figure 1 gives the feasible region for the problem

$$\begin{array}{ll} \text{Maximize:} & z = 75x + 50y \\ \text{Subject to:} & \frac{6}{5}x + y \leq 72 \\ & x + \frac{1}{2}y \leq 50 \\ & \frac{3}{2}x + 2y \leq 120 \\ & x, y \geq 0 \end{array} \tag{1}$$

given in Example 2 of the previous section. We want to find the point (or points) in this region that make the objective function as large as possible.

For each fixed value of z, the objective function $z = 75x+50y$ corresponds to a straight line. For instance, if $z = 1200$, then we have the line $75x+50y = 1200$ shown in Figure 1. As the value of z changes we get a family of parallel lines (called *level lines*). Level lines for $z = 2400$ and $z = 4125$ are also shown in Figure 1.

Note that as z gets larger (which is what we want) the level lines move upward to the right. The last of these upward moving lines that still intersects the feasible region passes through the corner point $(35, 30)$. Since this line gives us the largest possible z value, we find that $x = 35, y = 30$ is the solution to our problem with $z = 75 \cdot 35 + 50 \cdot 30 = 4125$ the maximal amount that the woodworker can make in a week.

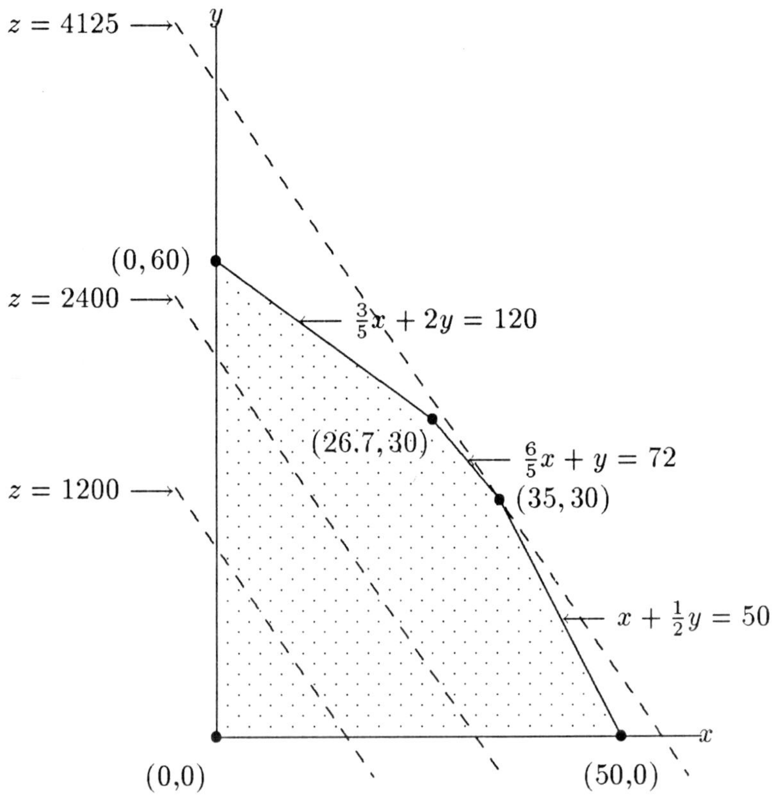

Figure 1

Generally you should be able to convince yourself (at least intuitively) that a family of level lines corresponding to different z values for an objective function $z = ax + by$ will "leave" the feasible region at a corner point. The coordinates of this corner point give the maximal (or minimal solution) to the problem. Thus, one way of solving an LP problem is to find all corner points of the feasible region, substitute their coordinates into the objective function, and determine which corner point (or points) gives the maximal or minimal value. This, however, is not a very efficient method. Alternatively, for a problem with two variables, you can graph the level lines and determine geometrically which corner point corresponds to the largest (or smallest) z value. This method is also not efficient. In large problems there may be hundreds of difficult-to-determine corner points. The next section introduces a surprisingly efficient method for solving LP problems that does not rely on geometry.

Using *GRAPH* to Solve Two-Variable LP Problems

You can use the *GRAPH* program of CALIPSO to solve two-variable LP problems. *GRAPH* can evaluate the objective function z at corner points and it can plot level lines. Example 1 shows the way. You should carry out the steps described in Example 1 with CALIPSO.

Example 1: Using *GRAPH* to Solve (1)

We start by entering the objective function and constraints of (1). Note that *GRAPH* does not accept fractions. You must enter a fraction like $\frac{6}{5}$ as its decimal equaivalent 1.2 .

Note: You do not need to memorize the "*F*" keys used below because by pressing [*F1*] you can view on the screen a list of the functions performed by each "*F*" key.

To solve the problem by calculating z at each corner point, first select with the [*F4*] key one of the constraints corresponding to an edge of the feasible region. Select a constraint by moving the cursor to it and pressing the [*F4*] key. Then move the cursor to an adjacent edge. The coordinates of the corner point and the value of z appear in the Message Box. Simply record the value of z for each corner point. You can use Figure 1 to check your corner points. The largest z value is the maximum and the smallest is the minimum.

To solve the problem using level lines, press [*F5*]. A sequence of level lines is drawn on the graph in the direction of increasing values of z. Spot the corner point at which z is largest. You can find the coordinates of this corner point and the value of z there by first selecting one edge that passes through the desired corner point and then moving the cursor to another edge that passes through the corner point.

4.3A Exercises: Solving Two-Variable LP Problems

Solve the following problems by hand or with CALIPSO's *GRAPH* program.

1 – 4. Solve the LP problems described in Problems 1 – 4 of Section 4.2. The solutions to 1 and 3 are given in the answer section.

5.§ Maximize : $z = 120x + 100y$
Subject to :
$$\begin{array}{rcrcr} x & + & y & \leq & 4 \\ 5x & + & 3y & \leq & 15 \end{array}$$
$x, y \geq 0$

6. Maximize : $z = 6x - 2y$
Subject to :
$$\begin{array}{rcrcr} -2x & + & y & \leq & 2 \\ 3x & + & 5y & \leq & 15 \end{array}$$
$x, y \geq 0$

7.§ Maximize : $z = 2x + 9y$
Subject to :
$$\begin{array}{rcrcr} x & - & y & \leq & 6 \\ -x & + & 3y & \leq & 12 \\ x & + & y & \leq & 8 \end{array}$$
$x, y \geq 0$

8. Maximize : $z = 4x - 7y$
Subject to :
$$\begin{array}{rcrcr} x & + & y & \geq & 3 \\ -x & + & y & \leq & 3 \\ 2x & + & y & \leq & 8 \end{array}$$
$x, y \geq 0$

9.§ Maximize : $z = 7x + 5y$
Subject to :
$$\begin{array}{rcrcr} x & - & y & \leq & 2 \\ x & + & y & \leq & 12 \\ 2x & + & 3y & \leq & 12 \\ 2x & - & y & \geq & 1 \end{array}$$
$x, y \geq 0$

10. Maximize : $z = 5x + 2y$
Subject to :
$$\begin{array}{rcrcr} 2x & - & y & \geq & 2 \\ x & + & 3y & \geq & 5 \\ x & + & 3y & \leq & 9 \\ x & + & y & \leq & 6 \end{array}$$
$x, y \geq 0$

11.§ Maximize : $z = 120x - 310y$
Subject to :
$$\begin{array}{rcrcr} x & - & y & \leq & 12 \\ x & + & y & \geq & 5 \\ 2x & + & 3y & \geq & 12 \\ 2x & + & 4y & \geq & 15 \\ 3x & + & 7y & \geq & 15 \\ 2x & + & 3y & \leq & 36 \end{array}$$
$x, y \geq 0$

12. Maximize : $z = 15x + 24y$
Subject to :
$$\begin{array}{rcrcr} x & + & y & \leq & 120 \\ 2x & + & 3y & \geq & 80 \\ x & + & 3y & \leq & 100 \\ 2x & + & y & \leq & 240 \\ 3x & + & 5y & \geq & 200 \\ x & - & y & \geq & 50 \end{array}$$
$x, y \geq 0$

13. Graph the feasible region and objective function of the following LP problem. Use level lines to explain why the problem has more than one optimal solution.

$$\begin{array}{llcccl}
\text{Maximize}: z = & 20x & + & 20y & & \\
\text{Subject to}: & x & + & 2y & \leq & 24 \\
 & x & + & y & \leq & 15 \\
 & -x & + & 2y & \geq & 5 \\
 & x, y \geq 0 & & & &
\end{array}$$

4.4 THE SIMPLEX METHOD

While visually appealing, the geometric method discussed in the previous section is of limited use since most practical problems have more than two variables. The *simplex method*, which systematically finds solutions to equivalent systems of linear equations, provides a remarkably efficient means for obtaining optimal solutions. To illustrate how the simplex method works we solve the LP problem

$$\begin{array}{llcccl}
\text{Maximize}: z = & 2x & + & 9y & & \\
\text{Subject to}: & x & - & y & \leq & 6 \\
 & -x & + & 3y & \leq & 12 \\
 & x & + & y & \leq & 8 \\
 & x, y \geq 0 & & & &
\end{array} \qquad (1)$$

The first step in applying the simplex method to this problem is to introduce *slack variables* to convert each $\leq$ constraint to an $=$ constraint. As the name suggests, these variables take up any "slack" between the left-hand and right-hand sides of the constraints in (1). If we call these variables s_1, s_2, and s_3 we obtain the system of linear equations

$$\begin{array}{rcrcccccl}
x & - & y & + & s_1 & & & = & 6 \\
-x & + & 3y & & & + s_2 & & = & 12 \\
x & + & y & & & & + s_3 & = & 8 \\
\multicolumn{9}{l}{x \geq 0, y \geq 0; s_i \geq 0, 1 \leq i \leq 3}
\end{array} \qquad (2)$$

Note that all variables in this system are nonnegative.

Definition. A system of linear equations with m equations and n variables

$$\begin{array}{ccccccccc} a_{11}x_1 & + & a_{12}x_2 & + & \cdots & + & a_{1n}x_n & = & b_1 \\ a_{21}x_1 & + & a_{22}x_2 & + & \cdots & + & a_{2n}x_n & = & b_2 \\ \vdots & & & & & & & & \\ a_{m1}x_1 & + & a_{m2}x_2 & + & \cdots & + & a_{mn}x_n & = & b_m \end{array}$$

is in *canonical form* (or is a *canonical system* of linear equations) if

(a) $m \leq n$ and

(b) there are m "distinguished" variables, called *basic variables*, such that each basic variable x_i has coefficient 1 in exactly one equation and coefficient 0 in the other equations, and such that each equation contains exactly one basic variable. The other variables are called *nonbasic variables.*

Example 1: Canonical Form

(a) In the system of linear equations

$$\begin{array}{lllllllll} 3x_1 & & + & 4x_3 & & & + & x_6 & = 9 \\ 2x_1 & + x_2 & - & x_3 & & & & & = 14 \\ x_1 & & + & 2x_3 & + x_4 & & & & = 7 \\ & & & x_3 & & + x_5 & & & = 8 \end{array}$$

$x_i \geq 0, 1 \leq i \leq 6$

the variables x_2, x_4, x_5, x_6 are basic and the variables x_1 and x_3 are nonbasic.

(b) In (2) s_1, s_2, s_3 are basic variables and x, y are nonbasic variables. One of the useful features of a canonical system of equations is that you can always obtain a solution by setting each basic variable equal to the right-hand side of the equation in which it occurs and by setting all nonbasic variables equal to 0. This solution is called the *basic solution.* The basic solution to (2) is

$$s_1 = 6, s_2 = 12, s_3 = 8$$

Note that we do not usually give the values of the nonbasic variables with the basic solution since they always have the value 0. Also note that the value of the objective function $z = 2x + 9y$ corresponding to this solution is

$$z = 2 \cdot 0 + 9 \cdot 0 = 0$$

The main idea underlying the simplex method is to create a sequence of *equivalent* systems of equations whose basic variables provide better and better solutions to the original problem (better in the sense that the corresponding values of the objective function become increasingly larger). The nonbasic variables in these equivalent systems of equations will always be assigned the value 0.

Geometrically, the solutions to these equivalent systems of equations correspond to corner points of the feasible region.

For any given system of equations there are infinitely many equivalent systems. As indicated above, the simplex method selects those that produce increasingly better objective function values. To see how to make these selections for problem (1) first consider the objective function

$$z = 2x + 9y$$

In (2) the variables x and y are viewed as nonbasic variables with assigned values 0, and so $z = 0$. In order for z to increase toward a maximal value we need to try to make either x or y a basic variable with a value greater than 0. Since the coefficient of y is larger than the coefficient of x it is reasonable to make y a basic variable rather than x (a unit increase in y will increase z by 9, while a unit increase in x will increase z by 2). The variable y is called the *entering basic variable.*

Entering Basic Variable Rule: If the objective function is written in terms of the nonbasic variables, then the *entering basic variable* is the variable with the largest positive coefficient. Ties for the largest coefficient can be broken arbitrarily.

Since the system of equations (2) consists of three equations it can only have three basic variables. So if y is to become a basic variable, then we must make one of the current basic variables s_1, s_2, s_3 a nonbasic variable. To see which of these becomes nonbasic recall that $x = 0$, and therefore from (2) we can write

$$\begin{aligned} s_1 &= 6 + y \\ s_2 &= 12 - 3y \\ s_3 &= 8 - y \end{aligned} \tag{3}$$

Since s_1, s_2, and s_3 are all nonnegative, it follows from the middle equation that y cannot be larger than 4. (The first equation places no restrictions on y and the third equation forces y to be less than or equal to 8.) Finally, observe that if $y = 4$, then $s_2 = 0$ and so s_2 becomes a nonbasic variable. This variable is called the *departing basic variable.*

There is an equivalent but easier way to find the departing basic variable, which we describe in the following rule (see Problem 9 in Exercises 4.4A).

Departing Basic Variable Rule: Calculate the ratios of the values on the right-hand side to the corresponding coefficients of the entering basic variable. Consider only those ratios that are ≥ 0. The smallest of these ratios corresponds to the *departing basic variable.*

In (2) the ratios to be considered are $\frac{12}{3}$ and $\frac{8}{1}$; the smallest of these is $\frac{12}{3}$ and so s_2 is the departing basic variable.

To make y a basic variable and s_2 a nonbasic variable we perform a *pivot operation* on the system of equations (2). The ıpivot element is the coefficient of the entering basic variable y in the equation containing the departing basic variable s_2. Thus, the coefficient 3 is the pivot element. We use elementary row operations to convert the coefficient 3 to 1 and to convert the coefficients above and below 3 to 0. This gives us the equivalent system of equations

$$\begin{array}{rcrcrcrcrcr} \frac{2}{3}x & & & + & s_1 & + & \frac{1}{3}s_2 & & & = & 10 \\ -\frac{1}{3}x & + & y & & & + & \frac{1}{3}s_2 & & & = & 4 \\ \frac{4}{3}x & & & & & - & \frac{1}{3}s_2 & + & s_3 & = & 4 \end{array}$$

Observe that y, s_1, and s_3 are now the basic variables, and the basic solution is $s_1 = 10, y = 4, s_3 = 4$. Note too that the current value of the objective function z is

$$z = 2x + 9y = 2 \cdot 0 + 9 \cdot 4 = 36$$

We now repeat the entire procedure to obtain a new (equivalent) system of equations that will yield a better z value. First it is necessary to rewrite z in terms of the current nonbasic variables x and s_2. To do this use the second equation from (2)

$$\begin{array}{rcrcrcr} 3y & = & 12 & + & x & - & s_2 \qquad \text{or, equivalently,} \\ y & = & 4 & + & \frac{1}{3}x & - & \frac{1}{3}s_2 \end{array}$$

Substituting the right-hand side of this equation for y in the objective function

$$z = 2x + 9y$$

gives the new objective function

$$z = 2x + 9(4 + \tfrac{1}{3}x - \tfrac{1}{3}s_2) = 5x - 3s_2 + 36$$

written in terms of the nonbasic variables. From this equation and the Entering Basic Variable Rule we see that the next entering basic variable is x. Observe from this equation that we should keep s_2 nonbasic and equal to 0, since if s_2 becomes positive z will become smaller rather than larger. Generally, nonbasic variables with negative coefficients in the objective function are never candidates to become an entering basic variable. If they took on positive (rather than zero) values they would decrease the value of the objective function. In fact, if an objective function is written in terms of nonbasic variables and if the coefficients of all of these variables are all either 0 or negative, then we have reached a maximal solution since it is impossible to make the value of z any larger.

To carry out the steps just discussed (and succeeding ones) it is convenient to work with arrays of numbers called *tableaus.* The initial tableau corresponding to the equations (2) and the objective function

$$z = 2x + 9y \tag{4}$$

rewritten in the form

$$z - 2x - 9y = 0 \tag{5}$$

is

Tableau 1

	z	x	y	s_1	s_2	s_3	RHS
s_1	0	1	−1	1	0	0	6
s_2	0	−1	3	0	1	0	12
s_3	0	1	1	0	0	1	8
OBJ	1	−2	−9	0	0	0	0

The variables listed on the left-hand side of the tableau are the current (initial) basic variables. The bottom row of the tableau (excluding the rightmost entry, which is the current value of z) is called the *objective row*, and the rightmost column (again excluding the entry in the lower right-hand corner) is the *right-hand side* (RHS) of the tableau.

Note that the entering basic variable corresponds to the most *negative* entry in the objective row, since we have rewritten the objective function (4) in the form (5). In general, for simplex tableaus we have the following criterion for selecting the entering basic variable. This is the tableau equivalent of the entering basic variable rule given above.

Entering Basic Variable Rule for Tableaus: The *entering basic variable* is the variable heading the column in the tableau that contains the most negative value in the objective row. Ties may be broken arbitrarily. The column containing the entering basic variable is called the *pivot column.*

Departing Basic Variable Rule for Tableaus: To determine the *departing basic variable*, form the ratios of the entries in the RHS column to the corresponding positive-valued entries in the pivot column (the column headed by the entering basic variable). Ties may be broken arbitrarily. The departing basic variable is the variable corresponding to the smallest of these ratios. The row associated with the departing basic variable is called the *pivot row.* (Note: It may happen that there are no positive entries in the pivot column. In this case, the problem is unbounded. We discuss unbounded problems in Section 5.3.)

For the entering basic variable y in Tableau 1 the only nonnegative ratio is 12/3, which tells us that s_2 is the departing basic variable.

As we saw earlier, in order that y become a basic variable in place of s_2 in our example, we need to convert the entry 3 to 1, and all other entries in the y column to 0. This is called a *pivot operation* and the entry 3 is called the *pivot entry.* Using elementary row operations to perform this pivot operation we get the next tableau

Tableau 2

	z	x	y	s_1	s_2	s_3	RHS
s_1	0	$\frac{2}{3}$	0	1	$\frac{1}{3}$	0	10
y	0	$-\frac{1}{3}$	1	0	$\frac{1}{3}$	0	4
s_3	0	$\frac{4}{3}$	0	0	$-\frac{1}{3}$	1	4
OBJ	1	-5	0	0	3	0	36

Note that we also performed a row operation on the objective row to obtain a 0 in its pivot column. This makes the objective row correspond to an equation that expresses z in terms of the current *nonbasic* variables x and s_2.

From Tableau 2 we see that the current basic variables are s_1, y, and s_3. The solution to (2) given by this tableau is $x = 0, y = 4, s_1 = 10, s_2 =$

$0, s_3 = 4$, and $z = 36$. Geometrically this solution corresponds to the corner point (0,4) of the feasible region.

Since there is still a negative entry in the objective row we can improve the value of z. The negative entry -5 corresponds to the variable x, and this becomes the entering basic variable. To determine the departing basic variable consider the ratios $\frac{10}{\frac{2}{3}} = 10 \cdot \frac{3}{2} = 15$ and $\frac{4}{\frac{4}{3}} = 4 \cdot \frac{3}{4} = 3$ of the right-hand side entries of the tableau to the corresponding positive entries in the x column. The smaller of these ratios tells us that s_3 is the departing variable, and the pivot entry is $\frac{4}{3}$. Pivoting on $\frac{4}{3}$ gives us the next tableau

	z	x	y	s_1	s_2	s_3	RHS
s_1	0	0	0	1	$\frac{1}{2}$	$-\frac{1}{2}$	8
y	0	0	1	0	$\frac{1}{4}$	$\frac{1}{4}$	5
x	0	1	0	0	$-\frac{1}{4}$	$\frac{3}{4}$	3
OBJ	1	0	0	0	$\frac{7}{4}$	$\frac{15}{4}$	51

Since all entries in the objective row are nonnegative, we cannot improve on the value of z, and so we are done. To see this, note that the objective row in this tableau represents the equation

$$z + 0x + 0y + 0s_1 + \tfrac{7}{4}s_2 + \tfrac{15}{4}s_3 = 51 \text{ or, equivalently,}$$
$$z = -0x - 0y - 0s_1 - \tfrac{7}{4}s_2 - \tfrac{15}{4}s_3 + 51$$

Assigning positive values to either s_2 or s_3 (the current nonbasic variables) will only serve to decrease the value of z. In general, we have the following termination rule for simplex tableaus.

Termination Rule: If there are no negative entries in the objective row of a tableau, then an optimal solution has been obtained.

Thus the optimal solution for the system of equations (2) is $x = 3, y = 5, s_1 = 8, s_2 = 0, s_3 = 0; z = 51$. This solution corresponds to the corner point (3,5) of the feasible region. It is important to note, however, that in carrying out the simplex method we made no use of the graph of that region.

Definition. An LP problem is in *standard form* if it is in the form

$$\begin{array}{llcccccccl}
\text{Maximize: } z = & c_1x_1 & + & c_2x_2 & + & \cdots & + & c_nx_n & & \\
\text{Subject to:} & a_{11}x_1 & + & a_{12}x_2 & + & \cdots & + & a_{1n}x_n & \leq & b_1 \\
 & a_{21}x_1 & + & a_{22}x_2 & + & \cdots & + & a_{2n}x_n & \leq & b_2 \\
 & \vdots & & & & & & & & \\
 & a_{m1}x_1 & + & a_{m2}x_2 & + & \cdots & + & a_{mn}x_n & \leq & b_m \\
 & x_i \geq 0, 1 \leq i \leq n & & & & & & & &
\end{array}$$

where all the RHS entries b_i are nonnegative. The procedure outlined above is applicable to any LP problem in standard form.

For LP problems in standard form the simplex method consists of the following steps.

1. Introduce slack variables to each $\leq$ constraint.

2. Form the initial tableau

	z	x_1	x_2	$\cdots$	x_n	s_1	s_2	$\cdots$	s_m	RHS
s_1	0	a_{11}	a_{12}	$\cdots$	a_{1n}	1	0	$\cdots$	0	b_1
s_2	0	a_{21}	a_{22}	$\cdots$	a_{2n}	0	1	$\cdots$	0	b_2
$\vdots$										$\vdots$
s_m	0	a_{m1}	a_{m2}	$\cdots$	a_{mn}	0	0	$\cdots$	1	b_m
OBJ	1	$-c_1$	$-c_2$	$\cdots$	$-c_n$	0	0	$\cdots$	0	0

3. Use the Entering Basic Variable Rule to determine the pivot column.

4. Use the Departing Basic Variable Rule to determine the pivot row.

5. Pivot on the pivot entry.

6. Repeat steps 3 – 5 until all entries in the objective row are nonnegative.

In the next chapter we will see how to deal with LP problems that are not in standard form.

4.4A Exercises: Solving LP Problems by Hand

Use the simplex method to solve the following problems.

1. §

$$\begin{array}{llrcrcl} \text{Maximize: } z = & x_1 & + & 3x_2 & & \\ \text{Subject to:} & 2x_1 & + & 3x_2 & \leq & 12 \\ & 6x_1 & + & x_2 & \leq & 9 \\ & & & x_1 \geq 0, x_2 \geq 0 & & \end{array}$$

2.

$$\begin{array}{llrcrcl} \text{Maximize: } z = & 4x_1 & + & 3x_2 & & \\ \text{Subject to:} & 4x_1 & + & 2x_2 & \leq & 120 \\ & x_1 & + & x_2 & \leq & 40 \\ & & & x_1 \geq 0, x_2 \geq 0 & & \end{array}$$

3. §

$$\begin{array}{llrcrcl} \text{Maximize: } z = & 100x_1 & + & 120x_2 & & \\ \text{Subject to:} & 2x_1 & + & 2x_2 & \leq & 8 \\ & 3x_1 & + & 5x_2 & \leq & 15 \\ & & & x_1 \geq 0, x_2 \geq 0 & & \end{array}$$

4.

$$\begin{array}{llrcrcl} \text{Maximize: } z = & 8x_1 & + & 3x_2 & & \\ \text{Subject to:} & 2x_1 & + & x_2 & \leq & 80 \\ & 3x_1 & + & 5x_2 & \leq & 120 \\ & x_1 & + & 4x_2 & \leq & 68 \\ & & & x_1 \geq 0, x_2 \geq 0 & & \end{array}$$

5. §

$$\begin{array}{llclclcl} \text{Maximize: } z = & 3x_1 & + & 2x_2 & + & 4x_3 & & \\ \text{Subject to:} & 2x_1 & + & x_2 & & & \leq & 12 \\ & & & x_2 & + & x_3 & \leq & 8 \\ & & & x_1 \geq 0, x_2 \geq 0, x_3 \geq 0 & & & & \end{array}$$

6.

$$\begin{array}{llclclcl} \text{Maximize: } z = & 32x_1 & + & 18x_2 & + & 19x_3 & & \\ \text{Subject to:} & 2x_1 & + & x_2 & + & x_3 & \leq & 12 \\ & 8x_1 & + & 10x_2 & + & 3x_3 & \leq & 90 \\ & 4x_1 & + & 4x_2 & + & 3x_3 & \leq & 30 \\ & & & x_1 \geq 0, x_2 \geq 0, x_3 \geq 0 & & & & \end{array}$$

7. §

$$\begin{array}{llcccccccl}
\text{Maximize: } z = & 24x_1 & + & 16x_2 & + & 10x_3 & & \\
\text{Subject to:} & 3x_1 & + & 2x_2 & + & x_3 & \leq & 12 \\
& 2x_1 & + & 2x_2 & + & x_3 & \leq & 20 \\
& 3x_1 & + & x_2 & + & 3x_3 & \leq & 30 \\
\end{array}$$
$$x_1 \geq 0, x_2 \geq 0, x_3 \geq 0$$

8.

$$\begin{array}{llcccccccl}
\text{Maximize: } z = & 320x_1 & + & 108x_2 & + & 120x_3 & & \\
\text{Subject to:} & x_1 & + & 2x_2 & + & 5x_3 & \leq & 45 \\
& 6x_1 & + & 8x_2 & + & 2x_3 & \leq & 54 \\
& 3x_1 & + & 4x_2 & + & 2x_3 & \leq & 32 \\
\end{array}$$
$$x_1 \geq 0, x_2 \geq 0, x_3 \geq 0$$

9. Explain why using (3) to find the departing basic variable is equivalent to using the Departing Basic Variable Rule.

Solving LP Problems with *PIVOTER*

You can perform the steps of the simplex method with the *PIVOTER* program of CALIPSO. *PIVOTER* is a good way to learn the simplex method since it does the arithmetic for you, but you must make all the logical decisions. We introduce *PIVOTER* by using it to solve Example 2.

Example 2: Solving a Problem with *PIVOTER*

We solve the three-variable LP problem

$$\begin{array}{llcccccccl}
\text{Maximize: } z = & 2x_1 & + & 3x_2 & + & 3x_3 & & \\
\text{Subject to:} & 3x_1 & + & 2x_2 & & & \leq & 60 \\
& -x_1 & + & x_2 & + & 4x_3 & \leq & 10 \\
& 2x_1 & - & 2x_2 & + & 5x_3 & \leq & 5 \\
\end{array}$$
$$x_i \geq 0, 1 \leq i \leq 3$$

The first step is to introduce slack variables s_1, s_2, s_3 to form an equivalent LP problem with equality constraints.

$$
\begin{array}{llcccccccl}
\text{Maximize: } z = & 2x_1 & + & 3x_2 & + & 3x_3 & & & & \\
\text{Subject to:} & 3x_1 & + & 2x_2 & & & + s_1 & & & = 60 \\
 & -x_1 & + & x_2 & + & 4x_3 & & + s_2 & & = 10 \\
 & 2x_1 & - & 2x_2 & + & 5x_3 & & & + s_3 & = 5 \\
\end{array}
$$

$$x_i \geq 0, 1 \leq i \leq 3; s_i \geq 0, 1 \leq i \leq 3$$

The corresponding tableau is

Tableau 3

	x_1	x_2	x_3	s_1	s_2	s_3	RHS
s_1	3	2	0	1	0	0	60
s_2	−1	1	4	0	1	0	10
s_3	2	−2	5	0	0	1	5
OBJ	−2	−3	−3	0	0	0	0

Notice that we did not introduce a column for the z variable as we did in the earlier example. If you study the tableaus we obtained in that example by applying the simplex method, you see that the z column will never change. We often omit the z column to save space and effort. When you do this you must be careful in interpreting the meaning of the objective row. In order to interpret the OBJ row as an equation, you must remember to insert a 1 in place of the missing z column entry.

Here is how to solve our LP problem with *PIVOTER*.

1. Choose *PIVOTER* from the CALIPSO menu. As usual, you are asked for a file name. Enter TAB3 or any unused name you prefer.

2. Since there is no file stored under TAB3, a screen appears asking for the dimensions of the tableau. Make 4 rows, one for each row of constraints and one for the OBJ row. You should make 7 columns, one for each variable and one for the RHS. When you do this and press *Enter*, a tableau with all entries set to 0 will appear.

3. Notice the menu at the bottom of the screen. You make selections from the menu by moving the cursor to a function and pressing *Enter*. Alternatively, you can use the "hot keys" shown in parentheses to perform the function. We use the hot keys in these instructions since their use is easier to describe. (Note that some older keyboards do not have keys for *F11* and *F12*. You have to make these selections by moving the cursor in the menu and pressing *Enter*.)

4. As advertised in the menu, you press [F2] to enter column labels. Do so and enter $x1, x2, x3, s1, s2, s3$. Note that the last column is already labelled RHS. Also note that subscripts must be typed "on-line" in the *PIVOTER* display, e.g., x_1 appears as $x1$.

5. Now press [F3] and enter the row labels s_1, s_2, s_3. Note that the order of entering these labels, first column labels and then row labels, is important because *PIVOTER* does some bookkeeping to associate row labels with column labels as the row labels are entered.

6. Now press [F4] and enter the rest of the data in Tableau 3. Press [Esc] when you are done.

7. To perform pivots press [F6]. The cursor appears on 3, the entry in the upper left-hand corner. You can move the cursor with arrow keys. To perform a pivot, you locate the cursor on the desired element and press [Enter].

   The Entering Variable Rule says to select either the x_2 or x_3 column as the pivot column. Let's select the x_2 column. Move the cursor anywhere in the x_2 column.

   The Departing Variable Rule requires us to form ratios of entries in the RHS column to entries in the x_2 column. As advertised in the instructions at the bottom of the screen, *PIVOTER* will display these ratios for you when you press [F10]. *PIVOTER* displays the ratios 30, 10, $-5/2$. The smallest nonnegative ratio is 10 and it occurs in the second row. To do the pivot, move the cursor to the second row and press [Enter].

   Notice what happens in the the row label column when you pivot. The departing variable s_2 is replaced by the entering variable x_2.

8. Continue performing pivots in this way until you have an optimal tableau; that is, until there are no more negative entries in the OBJ row. The values of the basic variables then appear in the constraint rows of the RHS column and the value of z appears in the OBJ row of the RHS column. Since the values of the nonbasic variables are all 0, the solution is

$$x_1 = 8, x_2 = 18, x_3 = 0, s_1 = 0, s_2 = 0, s_3 = 25; z = 70$$

9. You can print the tableau that appears on the screen (at any time) by pressing [F5]. You can even print the tableau each time you pivot.

10. You can save the file to disk for later use by pressing [Esc] to get to the menu and then [F7]. Note that when you save a tableau, it is the one that currently appears on the screen that is saved. If you have performed pivots and want to save the original tableau, press [F11] to restore the display to the original tableau and then save it. (If your keyboard does not have an [F11], you can select this function by moving the cursor to it and pressing [Enter]).

11. If you make a mistake or become confused while executing the simplex method, you can start over by restoring the tableau to its initial form by pressing [Esc] to get to the menu and then pressing the [F11] key.

12. You leave the *PIVOTER* program by selecting *Quit* from the menu.

4.4B Exercises: Solving LP Problems with *PIVOTER*

Use *PIVOTER* to solve the following problems.

1. §

$$\begin{array}{llcccl} \text{Maximize: } z = & 120x & + & 100y & & \\ \text{Subject to:} & 5x & + & 3y & \leq & 15 \\ & 2x & + & 2y & \leq & 8 \\ & \multicolumn{5}{c}{x \geq 0, y \geq 0} \end{array}$$

2.

$$\begin{array}{lcccccccl} \text{Maximize: } z = & -3x_1 & + & 2x_2 & + & 5x_3 & & \\ \text{Subject to:} & -3x_1 & + & 2x_2 & + & 6x_3 & \leq & 9 \\ & x_1 & + & 2x_2 & - & x_3 & \leq & 6 \\ & -2x_1 & + & 4x_2 & + & 2x_3 & \leq & 5 \\ & \multicolumn{7}{c}{x_i \geq 0, 1 \leq i \leq 3} \end{array}$$

3. §

$$\begin{array}{lcccccccccl} \text{Maximize: } z = & \frac{2}{3}x_1 & + & 2x_2 & + & \frac{1}{2}x_3 & + & x_4 & & \\ \text{Subject to:} & x_1 & + & 3x_2 & & & + & x_4 & \leq & 4 \\ & x_1 & + & \frac{1}{2}x_2 & & & & & \leq & \frac{3}{2} \\ & & & x_2 & + & 4x_3 & + & x_4 & \leq & 3 \\ & \multicolumn{9}{c}{x_i \geq 0, 1 \leq i \leq 4} \end{array}$$

4.

$$\begin{array}{llllllllll}
\text{Maximize: } z = & 3x_1 & + & x_2 & + & 3x_3 & + & 5x_4 & & \\
\text{Subject to:} & 2x_1 & + & x_2 & & & + & x_4 & \leq & 4 \\
& \frac{1}{2}x_1 & + & x_2 & & & - & 3x_4 & \leq & \frac{3}{2} \\
& & & 3x_2 & + & 4x_3 & + & 2x_4 & \leq & 5 \\
& 2x_1 & - & 3x_2 & + & x_3 & + & x_4 & \leq & 9 \\
\end{array}$$
$$x_i \geq 0, 1 \leq i \leq 4$$

5. §

$$\begin{array}{llllllllll}
\text{Maximize: } z = & -10x_1 & + & 32x_2 & + & 48x_3 & + & 54x_4 & & \\
\text{Subject to:} & 2x_1 & + & 3x_2 & + & 5x_3 & + & x_4 & \leq & 24 \\
& 5x_1 & + & 2x_2 & + & x_3 & + & 3x_4 & \leq & 32 \\
& 8x_1 & + & 5x_2 & + & 6x_3 & + & 10x_4 & \leq & 64 \\
& 3x_1 & + & 6x_2 & + & 9x_3 & + & 12x_4 & \leq & 81 \\
\end{array}$$
$$x_i \geq 0, 1 \leq i \leq 4$$

6.

$$\begin{array}{llllllllll}
\text{Maximize: } z = & 3x_1 & - & 2x_2 & + & x_3 & + & 4x_4 & & \\
\text{Subject to:} & 2x_1 & - & x_2 & + & 3x_3 & + & x_4 & \leq & 12 \\
& x_1 & + & 2x_2 & + & 4x_3 & + & 5x_4 & \leq & 42 \\
& 3x_1 & + & x_2 & - & x_3 & + & 4x_4 & \leq & 24 \\
\end{array}$$
$$x_i \geq 0, 1 \leq i \leq 4$$

7. § (Enter this problem in decimal mode.)

$$\begin{array}{llllllllll}
\text{Maximize: } z = & 120x_1 & - & 32x_2 & + & 48x_3 & + & 64x_4 & & \\
\text{Subject to:} & 3x_1 & + & 2x_2 & - & x_3 & + & 12x_4 & \leq & 42 \\
& -x_1 & + & 2x_2 & + & 3x_3 & + & 12x_4 & \leq & 36 \\
& 4x_1 & + & 3x_2 & + & 5x_3 & + & 21x_4 & \leq & 45 \\
& 8x_1 & + & 3x_2 & + & 4x_3 & + & 5x_4 & \leq & 28 \\
& 6x_1 & + & 2x_2 & + & 4x_3 & + & x_4 & \leq & 14 \\
\end{array}$$
$$x_i \geq 0, 1 \leq i \leq 4$$

Solving LP Problems with *LP SOLVER*

The *LP SOLVER* program allows you to enter an LP problem as a system of inequalities and solves the problem automatically. We carry through an example to demonstrate its use.

1. As usual you start the program by selecting *LP SOLVER* from the CALIPSO menu and entering a file name. When you do this, a screen appears in which you can type inequalities. We will enter the LP problem

$$\begin{array}{llcccl} \text{Maximize: } z = & 120x & + & 100y & & \\ \text{Subject to:} & 5x & + & 3y & \leq & 15 \\ & 2x & + & 2y & \leq & 8 \\ & \multicolumn{5}{l}{x \geq 0, y \geq 0} \end{array}$$

2. You enter the objective function by typing

$$\text{Max } 120x + 100y;$$

   You can omit Max before the objective function and then *LP SOLVER* assumes you are solving a maximum problem. To solve a minimum problem, put Min before the objective function (this causes each coefficient in the objective function to be replaced by its negative).

   Note: Each objective function or constraint must end with a semicolon. This is the way *LP SOLVER* identifies an expression.

3. You enter the constraints by typing

$$\begin{array}{ccccc} 5x & + & 3y & <= & 15; \\ 2x & + & 2y & <= & 8; \end{array}$$

   You move to a new line on the screen with *Enter* or by using arrow keys. Actually you can type several expressions on the same line as long as you end each expression with a semicolon.

4. You should *not* enter the inequalities $x \geq 0$ and $y \geq 0$. *LP SOLVER* assumes that all variables are nonnegative.

5. *LP SOLVER* has many of the editing features of a word processor. You can correct typing mistakes in several ways. You can back up and erase with the *Backspace* key. You can delete a character by locating the cursor on it and pressing the *Delete* key. You can insert characters by pressing the *Insert* key and typing characters (to turn off the insert mode, press *Insert* again). You can also insert lines between two existing lines, remove lines, copy lines, and do other things. You do not need to remember all of the word processing features because

you can view a list of them by pressing [F2] as advertised at the top of the screen.

6. When you have the example entered correctly, press [F1] to solve the problem. The solution appears on the screen. You can, if you want, return to the inequalities by pressing [PgUp]. You can change the problem and solve the new problem.

7. You can print the problem and its solution with the [F8] key.

8. When you quit the program with the [Esc] key you will be given a chance to save the inequalities to disk for use again later.

9. *LP SOLVER* has many other features that we will consider in later sections.

4.4C Exercises: Using *LP SOLVER*

Solve the problems in Exercises 4.4B with *LP SOLVER*.

---

5 Intermediate Linear Programming*

5.1 NONSTANDARD FORM LP PROBLEMS

The simplex method is not immediately applicable to LP problems that are not in standard form. For instance, consider the following LP problem:

$$
\begin{array}{lrcrcrcl}
\text{Maximize}: z = & 2x_1 & + & x_2 & - & 3x_3 & & \\
\text{Subject to}: & x_1 & + & 2x_2 & + & 3x_3 & \leq & 5 \\
& 3x_1 & - & 2x_2 & + & x_3 & \leq & 8 \\
& 2x_1 & + & x_2 & + & 4x_3 & \geq & 6 \\
& x_1 & - & 2x_2 & + & 2x_3 & \geq & 4 \\
& \multicolumn{7}{l}{x_i \geq 0, 1 \leq i \leq 3}
\end{array}
\tag{1}
$$

The difficulty lies in the third and fourth inequalities. We need to convert these two inequalities to equalities in a way that produces an initial basic feasible solution. But if, for instance, we introduce a variable s such that $2x_1+x_2+4x_3+s = 6$, then s will be nonpositive (≤ 0); however, the simplex method requires that all variables be nonnegative (≥ 0).

Linear programming texts such as [CV] present a number of methods for dealing with this kind of problem. The most commonly used method, the two-phase method, is beyond the scope of this text. However, we will present another method that, though not particularly efficient, can be applied to small problems. In Section 5.2 we also describe the Big-M method, which is occasionally used to solve large LP problems that are not in standard form.

We can solve (1) using a procedure we call the *Modified Simplex Method* (MSM). First, multiply both sides of the third and fourth inequalities by -1 to obtain

$$
\begin{aligned}
-2x_1 - x_2 - 4x_3 &\leq -6 \\
-x_1 + 2x_2 - 2x_3 &\leq -4
\end{aligned}
$$

Next introduce nonnegative slack variables to all of the constraints and form the resulting tableau

Tableau 1

	x_1	x_2	x_3	s_1	s_2	s_3	s_4	RHS
s_1	1	2	3	1	0	0	0	5
s_2	3	−2	1	0	1	0	0	8
s_3	−2	−1	−4	0	0	1	0	−6
s_4	−1	2	−2	0	0	0	1	−4
OBJ	−2	−1	3	0	0	0	0	0

The problem now is that there are negative entries in the RHS; the basic solution corresponding to this tableau, $s_1 = 5, s_2 = 8, s_3 = -6, s_4 = -4$ is not a feasible solution, since variables are not allowed to take on negative values. We will use pivot operations to remove these negative entries. The pivot rules of the MSM are

1. Determine a *pivot column*. This is any column containing a negative entry from a row with a negative entry in the RHS. (Note: If there is no negative entry, then the problem is infeasible; that is, there is no solution. We discuss infeasibility in Section 5.3.)

2. Determine the *pivot row*. This is the row corresponding to the smallest nonnegative ratio of entries in the RHS to entries in the pivot column. Break ties arbitrarily.

3. Pivot on the *pivot entry*, which is the entry common to the pivot row and the pivot column.

4. Repeat Steps 1 – 3 until there are no negative entries in the RHS.

5. Proceed with the simplex method until an optimal solution is found.

CALIPSO

Executing the MSM with *PIVOTER* is very similar to executing the ordinary simplex method. By pressing *F10* you can view the ratios of the RHS column to the column that contains the cursor.

Example 1: Solving an LP Problem Not in Standard Form

We apply the MSM method to the LP problem (1). Since −6 occurs in the RHS of the third row of the initial tableau, we select any negative entry in this row, say −4, to determine the pivot column. The positive ratios of entries in the RHS to entries in the pivot column (the x_3 column) are $\frac{5}{3}, \frac{8}{1}, \frac{-6}{-4}, \frac{-4}{-2}$. Since the smallest of these ratios is $\frac{-6}{-4}$, the third row is the pivot row and the pivot entry is -4. Using CALIPSO to pivot, we get the next tableau

Tableau 2

	x_1	x_2	x_3	s_1	s_2	s_3	s_4	RHS
s_1	$-\frac{1}{2}$	$\frac{5}{4}$	0	1	0	$\frac{3}{4}$	0	$\frac{1}{2}$
s_2	$\frac{5}{2}$	$-\frac{9}{4}$	0	0	1	$\frac{1}{4}$	0	$\frac{13}{2}$
x_3	$\frac{1}{2}$	$\frac{1}{4}$	1	0	0	$-\frac{1}{4}$	0	$\frac{3}{2}$
s_4	0	$\frac{5}{2}$	0	0	0	$-\frac{1}{2}$	1	-1
OBJ	$-\frac{7}{2}$	$-\frac{7}{4}$	0	0	0	$\frac{3}{4}$	0	$-\frac{9}{2}$

Tableau 2 still has a negative entry in the right-hand side. The pivot column is the s_3 column (since $-\frac{1}{2}$ is the only negative entry in the row containing -1). The positive ratios of entries in the RHS to entries in the pivot column are $\frac{1/2}{3/4}, \frac{13/2}{1/4}, \frac{-1}{-1/2}$ The smallest of these ratios is $\frac{1/2}{3/4}$ so the first row is the pivot row and $\frac{3}{4}$ is the pivot entry. Pivoting we get Tableau 3.

Tableau 3

	x_1	x_2	x_3	s_1	s_2	s_3	s_4	RHS
s_3	$-\frac{2}{3}$	$\frac{5}{3}$	0	$\frac{4}{3}$	0	1	0	$\frac{2}{3}$
s_2	$\frac{8}{3}$	$-\frac{8}{3}$	0	$-\frac{1}{3}$	1	0	0	$\frac{19}{3}$
x_3	$\frac{1}{3}$	$\frac{2}{3}$	1	$\frac{1}{3}$	0	0	0	$\frac{5}{3}$
s_4	$-\frac{1}{3}$	$\frac{10}{3}$	0	$\frac{2}{3}$	0	0	1	$-\frac{2}{3}$
OBJ	-3	-3	0	-1	0	0	0	-5

Tableau 3 still has a negative entry in the right-hand side. The pivot column is the x_1 column (since $-\frac{1}{3}$ is the only negative entry in the row containing $-\frac{2}{3}$). The positive ratios of entries in the RHS to entries in the pivot column are $\frac{19/3}{8/3}, \frac{5/3}{1/3}, \frac{-2/3}{-1/3}$. The smallest of these ratios is $\frac{-2/3}{-1/3,}$ so the fourth row is the pivot row and $-\frac{1}{3}$ is the pivot entry. Pivoting we get Tableau 4.

Tableau 4

	x_1	x_2	x_3	s_1	s_2	s_3	s_4	RHS
s_3	0	−5	0	0	0	1	−2	2
s_2	0	24	0	5	1	0	8	1
x_3	0	4	1	1	0	0	1	1
x_1	1	−10	0	−2	0	0	−3	2
OBJ	0	−33	0	−7	0	0	−9	1

Tableau 4 has all nonnegative entries in its RHS column and so is in the form to start the simplex method. If you carry out two pivots of the simplex method, you will discover that the optimal solution is

$$x_1 = \frac{12}{5}, x_2 = 0, x_3 = \frac{4}{5}, s_1 = \frac{1}{5}, s_2 = 0, s_3 = 2, s_4 = 0; z = \frac{12}{5}$$

5.1A Exercises: The MSM Method

Solve the following exercises by hand and then check your answers with CALIPSO.

1. §

$$\begin{array}{lrcrcrcr} \text{Maximize:} & z & = & x_1 & - & 2x_2 & & \\ \text{Subject to:} & 2x_1 & + & 5x_2 & \le & 60 \\ & x_1 & + & x_2 & \ge & 10 \\ & & & x_1, x_2 \ge 0 \end{array}$$

2.

$$\begin{array}{lrcrcrcr} \text{Maximize:} & z & = & -3x_1 & + & 5x_2 \\ \text{Subject to:} & x_1 & - & x_2 & \le & 6 \\ & x_1 & + & x_2 & \ge & 1 \\ & 2x_1 & + & x_2 & \le & 8 \\ & & & x_1, x_2 \ge 0 \end{array}$$

3. §

$$\begin{array}{lrcrcrcr} \text{Maximize:} & z & = & 7x_1 & + & 5x_2 \\ \text{Subject to:} & 3x_1 & + & 2x_2 & \le & 6 \\ & x_1 & + & x_2 & \ge & 2 \\ & & & x_1, x_2 \ge 0 \end{array}$$

4.

$$\begin{array}{lrcrcl} \text{Maximize:} & z & = & -3x_1 & - & 5x_2 \\ \text{Subject to:} & x_1 & + & x_2 & \geq & 12 \\ & x_1 & + & 2x_2 & \geq & 30 \\ & 2x_1 & + & x_2 & \geq & 20 \\ & & & x_1, x_2 \geq 0 & & \end{array}$$

Minimization Problems

Frequently it is necessary to solve a minimization problem rather than a maximization problem. For example, the LP problem may be

$$\text{Minimize}: z = -2x_1 + 3x_2 - 4x_3$$

subject to various constraints. It is easy to convert this to a maximization problem. You only need to observe that the values of x_1, x_2, x_3 that make z a minimum are precisely the same values that maximize

$$w = -z = 2x_1 - 3x_2 + 4x_3$$

Thus, to solve a minimization problem you convert it to a maximization problem by multiplying the objective function by -1 and then use any of the methods we have discussed to solve the resulting maximization problem. Of course, you must remember to change the sign of the value of z in the answer produced by the computer.

Problems with Equality Constraints

Recall that two numbers a and b satisfy $a \leq b$ and $a \geq b$ if and only if $a = b$. You can use this fact to replace an equality constraint in an LP problem with two inequality constraints. For instance, you can replace the equality constraint

$$2x_1 - 3x_2 + x_3 = 8$$

by the two constraints

$$\begin{array}{rcrcrcl} 2x_1 & - & 3x_2 & + & x_3 & \leq & 8 \\ 2x_1 & - & 3x_2 & + & x_3 & \geq & 8 \end{array}$$

You can then multiply the $\geq$ constraint by -1 and solve the resulting problem by the MSM method.

Example 2: Removing Equality Constraints

The minimization problem

$$\begin{array}{llllllll}
\text{Minimize: } z = & -80x_1 & - & 42x_2 & - & 30x_3 & & \\
\text{Subject to:} & 3x_1 & + & 2x_2 & + & x_3 & \leq & 12 \\
 & 3x_1 & - & 5x_2 & + & 6x_3 & \geq & 15 \\
 & 2x_1 & - & 3x_2 & + & x_3 & = & 8 \\
 & \multicolumn{7}{l}{x_i \geq 0, 1 \leq i \leq 3}
\end{array}$$

is equivalent to the problem with slack variables

$$\begin{array}{llllllllllll}
\text{Maximize: } z = & 80x_1 & + & 42x_2 & + & 30x_3 & & & & & & \\
\text{Subject to:} & 3x_1 & + & 2x_2 & + & x_3 & + s_1 & & & & = & 12 \\
 & -3x_1 & + & 5x_2 & - & 6x_3 & & + s_2 & & & = & -15 \\
 & 2x_1 & - & 3x_2 & + & x_3 & & & + s_3 & & = & 8 \\
 & -2x_1 & + & 3x_2 & - & x_3 & & & & + s_4 & = & -8 \\
 & \multicolumn{11}{l}{x_i \geq 0, 1 \leq i \leq 3; s_i \geq 0, 1 \leq i \leq 4}
\end{array}$$

5.1B Exercises: Nonstandard LP Problems

Solve Problems 1 and 2 by hand and the rest with *PIVOTER*.

1. §

$$\begin{array}{llllll}
\text{Minimize:} & z & = & 2x_1 & - & 4x_2 \\
\text{Subject to:} & 2x_1 & + & 3x_2 & \leq & 5 \\
 & x_1 & - & 3x_2 & \geq & 2 \\
 & \multicolumn{5}{l}{x_1, x_2 \geq 0}
\end{array}$$

Notice this is a minimization problem!

2.

$$\begin{array}{llllllll}
\text{Maximize: } z = & -2x_1 & - & 2x_2 & + & 5x_3 & & \\
\text{Subject to:} & 3x_1 & + & 2x_2 & - & 4x_3 & \geq & 7 \\
 & x_1 & + & x_2 & + & 3x_3 & = & 12 \\
 & \multicolumn{7}{l}{x_i \geq 0, 1 \leq i \leq 3}
\end{array}$$

3. §

$$\begin{array}{llllllll}
\text{Maximize: } z = & 2x_1 & - & x_2 & + & 3x_3 & & \\
\text{Subject to:} & 2x_1 & + & x_2 & + & 3x_3 & \leq & 10 \\
 & 2x_1 & - & 2x_2 & + & 3x_3 & \leq & 9 \\
 & x_1 & + & 3x_2 & + & 2x_3 & \geq & 4 \\
 & 2x_1 & - & x_2 & + & 3x_3 & \geq & 5 \\
 & \multicolumn{7}{l}{x_i \geq 0, 1 \leq i \leq 3}
\end{array}$$

4.

$$\begin{array}{llrcrcl}
\text{Maximize: } z = & 2x_1 & + & 3x_2 & & \\
\text{Subject to:} & -4x_1 & + & 3x_2 & \leq & 12 \\
& 2x_1 & + & x_2 & \leq & 6 \\
& x_1 & + & x_2 & \geq & 3 \\
& 5x_1 & + & x_2 & \geq & 5 \\
& x_1, x_2 \geq 0 & & & &
\end{array}$$

5. §

$$\begin{array}{llrcrcrcrcl}
\text{Minimize: } z = & x_1 & + & 2x_2 & + & 3x_3 & + & 4x_4 & & \\
\text{Subject to:} & x_1 & & & + & x_3 & - & 4x_4 & \geq & 2 \\
& x_1 & - & x_2 & + & 3x_3 & + & x_4 & \leq & 9 \\
& x_1 & + & x_2 & - & x_3 & + & x_4 & \geq & 3 \\
& x_i \geq 0, 1 \leq i \leq 4 & & & & & & & &
\end{array}$$

6. Solve the LP problem given in Example 2.

7. §

$$\begin{array}{llrcrcrcl}
\text{Maximize: } z = & 20x_1 & + & 34x_2 & + & 36x_3 & & \\
\text{Subject to:} & 5x_1 & - & 2x_2 & + & 3x_3 & \geq & 8 \\
& x_1 & - & x_2 & + & 2x_3 & = & 5 \\
& 2x_1 & + & x_2 & + & x_3 & \leq & 12 \\
& x_i \geq 0, 1 \leq i \leq 3 & & & & & &
\end{array}$$

8.

$$\begin{array}{llrcrcrcl}
\text{Minimize: } z = & x_1 & + & 2x_2 & + & x_3 & & \\
\text{Subject to:} & 3x_1 & - & x_2 & + & 3x_3 & \geq & 5 \\
& -x_1 & + & 2x_2 & + & 5x_3 & = & 6 \\
& 2x_1 & + & x_2 & - & x_3 & \geq & 2 \\
& x_i \geq 0, 1 \leq i \leq 3 & & & & & &
\end{array}$$

Solving Nonstandard Problems with *LP SOLVER*

You can enter an LP problem with *mixed constraints* ($\leq, \geq, =$) in the *LP SOLVER* program of CALIPSO. You do not need to change $\geq$ or $=$ constraints to $\leq$ constraints; *LP SOLVER* will give you the solution automatically. *LP SOLVER* uses the two-phase method, a method that is beyond the scope of this book; however, since you only see the solution, you need not be concerned about how *LP SOLVER* found it.

5.1C Exercises: Using *LP SOLVER*

Solve the problems in Exercises 5.1B with *LP SOLVER*.

5.2 THE BIG-M METHOD *

In this section we look at another method for solving LP problems in nonstandard form. This method, known as the *Big-M method*, is generally more efficient than the MSM method described in Section 5.1 though it is subject to computer rounding errors.

We use the following LP problem to illustrate the Big-M procedure.

$$\begin{array}{llcccccccc} \text{Maximize}: z = & 2x_1 & + & 3x_2 & - & x_3 & & \\ \text{Subject to}: & x_1 & + & 2x_2 & + & 3x_3 & \leq & 12 \\ & x_1 & + & 5x_2 & - & 6x_3 & \geq & 7 \\ & 3x_1 & - & x_2 & + & 4x_3 & \geq & 6 \\ & 2x_1 & - & 3x_2 & + & 4x_3 & = & 8 \\ & x_1, x_2, x_3 \geq 0 \end{array} \tag{1}$$

As we indicated previously we cannot directly apply the simplex method to this problem because of the difficulty in finding an initial basic feasible solution. To resolve this difficulty we introduce three kinds of nonnegative variables – *slack variables*, *surplus variables*, and *artificial variables*. Introducing these variables results in a system of linear equations to which we can apply the simplex method. Specifically,

(1) To each $\leq$ constraint add (as before) a nonnegative *slack variable*; thus, in our example

$$x_1 + 2x_2 + 3x_3 + s_1 = 12$$

(2) To each $\geq$ constraint, first subtract a nonnegative *surplus variable*; thus, in our example

$$\begin{array}{ccccccccccc} x_1 & + & 5x_2 & - & 6x_3 & - & s_2 & & & = & 7 \\ 3x_1 & - & x_2 & + & 4x_3 & & & - & s_3 & = & 6 \end{array}$$

Then, in order to create an initial basic solution, add a nonnegative *artificial variable* to each of these equalities; thus

$$\begin{array}{lllllllllllll} x_1 & + & 5x_2 & - & 6x_3 & - & s_2 & & & + & y_1 & & = 7 \\ 3x_1 & - & x_2 & + & 4x_3 & & & - & s_3 & & & + \; y_2 & = 6 \end{array}$$

(Eventually the artificial variables will have value 0.)

(3) To each = constraint, add an artificial variable; thus, in our example

$$2x_1 - 3x_2 + 4x_3 + y_3 = 8$$

We emphasize again that all of these new variables are nonnegative. The artificial variables serve to give us a basic feasible solution. If the problem has feasible solutions, the artificial variables will have value 0 in the optimal solution. If any of them appear with positive values in the optimal solution, then the problem is infeasible. In our example we now have

$$\begin{array}{l} \text{Maximize}: z = 2x_1 + 3x_2 - x_3 \\ \text{Subject to}: \\ \begin{array}{lllllllllllllll} x_1 & + & 2x_2 & + & 3x_3 & + & s_1 & & & & & & & = & 12 \\ x_1 & + & 5x_2 & - & 6x_3 & - & s_2 & + & y_1 & & & & & = & 7 \\ 3x_1 & - & x_2 & + & 4x_3 & - & s_3 & & & + & y_2 & & & = & 6 \\ 2x_1 & - & 3x_2 & + & 4x_3 & & & & & & & + & y_3 & = & 8 \end{array} \\ \text{All variables} \geq 0 \end{array} \qquad (2)$$

Note that the current basic feasible solution is $s_1 = 12, y_1 = 7, y_2 = 6, y_3 = 8$. Our goal is to obtain an optimal solution for which all of the artificial variables are equal to 0. To do this we maximize a new objective function

$$z' = 2x_1 + 3x_2 - x_3 - M(y_1 + y_2 + y_3) \qquad (3)$$

where M is a large number. If M is large enough, then since the artificial variables are nonnegative and since we are trying to maximize z', it is reasonable to expect that at the optimal solution the artificial variables y_1, y_2, y_3 will all be equal to 0.

To apply the simplex method we first must rewrite z' in terms of the nonbasic variables. Solving for y_1, y_2, and y_3 in the equations in (2) involving these artificial variables we find

$$\begin{aligned} y_1 &= 7 - x_1 - 5x_2 + 6x_3 + s_2 \\ y_2 &= 6 - 3x_1 + x_2 - 4x_3 + s_3 \\ y_3 &= 8 - 2x_1 + 3x_2 - 4x_3 \end{aligned}$$

Adding these equations we get

$$y_1 + y_2 + y_3 = 21 - 6x_1 - x_2 - 2x_3 + s_2 + s_3$$

Substituting in (3) and combining terms we obtain

$$\begin{aligned} z' &= 2x_1 + 3x_2 - x_3 - M(21 - 6x_1 - x_2 - 2x_3 + s_2 + s_3) \\ &= (2 + 6M)x_1 + (3 + M)x_2 + (-1 + 2M)x_3 - Ms_2 - Ms_3 - 21M \end{aligned}$$

The LP problem in standard form is

Maximize:
$z' = (2 + 6M)x_1 + (3 + M)x_2 + (-1 + 2M)x_3 - Ms_2 - Ms_3 - 21M$
Subject to:

$$\begin{array}{lllllllllll} x_1 & + 2x_2 & + 3x_3 & + s_1 & & & & & & = & 12 \\ x_1 & + 5x_2 & - 6x_3 & & - s_2 & & + y_1 & & & = & 7 \\ 3x_1 & - x_2 & + 4x_3 & & & - s_3 & & + y_2 & & = & 6 \\ 2x_1 & - 3x_2 & + 4x_3 & & & & & & + y_3 & = & 8 \end{array}$$

All variables ≥ 0

The corresponding simplex tableau is

	x_1	x_2	x_3	s_1	s_2	s_3	y_1	y_2	y_3	RHS
s_1	1	2	3	1	0	0	0	0	0	12
y_1	1	5	-6	0	-1	0	1	0	0	7
y_2	3	-1	4	0	0	-1	0	1	0	6
y_3	2	-3	4	0	0	0	0	0	1	8
OBJ	$-2-6M$	$-3-M$	$1-2M$	0	M	M	0	0	0	$-21M$

Now choose a large value for M, say $M = 100$, and then proceed with the simplex method. Using *PIVOTER* you will find that the optimal solution is $x_1 = 7.429, x_2 = 2.286, x_3 = 0, s_1 = 0, s_2 = 11.857, s_3 = 14, y_1 = 0, y_2 = 0, y_3 = 0; z = 21.714$. Why did we choose $M = 100$? This is one of the problems with the Big-M method. In order to avoid computer rounding errors you should try to choose M as small as possible; but M must be large enough to "drive" the artificial variables to 0. There is no obvious way for selecting M.

Example 1: Converting an LP Problem to Big-M Form

Consider the LP problem

$$
\begin{array}{lrcrcrcrcl}
\text{Maximize: } z = & 2x_1 & - & x_2 & + & x_3 & + & 6x_4 & & \\
\text{Subject to:} & x_1 & + & 2x_2 & - & 5x_3 & + & 4x_4 & \le & 9 \\
& 3x_1 & - & 3x_2 & + & 6x_3 & + & 2x_4 & \le & 7 \\
& x_1 & + & 5x_2 & - & x_3 & + & 2x_4 & \ge & 10 \\
& 4x_1 & - & 2x_2 & + & 4x_3 & & & = & 8 \\
& & & x_i \ge 0, 1 \le i \le 4 & & & & & &
\end{array}
$$

Introducing slack, surplus, and artificial variables we obtain the constraints

$$
\begin{array}{rcrcrcrcccccccl}
x_1 & + & 2x_2 & - & 5x_3 & + & 4x_4 & + & s_1 & & & & & = & 9 \\
3x_1 & - & 3x_2 & + & 6x_3 & + & 2x_4 & & & + s_2 & & & & = & 7 \\
x_1 & + & 5x_2 & - & x_3 & + & 2x_4 & & & & - s_3 & + y_1 & & = & 10 \\
4x_1 & - & 2x_2 & + & 4x_3 & & & & & & & & + y_2 & = & 8
\end{array}
$$

All variables ≥ 0

Since

$$
\begin{aligned}
y_1 + y_2 &= (10 - x_1 - 5x_2 + x_3 - 2x_4 + s_3) + (8 - 4x_1 + 2x_2 - 4x_3) \\
&= 18 - 5x_1 - 3x_2 - 3x_3 - 2x_4 + s_3
\end{aligned}
$$

it follows that

$$
\begin{aligned}
z' &= 2x_1 - x_2 + x_3 + 6x_4 - M(y_1 + y_2) \\
&= (2 + 5M)x_1 + (-1 + 3M)x_2 + (1 + 3M)x_3 \\
&\quad +(6 + 2M)x_4 - Ms_3 - 18M
\end{aligned}
$$

You can assign a value (say $M = 100$) and enter these data in *PIVOTER* to solve the original problem. The solution is $x_1 = 2.460, x_2 = 1.044, x_3 = 0.0619, x_4 = 1.190$ with all other variables 0. The maximum value of the objective function is $z = 11.080$.

5.2A Exercises: Using the Big-M Method with *PIVOTER*

Solve the problems in Exercises 5.1B by the Big-M method with M replaced by a large value, say 100.

5.3 INFEASIBLE, UNBOUNDED, AND NONUNIQUE SOLUTIONS

There are three situations in which an LP problem can fail to have exactly one solution:

(1) The problem has no solution because there is no way of simultaneously satisfying all the constraints. In this case we say the problem is *infeasible.*

(2) For a maximization problem, there are feasible solutions for which the objective function attains arbitrarily large positive values; or, for a minimization problem, there are feasible solutions for which the objective function attains arbitrarily large negative values. In either of these cases, we say that the LP problem is *unbounded on the feasible set* or, more simply, *unbounded.*

(3) The solution is not unique; that is, it has more than one solution that optimizes the objective function. In this case we say the problem has *nonunique* or *alternative optimal solutions.*

We consider these situations separately in Parts 1, 2, and 3 below.

Part 1: Infeasible Problems

The LP problem

$$\begin{array}{llcccl}
\text{Maximize: } z = & 60x & + & 80y & & \\
\text{Subject to:} & x & + & y & \leq & 2 \\
 & 2x & + & 2y & \geq & 12 \\
 & x, y \geq 0 & & & &
\end{array}$$

obviously has no feasible solutions since $x + y \leq 2$ implies that $2x + 2y \leq 4$, and so there can be no solutions that also satisfy the constraint $2x+2y \geq 12$. However, it is not immediately obvious that the problem in Example 1 has no feasible solutions.

Example 1: A Problem with No Feasible Solution

The LP problem

$$
\begin{array}{llcccl}
\text{Maximize}: z = & -2x & + & 3y & & \\
\text{Subject to}: & -x & + & y & \geq & 3 \\
& 3x & + & y & \leq & 5 \\
& x & + & y & \geq & 6 \\
& x, y \geq 0 & & & &
\end{array} \tag{1}
$$

does not have any feasible solutions. You can see why this is so from the graph in Figure 1.

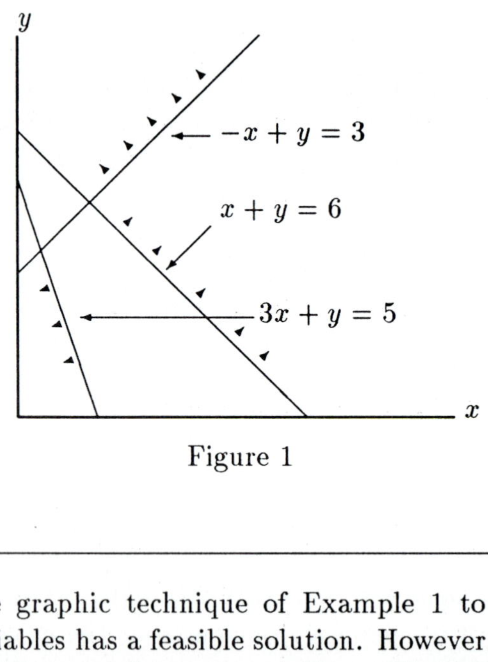

Figure 1

We could use the graphic technique of Example 1 to determine if a problem with two variables has a feasible solution. However, such a method will not work for problems with many variables. We need to develop an algebraic way of discovering that a problem has no feasible solutions. To this end, first note that any problem in standard form (see Section 4.4) has feasible solutions. In particular, setting all of the variables equal to 0 gives a feasible solution. The difficulty in discovering infeasibility occurs only with nonstandard problems. We will consider how infeasibility can be detected with the MSM method. You can find other ways of showing that a problem is infeasible in linear programming books such as [CV].

Infeasibility Termination Rule for MSM Method

If at some step of the MSM method, there is a row with a negative entry in the RHS column and all nonnegative entries in other columns, then the LP problem is infeasible.

Example 2 illustrates the general rule for testing an LP problem for infeasibility when this method is used.

Example 2: Detecting Infeasibility with the MSM Method

In Section 5.1 we used the MSM method to solve problems like (1) by multiplying the $\geq$ constraints by -1 and adding slack variables to obtain an equivalent problem. For problem (1), this equivalent problem is

$$\begin{array}{llcrcrcrcrcr} \text{Maximize}: z = & -2x & + & 3y & & & & & & & & \\ \text{Subject to}: & x & - & y & + & s_1 & & & & & = & -3 \\ & 3x & + & y & & & + & s_2 & & & = & 5 \\ & -x & - & y & & & & & + & s_3 & = & -6 \\ & \multicolumn{11}{l}{x, y, s_1, s_2, s_3 \geq 0} \end{array} \qquad (2)$$

From (2) we form the tableau

	x	y	s_1	s_2	s_3	RHS
s_1	1	−1	1	0	0	−3
s_2	3	1	0	1	0	5
s_3	−1	−1	0	0	1	−6
OBJ	2	−3	0	0	0	0

(3)

Since -3 is a negative RHS entry we can pivot on the -1 in the y column. We form the ratios $\{\frac{-3}{-1}, \frac{5}{1}, \frac{-6}{-1}\}$. Since the first ratio is smallest, we pivot on the -1 in the y column and s_1 row to obtain tableau (4).

	x	y	s_1	s_2	s_3	RHS
y	−1	1	−1	0	0	3
s_2	4	0	1	1	0	2
s_3	−2	0	−1	0	1	−3
OBJ	−1	0	−3	0	0	9

(4)

Next we pivot on the -1 in the s_3 row and s_1 column to obtain tableau (5).

	x	y	s_1	s_2	s_3	RHS
y	1	1	0	0	-1	6
s_2	2	0	0	1	1	-1
s_1	2	0	1	0	-1	3
OBJ	5	0	0	0	-3	18

(5)

The second row has RHS -1. However, all of the other entries in this row are nonnegative. The equation corresponding to the second row is

$$2x + s_2 + s_3 = -1$$

It is impossible to satisfy this equation by assigning nonnegative values to x, s_2, and s_3 because the left side will be nonnegative and the right side is negative. It follows that the system of equations (2) is infeasible. Problem (1) is also infeasible because (2) is equivalent to the original system (1).

5.3A Exercises: Infeasible Problems

Use the MSM method to show that Problems 1–5 are infeasible.

1.

$$\begin{array}{llrcrcl} \text{Maximize: } z = & 4x & + & 2y & & \\ \text{Subject to:} & x & - & 2y & \leq & 5 \\ & 3x & + & y & \leq & 6 \\ & 2x & - & y & \geq & 5 \\ & x, y \geq 0 \end{array}$$

2. §

$$\begin{array}{llrcrcl} \text{Maximize: } z = & 3x & - & y & & \\ \text{Subject to:} & 2x & + & 2y & \geq & 20 \\ & 3x & - & 2y & \leq & 20 \\ & 2x & + & y & \leq & 5 \\ & x, y \geq 0 \end{array}$$

3.

$$\begin{array}{llrcrcl} \text{Maximize: } z = & 2x & + & 3y & & \\ \text{Subject to:} & x & + & 4y & \leq & 4 \\ & x & - & 2y & \leq & 3 \\ & x & - & y & \geq & 4 \\ & x, y \geq 0 \end{array}$$

4. §

$$
\begin{array}{lrcrcl}
\text{Maximize: } z = & 3x & + & 2y & & \\
\text{Subject to:} & 3x & + & 4y & \leq & 5 \\
& 2x & + & y & \geq & 2 \\
& -x & + & y & \geq & 1 \\
& x, y \geq 0 & & & &
\end{array}
$$

5.

$$
\begin{array}{lrcrcrcl}
\text{Maximize: } z = & x_1 & + & x_2 & + & x_3 & & \\
\text{Subject to:} & x_1 & + & 2x_2 & + & 3x_3 & \leq & 27 \\
& 2x_1 & + & x_2 & + & x_3 & \geq & 54 \\
& x_1 & - & x_2 & + & x_3 & = & 9 \\
& x_i \geq 0, 1 \leq i \leq 3 & & & & & &
\end{array}
$$

6. §

$$
\begin{array}{lrcrcrcl}
\text{Minimize: } z = & 2x_1 & - & 3x_2 & - & 5x_3 & & \\
\text{Subject to:} & 3x_1 & + & 4x_2 & + & x_3 & \leq & 5 \\
& -2x_1 & + & 3x_2 & - & 4x_3 & \geq & 7 \\
& 5x_1 & + & x_2 & + & 5x_3 & \geq & 12 \\
& x_i \geq 0, 1 \leq i \leq 3 & & & & & &
\end{array}
$$

Part 2: Unbounded LP Problems

The test for unboundedness of the objective function on the feasible region is the same for problems solved by the simplex, MSM, and Big-M methods.

We use the LP problem

$$
\begin{array}{lrcrcl}
\text{Maximize : } z = & 3x & - & 2y & & \\
\text{Subject to :} & -x & + & y & \leq & 3 \\
& x & + & y & \geq & 6 \\
& x, y \geq 0 & & & &
\end{array}
\tag{6}
$$

to illustrate the ideas. Figure 2 gives the graph of (6).

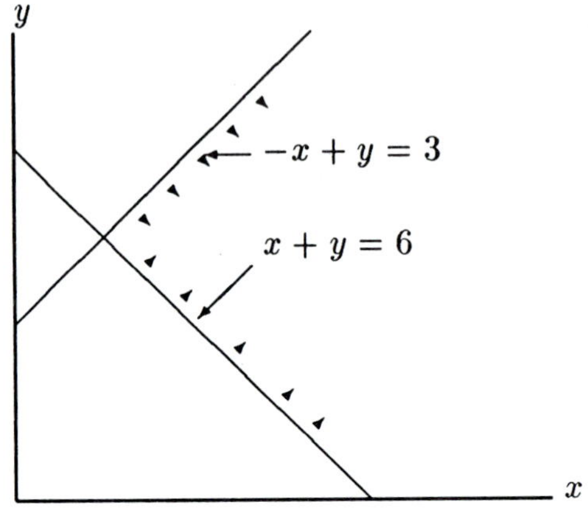

Figure 2

You can see from Figure 2 that the feasible set of (6) is *unbounded*; that is, there are feasible points at arbitrarily large distances from the origin. When the feasible region is unbounded it is possible (but not necessarily certain) that the objective function can take on arbitrarily large values. To solve (6) by the MSM method we form the initial tableau

	x	y	s_1	s_2	RHS
s_1	−1	1	1	0	3
s_2	−1	−1	0	1	−6
OBJ	−3	2	0	0	0

(7)

The MSM method requires us to pivot in either column x or y of (7). We choose to pivot in column x and row s_2 and obtain tableau (8)

	x	y	s_1	s_2	RHS
s_1	0	2	1	−1	9
x	1	1	0	−1	6
OBJ	0	5	0	−3	18

(8)

Notice that column s_2 of (8) has all entries ≤ 0. We will see that this situation indicates that the objective function is unbounded. The LP problem that corresponds to (8) is

$$
\begin{array}{llcccccccl}
\text{Maximize}: & z = -5y + 3s_2 + 18 \\
\text{Subject to}: & & & 2y & + & s_1 & - & s_2 & = & 9 \\
& x & + & y & & & - & s_2 & = & 6 \\
& x, y, s_1, s_2 \geq 0
\end{array}
\tag{9}
$$

If we set the nonbasic variable y equal to 0 in the two constraint equations of (9) and solve the first equation for the basic variable s_1 and the second for the basic variable x we obtain

$$
\begin{aligned}
s_1 &= s_2 + 9 \\
x &= s_2 + 6
\end{aligned}
\tag{10}
$$

From (10) you see that we can find feasible solutions for arbitrarily large assignments to s_2. Since s_2 has a positive coefficient in the objective function of (9), we can make that objective function as large as we want. Thus, the original LP problem (6) is unbounded on the feasible set. This example characterizes a general test for unboundedness.

The Unbounded Termination Rule

If any tableau obtained while applying the simplex method has all nonnegative entries in the RHS column (except, possibly, for the entry in the OBJ row) and has a column whose entries are all ≤ 0 (including the OBJ row entry), then the corresponding LP problem is unbounded. This rule applies to problems solved by the ordinary simplex method, the modified simplex method, or the Big-M method (as well as to some other methods we have not considered).

Notice that tableau (7) has all negative entries in its first column. This does not demonstrate that the objective function is unbounded because an entry in the RHS column is negative.

Unboundedness of the feasible set of a problem is a necessary condition for unboundedness of the objective function, but it is not sufficient. Example 5 shows that changing (6) to a minimization problem results in a problem with a bounded solution. Thus, a problem with an unbounded feasible region need *not* have an unbounded objective function.

Example 5: A Bounded Objective Function on an Unbounded Feasible Region

The LP problem

$$\begin{array}{llcll} \text{Minimize} : z = & 3x & - & 2y & \\ \text{Subject to} : & -x & + & y & \leq 3 \\ & x & + & y & \geq 6 \\ & x, y \geq 0 & & & \end{array} \tag{11}$$

has the initial tableau

(12)

	x	y	s_1	s_2	RHS
s_1	-1	1	1	0	3
s_2	-1	-1	0	1	-6
OBJ	3	-2	0	0	0

Following the rules of the MSM we pivot on the -1 in the s_2 row and x column of (12) to obtain (13)

(13)

	x	y	s_1	s_2	RHS
s_1	0	2	1	-1	9
x	1	1	0	-1	6
OBJ	0	-5	0	3	-18

We now follow the rules of the ordinary simplex method and pivot on the 2 in row s_1 and column y of (13) to obtain (14)

(14)

	x	y	s_1	s_2	RHS
y	0	1	$\frac{1}{2}$	$-\frac{1}{2}$	$\frac{9}{2}$
x	1	0	$-\frac{1}{2}$	$-\frac{1}{2}$	$\frac{3}{2}$
OBJ	0	0	$\frac{5}{2}$	$\frac{1}{2}$	$\frac{9}{2}$

Tableau (14) shows that problem (11) has a finite solution.

5.3B Exercises: Unbounded Objective Functions

Show that the following problems are unbounded by using either the ordinary simplex method or the MSM method, as is appropriate.

1.

$$\begin{array}{lrcrcl}
\text{Maximize: } z = & 2x & + & 3y & & \\
\text{Subject to:} & x & - & 2y & \leq & 2 \\
 & -5x & + & 3y & \leq & 3 \\
 & x, y \geq 0 & & & &
\end{array}$$

2. §

$$\begin{array}{lrcrcl}
\text{Maximize: } z = & 2x & + & 3y & & \\
\text{Subject to:} & x & - & y & \leq & 1 \\
 & x & + & y & \geq & 2 \\
 & x, y \geq 0 & & & &
\end{array}$$

3.

$$\begin{array}{lrcrcl}
\text{Maximize: } z = & x & + & y & & \\
\text{Subject to:} & 2x & - & 3y & \leq & 10 \\
 & 3x & - & 2y & \leq & 20 \\
 & x, y \geq 0 & & & &
\end{array}$$

4. §

$$\begin{array}{lrcrcl}
\text{Maximize: } z = & 3x & - & y & & \\
\text{Subject to:} & 2x & + & 2y & \geq & 20 \\
 & 3x & - & 2y & \leq & 20 \\
 & x, y \geq 0 & & & &
\end{array}$$

5. §

$$\begin{array}{lrcrcrcl}
\text{Minimize: } z = & 6x_1 & - & 2x_2 & - & 2x_3 & & \\
\text{Subject to:} & 3x_1 & - & 2x_2 & - & 2x_3 & \leq & 4 \\
 & -2x_1 & + & x_2 & & & \geq & 1 \\
 & -2x_1 & + & 5x_2 & & & \leq & 8 \\
 & x_i \geq 0, 1 \leq i \leq 3 & & & & & &
\end{array}$$

6.

$$\begin{array}{lrcrcrcl}
\text{Maximize: } z = & 10x_1 & + & 42x_2 & - & 3x_3 & & \\
\text{Subject to:} & -14x_1 & - & 80x_2 & + & 6x_3 & \leq & 88 \\
 & 28x_1 & + & 7x_2 & - & 21x_3 & \leq & 112 \\
 & -7x_1 & + & 14x_2 & - & 24x_3 & \leq & 107 \\
 & x_i \geq 0, 1 \leq i \leq 3 & & & & & &
\end{array}$$

7.

$$\begin{array}{lrcrcrcl}
\text{Maximize: } z = & 2x_1 & + & x_2 & - & 3x_3 & & \\
\text{Subject to:} & 3x_1 & + & 2x_2 & - & 4x_3 & \geq & 7 \\
 & x_1 & - & x_2 & + & 3x_3 & \geq & 2 \\
 & x_i \geq 0, 1 \leq i \leq 3 & & & & & &
\end{array}$$

8.

$$\begin{array}{llrcrcrcrcl}
\text{Minimize: } z = & & 13x_1 & - & 6x_2 & + & x_3 & & & & \\
\text{Subject to:} & & -2x_1 & + & 6x_2 & + & 2x_3 & - & 3x_4 & \leq & 20 \\
& & -4x_1 & + & 3x_2 & + & x_3 & - & x_4 & \leq & 10 \\
& & & & -5x_2 & + & 3x_3 & - & x_4 & \leq & 60 \\
& & x_i \geq 0, 1 \leq i \leq 4 & & & & & & & &
\end{array}$$

Part 3: Nonunique Optimal Solutions

There are two circumstances in which the final tableau of an LP problem indicates the possible existence of nonunique (alternative) optimal solutions. One is the presence of a 0 in the OBJ row of a column corresponding to a nonbasic variable. The other is a final value of 0 for a basic variable. Example 6 illustrates the first circumstance and Example 7 illustrates the second.

Example 6: A Zero Coefficient in a Nonbasic Column of the Objective Function

You can verify with *PIVOTER* that the LP problem

$$\begin{array}{lrcrcrcl}
\text{Maximize}: z = & 60x_1 & + & 35x_2 & + & 20x_3 & & \\
\text{Subject to}: & 8x_1 & + & 6x_2 & + & x_3 & \leq & 48 \\
& 4x_1 & + & 2x_2 & + & \frac{3}{2}x_3 & \leq & 20 \\
& 2x_1 & + & \frac{3}{2}x_2 & + & \frac{1}{2}x_3 & \leq & 8 \\
& x_1, x_2, x_3 \geq 0 & & & & & &
\end{array} \tag{15}$$

has the final tableau

(16)

	x_1	x_2	x_3	s_1	s_2	s_3	RHS
s_1	0	-2	0	1	2	-8	24
x_3	0	-2	1	0	2	-4	8
x_1	1	$\frac{5}{4}$	0	0	$-\frac{1}{2}$	$\frac{3}{2}$	2
OBJ	0	0	0	0	10	10	280

Notice that the nonbasic variable x_2 has a 0 coefficient in the objective function of the LP problem corresponding to (16). Pivoting on any entry in the x_2 column will not change the value of the objective function; however, pivoting on some entries might make entries in the RHS column negative. You can avoid this by choosing the pivot entry by the Departing Variable Rule of the ordinary simplex method. In (16) this rule selects $\frac{5}{4}$ as the pivot element. Pivoting on $\frac{5}{4}$ we get tableau (17).

	x_1	x_2	x_3	s_1	s_2	s_3	RHS
s_1	$\frac{8}{5}$	0	0	1	$\frac{6}{5}$	$-\frac{28}{5}$	$\frac{136}{5}$
x_3	$\frac{8}{5}$	0	1	0	$\frac{6}{5}$	$-\frac{8}{5}$	$\frac{56}{5}$
x_2	$\frac{4}{5}$	1	0	0	$-\frac{2}{5}$	$\frac{6}{5}$	$\frac{8}{5}$
OBJ	0	0	0	0	10	10	280

(17)

Tableau (16) gives the optimal solution

$$x_1 = 2, x_2 = 0, x_3 = 8, s_1 = 24, s_2 = 0; z = 280$$

and tableau (17) gives the optimal solution

$$x_1 = 0, x_2 = \tfrac{8}{5}, x_3 = \tfrac{56}{3}, s_1 = \tfrac{136}{5}, s_2 = 0; z = 280$$

Example 7: A Zero in the RHS Column

You can verify using the MSM method with *PIVOTER* that the LP problem

$$\begin{array}{lrcrcl} \text{Minimize}: z = & 2x & + & y & & \\ \text{Subject to}: & x & + & y & \geq & 5 \\ & x & + & 2y & \geq & 8 \\ & -x & + & y & \leq & 1 \\ & x, y \geq 0 & & & & \end{array} \tag{18}$$

has the final tableau

	x	y	s_1	s_2	s_3	RHS
s_1	0	0	1	$-\frac{2}{3}$	$-\frac{1}{3}$	0
x	1	0	0	$-\frac{1}{3}$	$-\frac{2}{3}$	2
y	0	1	0	$-\frac{1}{3}$	$\frac{1}{3}$	3
OBJ	0	0	0	1	1	-7

(19)

You can also get other final tableaus if you make different choices than we did in executing the MSM. Since the first row of (19) has a 0 in its RHS column, pivoting on any entry in the first row will not change the value -7 of the objective function. However, pivoting on some entries in this row might cause negative numbers to appear in the OBJ row. You can determine which entry to pivot on by forming ratios of the OBJ row entries to negative entries in the pivot row. For (19) these ratios are

$$\frac{1}{-2/3} \quad \text{and} \quad \frac{1}{-1/3}$$

Since the larger ratio is $\frac{1}{-2/3}$ we pivot on the $-\frac{2}{3}$ in the pivot row to obtain another optimal tableau (20). (See the Entering Variable Rule of the Dual Simplex Method in Section 6.1 for a more thorough explanation of this rule.)

	x	y	s_1	s_2	s_3	RHS
s_2	0	0	$-\frac{3}{2}$	1	$\frac{1}{2}$	0
x	1	0	$-\frac{1}{2}$	0	$-\frac{1}{2}$	2
y	0	1	$-\frac{1}{2}$	0	$\frac{1}{2}$	3
OBJ	0	0	$\frac{3}{2}$	0	$\frac{1}{2}$	-7

(20)

Tableau (19) gives the optimal solution $x = 2, y = 3, s_1 = 0, s_2 = 0; z = 7$ and Tableau (20) gives an alternative optimal solution $x = 2, y = 3, s_1 = 0, s_2 = 0; z = 7$.

In general, we have the following rule.

Finding Alternative Optimal Solutions

(1) Look in the final tableau for a 0 in the OBJ row of a nonbasic variable. If the column containing such a 0 contains any positive entries, choose a pivot element by the Departing Basic Variable Rule of the ordinary simplex method. Performing the pivot will yield another optimal tableau giving an alternative solution.

(2) Look in the final tableau for a 0 in the RHS column. If a row with 0 in its RHS has any negative entries, form the ratios of the OBJ row entries to negative entries in the pivot row. Choose the largest (or one of the largest) of these ratios. The column from which a largest ratio comes is the pivot column. Pivoting on such an entry will yield an alternative optimal solution.

5.3C Exercises: Alternative Optimal Solutions

Show that the following problems have alternative optimal solutions by solving them by hand or with *PIVOTER*.

1.

$$\begin{array}{llcll} \text{Maximize: } z = & 6x_1 & + & 9x_2 & \\ \text{Subject to:} & 2x_1 & + & x_2 & \leq 8 \\ & 2x_1 & + & 3x_2 & \leq 12 \\ & -2x_1 & + & x_2 & \leq 4 \\ & \multicolumn{4}{l}{x_i \geq 0, 1 \leq i \leq 3} \end{array}$$

2. §

$$\begin{array}{llcll} \text{Maximize: } z = & x_1 & + & x_2 & \\ \text{Subject to:} & x_1 & + & 2x_2 & \leq 24 \\ & x_1 & + & x_2 & \leq 14 \\ & 5x_1 & + & 3x_2 & \leq 60 \\ & \multicolumn{4}{l}{x_i \geq 0, 1 \leq i \leq 3} \end{array}$$

3.

$$\begin{array}{llclcll} \text{Maximize: } z = & -1320x_1 & + & 390x_2 & - & 5x_3 & \\ \text{Subject to:} & 8x_1 & - & 2x_2 & + & 4x_3 & \leq 12 \\ & 6x_1 & + & 4x_2 & + & 2x_3 & \leq 8 \\ & 6x_1 & + & 4x_2 & + & x_3 & \leq 8 \\ & \multicolumn{6}{l}{x_i \geq 0, 1 \leq i \leq 3} \end{array}$$

4. §

$$\begin{array}{llclcll} \text{Minimize: } z = & -10x_1 & + & 15x_2 & - & 5x_3 & \\ \text{Subject to:} & 2x_1 & - & 3x_2 & + & x_3 & \leq 20 \\ & x_1 & - & 2x_2 & + & 3x_3 & \leq 53 \\ & 6x_1 & + & 4x_2 & + & 2x_3 & \leq 44 \\ & \multicolumn{6}{l}{x_i \geq 0, 1 \leq i \leq 3} \end{array}$$

5.

$$\begin{array}{llclcll} \text{Maximize: } z = & 10x_1 & + & 20x_2 & + & 10x_3 & \\ \text{Subject to:} & 2x_1 & + & 3x_2 & + & x_3 & \leq 12 \\ & x_1 & + & 2x_2 & + & x_3 & \leq 5 \\ & -2x_1 & + & 3x_2 & + & 4x_3 & \leq 6 \\ & \multicolumn{6}{l}{x_i \geq 0, 1 \leq i \leq 3} \end{array}$$

6.

$$\begin{array}{lrcrcrcr} \text{Maximize: } z = & 10x_1 & - & 20x_2 & + & 30x_3 & & \\ \text{Subject to:} & 2x_1 & - & 3x_2 & + & x_3 & \leq & 100 \\ & x_1 & + & 2x_2 & + & 3x_3 & \leq & 30 \\ & 4x_1 & - & 7x_2 & + & 6x_3 & \leq & 60 \\ & \multicolumn{7}{l}{x_i \geq 0, 1 \leq i \leq 3} \end{array}$$

The next two problems could result from transportation problems, which are discussed in the next section. They both have alternative optimal solutions. Formulate them for solution by the Big-M method and use *PIVOTER* to verify that they have nonunique solutions.

7.

$$\begin{array}{lrcrcrcr} \multicolumn{8}{l}{\text{Minimize: } z = 4x_{11} + 2x_{12} + 3x_{13} + 2x_{21} + 4x_{22} + 3x_{23}} \\ \text{Subject to:} & x_{11} & + & x_{12} & + & x_{13} & \leq & 40 \\ & x_{21} & + & x_{22} & + & x_{23} & \leq & 30 \\ & x_{11} & + & x_{21} & = & 10 & & \\ & x_{12} & + & x_{22} & = & 45 & & \\ & x_{13} & + & x_{23} & = & 15 & & \\ & \multicolumn{7}{l}{x_{ij} \geq 0, 1 \leq i \leq, 1 \leq j \leq 3} \end{array}$$

8. §

$$\begin{array}{lrcrcrcr} \multicolumn{8}{l}{\text{Minimize: } z = 4x_{11} + 2x_{12} + 3x_{13} + 2x_{21} + 4x_{22} + 3x_{23}} \\ \text{Subject to:} & x_{11} & + & x_{12} & + & x_{13} & \leq & 60 \\ & x_{21} & + & x_{22} & + & x_{23} & \leq & 20 \\ & x_{11} & + & x_{21} & = & 15 & & \\ & x_{12} & + & x_{22} & = & 30 & & \\ & x_{13} & + & x_{23} & = & 15 & & \\ & \multicolumn{7}{l}{x_{ij} \geq 0, 1 \leq i \leq 3, 1 \leq j \leq 3} \end{array}$$

5.4 SOLVING APPLIED PROBLEMS

This section consists primarily of a number of word problems. These problems involve three or more variables and are somewhat more difficult than those encountered thus far. Some of the exercises found in later sections on duality and sensitivity analysis will refer to these problems.

Note: In applying the simplex method you frequently will obtain optimal solutions that though mathematically correct, make no practical sense. In particular, optimal solutions may need to be integer valued. Suppose, for instance, in determining the number of cars to manufacture in order to maximize profits you obtain the optimal solution: Model A, 17.6591; Model B, .0023, and Model C, 300.21. The market value of .0023 of a car or .6591 of a car is questionable. To find an optimal integer solution to this problem you might try to round the noninteger solution. However, as illustrated in Problem 14 in Exercises 5.4A, answers obtained from either rounding up or down (or both) may be far from the true integer-valued solution. You can resolve this problem by using techniques from Integer Programming, a topic discussed in [CV] but not in this text. For purposes of the exercises you can simply accept noninteger optimal solutions, but you should be aware that in practice more work (in fact, a different approach) may be called for.

The first example reviews a general approach to setting up and solving word problems.

Example 1: A Media Problem

A politician has $500,000 available for a media campaign. She finds that each minute of TV time costs $750, each minute of radio time costs $400, and each newspaper ad costs $300. Each minute of TV time will reach 80,000 people, each minute of radio time will reach 50,000 people, and each newspaper ad will reach 40,000 people. The candidate calculates that she needs to reach at least 2,000,000 people. She also wants to have at least twice as much radio time as TV time. Moreover, the amount spent on newspaper ads should be no more than the amount spent on radio and TV ads together. How should she organize her media campaign to realize her goals at a minimum cost?

Step 1. Determine the variables.

You are asked to find the number of TV and radio minutes as well as the number of newspaper ads. These are your variables. Let x_1 be the number of TV minutes purchased, x_2 the number of radio minutes, and x_3 the number of newspaper ads.

Step 2. Find the objective function.

You are trying to minimize total costs so the objective function should reflect the total cost of advertising. Since it costs $750 per TV minute,

\$400 per radio minute, and \$300 per newspaper ad, the objective function is

$$z = 750x_1 + 400x_2 + 300x_3$$

Step 3. Find the constraints.

Since the candidate must reach 2,000,000 people we have the constraint

$$80,000x_1 + 50,000x_2 + 40,000x_3 \geq 2,000,000$$

Numerically it is easier to work with the equivalent constraint

$$8x_1 + 5x_2 + 4x_3 \geq 200$$

The constraint

$$x_2 \geq 2x_1, \text{or, equivalently, } 2x_1 - x_2 \leq 0$$

ensures that there will be at least twice as much radio time as TV time. Finally, the constraint

$$300x_3 \leq 750x_1 + 400x_2, \text{ or, equivalently, } 75x_1 + 40x_2 - 30x_3 \geq 0$$

ensures that no more money will be spent on newspaper ads than on TV and radio spots.

Thus, one formulation of this problem as an LP problem is

$$\begin{array}{llcccccl}
\text{Minimize: } z = & 750x_1 & + & 400x_2 & + & 300x_3 & & \\
\text{Subject to:} & 8x_1 & + & 5x_2 & + & 4x_3 & \geq & 200 \\
 & 2x_1 & - & x_2 & & & \leq & 0 \\
 & 75x_1 & + & 40x_2 & - & 30x_3 & \geq & 0 \\
 & & & x_1, x_2, x_3 \geq 0 & & & &
\end{array}$$

You can enter this problem in *LP SOLVER* to get the solution

$$x_1 = 0, x_2 = 19.355, x_3 = 25.806; z = \$15,483.87$$

In many problems it is convenient to first set up a table to organize the given data. Example 2 illustrates this.

Example 2: A Production Problem

A small firm uses four machines A, B, C, D to produce three products P_1, P_2, P_3. Producing one unit of product P_1 requires 2 hours on machine A, 4 hours on machine B, 1 hour on machine C, and 2 hours on machine D. Machine times for producing one unit of product P_2 are 5 hours, 3 hours, 3 hours, and 2 hours on machines A, B, C, D, respectively. Machine times for producing one unit of product P_3 are 2 hours, 4 hours, 1 hour, and 6 hours on machines A, B, C, D, respectively. Machine A is available 40 hours per week; machine B, 50 hours per week; machine C, 35 hours per week; and machine D, 60 hours per week. The profits per unit produced of the products P_1, P_2, P_3 are \$150, \$200, and \$130, respectively. To maximize profits how many units of each product should be produced each week?

Before carrying out the steps outlined in Example 1, you should first set up a table similar to the following which gives the machine times required for each product, the available machine times, and the profits for each product.

	A	B	C	D	Profit(\$)
P_1	2	4	1	2	150
P_2	5	3	3	2	200
P_3	2	4	1	6	130
Amt. Avail.	40	50	35	60	

Letting x_1 be the number of units of P_1 produced, x_2 the number of units of P_2, and x_3 the number of units of P_3, we obtain from the table the LP problem

$$\begin{array}{llllllll}
\text{Maximize: } z = & 150x_1 & + & 200x_2 & + & 130x_3 & & \\
\text{Subject to:} & 2x_1 & + & 5x_2 & + & 2x_3 & \leq & 40 \\
 & 4x_1 & + & 3x_2 & + & 4x_3 & \leq & 50 \\
 & x_1 & + & 3x_2 & + & x_3 & \leq & 35 \\
 & 2x_1 & + & 2x_2 & + & 6x_3 & \leq & 60 \\
 & & & x_1, x_2, x_3 \geq 0 & & & &
\end{array}$$

The Transportation Problem

The *transportation problem* was one of the first applications of linear programming methods. In its simplest form this problem involves distributing a single product P from a number of *supply points* (say, warehouses) to a number of *demand points* (say, retail stores). You are given the costs of shipping one unit of product P from any given supply point to any given demand point. The problem is to meet (at minimal cost) all of the demand requirements subject to constraints on the amounts of supplies available at each supply point. The next example illustrates this type of problem.

Example 3: A Transportation Problem

The following *cost table* provides the information necessary to set up an LP transportation problem. In this problem there are 3 supply points S_1, S_2, S_3 and 4 demand points D_1, D_2, D_3, D_4. The numbers at the right of the table indicate the supplies available of a product P at each supply point. For instance, there are 400 units of P available at supply point S_2.

The values at the bottom of the table indicate the demands at each demand point. For instance, demand point D_3 requires 300 units of product P. Finally, the values in the body of the table give the costs of shipping 1 unit of P from a supply point to a demand point, Thus, for example, the cost is 12 to ship 1 unit of P from supply point S_3 to demand point D_2.

Supply points	Demand points				
	D_1	D_2	D_3	D_4	Supply
S_1	6	4	3	8	800
S_2	10	5	13	7	400
S_3	8	12	4	11	500
Demand	200	450	300	250	

To set up the corresponding LP problem you first must determine the variables. Let x_{ij} be the number of units shipped from supply point S_i to demand point D_j (for instance, x_{23} is the number units of P shipped from supply point S_2 to demand point D_3). From the costs found in the cost table we obtain the objective function

$$\begin{aligned} z = \; & 6x_{11} + 4x_{12} + 3x_{13} + 8x_{14} + 10x_{21} + 5x_{22} \\ & +13x_{23} + 7x_{24} + 8x_{31} + 12x_{32} + 4x_{33} + 11x_{34} \end{aligned} \tag{1}$$

that we want to minimize.

The transportation problem has two kinds of constraints: *supply constraints* and *demand constraints.* The supply constraints are determined by the available supplies. For instance, since there are 800 units of P available at supply point S_1 and since the total amount of P shipped from S_1 is

$x_{11} + x_{12} + x_{13} + x_{14}$ we have the first supply constraint

$$x_{11} + x_{12} + x_{13} + x_{14} \leq 800 \tag{2}$$

Similar reasoning gives the other supply constraints for supply points S_2 and S_3.

$$x_{21} + x_{22} + x_{23} + x_{24} \leq 400 \tag{3}$$

$$x_{31} + x_{32} + x_{33} + x_{34} \leq 500 \tag{4}$$

Demand constraints are found in an analogous manner. The demand for P at demand point D_1 is 200 units. Since $x_{11} + x_{21} + x_{31}$ represents the total amount of P shipped to D_1, we have the first demand constraint

$$x_{11} + x_{21} + x_{31} \geq 200 \tag{5}$$

The remaining demand constraints are

$$x_{12} + x_{22} + x_{32} \geq 450 \tag{6}$$

$$x_{13} + x_{23} + x_{33} \geq 300 \tag{7}$$

$$x_{14} + x_{24} + x_{34} \geq 250 \tag{8}$$

Thus, the LP problem to be solved has objective function (1) and constraints (2) – (8). The solution found by *LP SOLVER* is

$$x_{11} = 200, x_{12} = 300, x_{13} = 300, x_{22} = 150, x_{24} = 250; z = 5800$$

Example 4: Representing a Scheduling Problem as a Transportation Problem

Some scheduling problems can be viewed as transportation problems. We illustrate this with the following situation. Table 1 gives information needed to establish an optimal production schedule for a manufacturer operating on a quarterly basis.

Table 1

Quarter	Quarterly demand	Normal production capacity	Overtime production capacity	Unit storage costs	Unit normal costs	Overtime additional unit costs
1	240	190	60	5	25	15
2	150	170	45	6	30	10
3	200	230	0	5	35	15
4	350	300	40	3	30	20

The problem is to meet the demands at minimal cost. To set this problem up as a transportation problem we define the "supply points" to be the various quarters, where we make a distinction between normal and overtime production. Specifically we have

S_1:	Quarter 1, normal production	190
S_2:	Quarter 1, overtime production	60
S_3:	Quarter 2, normal production	170
S_4:	Quarter 2, overtime production	45
S_5:	Quarter 3, normal production	230
S_6:	Quarter 4, normal production	300
S_7:	Quarter 4, overtime production	40

The "demand points" are the four quarters

D_1:	Quarter 1	240	D_2:	Quarter 2	150
D_3:	Quarter 3	200	D_4:	Quarter 4	350

Table 2 is the corresponding cost table.

Table 2

Supply points	Demand points				
	D_1	D_2	D_3	D_4	Supply
S_1	25	30	36	41	190
S_2	40	45	51	56	60
S_3	∞	30	36	41	170
S_4	∞	40	46	51	45
S_5	∞	∞	35	40	230
S_6	∞	∞	∞	30	300
S_7	∞	∞	∞	50	40
Demand	240	150	200	350	

Notes:

- We obtain the S_1 row of Table 2 as follows. The entry (25) in the D_1 column is the Normal Unit Cost for Quarter 1. The entry in the D_2 column is $25 + 5 = 30$, the sum of the Normal Unit Cost to produce the product in the first quarter and the cost of storing it until the second quarter. The entry in the D_3 column is $25 + 5 + 6 = 36$, the production cost plus the cost of storage for the first quarter plus the cost of storage for the second quarter. Similarly, the entry in the D_4 column is $25 + 5 + 6 + 5 = 41$.

- We obtain rows S_3, S_5, and S_6 in a way similar to row S_1. You should check the entries in these rows.

- We obtain the S_2 row of Table 2 as follows. The entry in the D_1 column is $40 = 25 + 15$, the sum of the normal and overtime unit costs in Quarter 1. The entry in the D_2 column is $40 + 5 = 45$, the sum of the total unit cost plus the storage cost for the first quarter. We obtain the other columns in row S_2 by adding on more storage costs as we did for row S_1.

- The ∞ entries indicate that it is impossible to supply one quarter's demand with the production in a later quarter.

The transportation problem obtained from Table 2 (with 500 replacing the ∞ entries) is

Minimize:

$$\begin{array}{rrcrcrcrc}
z = & 25x_{11} & + & 30x_{12} & + & 36x_{13} & + & 41x_{14} \\
 & 40x_{21} & + & 45x_{22} & + & 51x_{23} & + & 56x_{24} \\
 & 500x_{31} & + & 30x_{32} & + & 36x_{33} & + & 41x_{34} \\
 & 500x_{41} & + & 40x_{42} & + & 46x_{43} & + & 51x_{44} \\
 & 500x_{51} & + & 500x_{52} & + & 35x_{53} & + & 40x_{54} \\
 & 500x_{61} & + & 500x_{62} & + & 500x_{63} & + & 30x_{64} \\
 & 500x_{71} & + & 500x_{72} & + & 500x_{73} & + & 50x_{74}
\end{array}$$

Subject to:

$$\begin{array}{ccccccccccccccc}
x_{11} & + & x_{12} & + & x_{13} & + & x_{14} & \leq & 190 \\
x_{21} & + & x_{22} & + & x_{23} & + & x_{24} & \leq & 60 \\
x_{31} & + & x_{32} & + & x_{33} & + & x_{34} & \leq & 170 \\
x_{41} & + & x_{42} & + & x_{43} & + & x_{44} & \leq & 45 \\
x_{51} & + & x_{52} & + & x_{53} & + & x_{54} & \leq & 230 \\
x_{61} & + & x_{62} & + & x_{63} & + & x_{64} & \leq & 300 \\
x_{71} & + & x_{72} & + & x_{73} & + & x_{74} & \leq & 40 \\
x_{11} & + & x_{21} & + & x_{31} & + & x_{41} & + & x_{51} & + & x_{61} & + & x_{71} & \geq & 240 \\
x_{12} & + & x_{22} & + & x_{32} & + & x_{42} & + & x_{52} & + & x_{62} & + & x_{72} & \geq & 150 \\
x_{13} & + & x_{23} & + & x_{33} & + & x_{43} & + & x_{53} & + & x_{63} & + & x_{73} & \geq & 200 \\
x_{14} & + & x_{24} & + & x_{34} & + & x_{44} & + & x_{54} & + & x_{64} & + & x_{74} & \geq & 350
\end{array}$$

$x_{ij} \geq 0, 1 \leq i \leq 7, 1 \leq j \leq 4$

Solving this problem with CALIPSO's *LP SOLVER* we get the optimal solution $z = 29,270$ with

$$x_{11} = 190, x_{21} = 50, x_{22} = 150, x_{33} = 20, x_{53} = 180, x_{54} = 50, x_{64} = 300$$

and the other decision variables are 0.

Note: The *LP SOLVER* program of CALIPSO is capable of generating systems of inequalities with commands that resemble those in the BASIC programming language. Problem 15 shows you how to use this facility.

5.4A Exercises: Applied Problems

In the these exercises you should, when appropriate, first set up a table to organize the data and then formulate the corresponding LP problem. You can solve the problems by hand or, more easily, with *PIVOTER* or *LP SOLVER*.

1. § Determine the lowest-cost diet, chosen from the nutrients in the following table, that supplies the minimum daily requirements of 80 g of protein, 50 mg of vitamin C, and 10 mg of iron.

Food	Protein (g/unit)	Vitamin C (mg/unit)	Iron (mg/unit)	Cost ($/unit)
Squash	0.5	3	0.4	0.05
Carrots	0.8	4	0.4	0.09
Apples	0.4	7	0.5	0.10
Bananas	1.2	10	.04	0.12
Beef	6.0	0	1.5	0.90

2. A farmer can choose between three feeds for his llamas. The following table gives the nutritional requirements and costs. The minimum daily requirements of nutrients A, B, and C are 72, 81, and 69 units, respectively. Determine the mixture of feeds that will supply the minimum nutritional requirements at least cost.

Feed	A (units/lb)	B (units/lb)	C (units/lb)	Cost ($/lb)
Feed 1	5	8	3	0.10
Feed 2	2	4	4	0.07
Feed 3	6	5	3	0.06

3. § A manufacturing firm has two plants P_1 and P_2. Each plant is capable of manufacturing four products: A, B, C, and D. The firm

is under contract to supply a distributor with 1000 units of Product A, 800 units of Product B, 900 units of Product C, and 1500 units of Product D. The following table gives the number of units of these products that plants P_1 and P_2 can produce per day.

Product	Plant P_1 (units/day)	Plant P_2 (units/day)
A	200	100
B	60	200
C	90	150
D	130	80

The cost of operating Plant P_1 is \$8000 per day and the cost of operating Plant P_2 is \$11,000 per day. How many days should each plant operate per week to meet the demands of the distributor at minimum cost?

4. A large manufacturing company has 5 factories F_1, F_2, F_3, F_4, F_5 that are capable of producing 4 kinds of appliances: refrigerators, freezers, dishwashers, and ovens. The following table lists the daily productions of each factory and the per-day costs of operating the factories:

	Refrigerators	Freezers	Dishwashers	Ovens	Costs(\$)
F_1	30	25	60	5	10,000
F_2	35	10	40	0	15,000
F_3	50	40	10	0	8,000
F_4	20	15	35	60	15,000
F_5	15	10	60	30	12,000

The plant has received orders for 1500 refrigerators, 1000 freezers, 1000 dishwashers, and 750 ovens. How many days should each factory operate to meet the demands but minimize the costs?

5. § Having won \$200,000 in the state lottery Jennifer decides to invest her money in stocks, bonds, real estate, and a savings account. She estimates her potential annual return from stocks to be 12%, from bonds 8%, from real estate 10%, and from the savings account, 6%. After a good deal of thought she has decided on the following restrictions on her investments.

(a) She will invest no more money in stocks and real estate than she puts in her savings account.

(b) The amount invested in bonds will be at least 2/3 of the amount placed in stocks.

(c) She will invest at least \$50,000 in stocks, bonds, and real estate (together).

To maximize her annual return how much should Jennifer invest in each of the four possibilities?

6. At a large university, precalculus courses are taught by graduate students, instructors, and professors. It is recommended that graduate students teach classes of size 30; instructors, classes of size 40; and professors, classes of size 70. The cost per hour of graduate students is \$10, of instructors, \$18, and professors, \$45. Instructors are to teach at least twice as many classes as graduate students and the number of classes taught by professors is to be at least the number of classes taught by graduate students and instructors combined. If there are 1500 students enrolled in precalculus math courses, how many classes should be taught by graduate students, instructors, and professors to minimize costs?

7. § Blended Fruits produces four kinds of drinks, each consisting of a blend of apple, orange, lime, cranberry, and peach juices. A quart (32 oz) of Drink 1 consists of 5 oz of apple, 8 oz of orange, 12 oz of lime, 3 oz of cranberry, and 4 oz of peach juice. Drink 2 consists of 8 oz of apple, 16 oz of cranberry, and 8 oz of peach juice. Drink 3 consists of 4 oz of apple, 12 oz of orange, 3 oz of lime, and 13 oz of cranberry juice. Drink 4 consists of 5 oz of apple, 6 oz of lime, 8 oz of cranberry, and 13 oz of peach juice. On a daily basis Blended Fruits has available 7500 quarts of apple juice, 5000 quarts of orange juice, 1200 quarts of lime juice, 4000 quarts of cranberry juice, and 1500 quarts of peach juice. The profit on the sale of a quart of Drink 1 is \$0.60; a quart of Drink 2, \$0.40; a quart of Drink 3, \$0.80; and a quart of Drink 4, \$0.50. How many quarts of each kind of drink should Blended Fruits produce daily to maximize profits?

8. A large industrial plant requires many security personnel. The following table gives the minimal personnel requirements for six 4-hour periods.

Period	Start-End	Personnel Requirements
1	12AM- 4AM	20
2	4AM- 8AM	24
3	8AM-12PM	32
4	12PM- 4PM	25
5	4PM- 8PM	20
6	8PM-12AM	16

The plant hires security personnel for 8-hour shifts that start at the beginning of each period. What is the minimum number of security personnel needed and how should they be scheduled? (Hint: let x_i be the number of people who start work at the beginning of period i. Then the requirement that 20 people work from 12AM to 4AM can be modeled with the inequality $x_1 + x_6 \geq 20$. We use $\geq$ because it may be impossible to meet the requirements exactly with people working 8-hour shifts.)

9. § The *stock cutting problem* arises in many contexts. This exercise presents a simple illustration of this kind of problem. A particular brand of pipe comes in 30-ft lengths. You are a contractor and need at least 150 7-ft pieces, 200 12-ft pieces, and 300 16-ft pieces. You can reasonably use any of the following cutting patterns to obtain the desired sizes:

Pattern	$7'$ pieces	$12'$ pieces	$16'$ pieces	Feet of Scrap
1	4	0	0	2
2	2	1	0	4
3	2	0	1	0
4	0	2	0	6
5	0	1	1	2

How many times should each cutting pattern be used to mimimize the amount of scrap? (Note that the simplex method will give a solution that uses some cutting patterns a noninteger number of times. A solution in integers requires the use of integer programming methods that are beyond the scope of this book.)

10. Suppose that pipe in Problem 9 comes in 25-ft lengths and you need 400 pieces of 7-ft pipe, 350 pieces of 8-ft pipe, and 200 pieces of 11-ft pipe. As in Problem 9 find all reasonable patterns and then determine how many times each pattern should be used to minimize the amount of scrap.

11. § Solve the transportation problem with cost table

Supply points	Demand points					
	D_1	D_2	D_3	D_4	D_5	Supply
S_1	8	6	13	11	7	420
S_2	3	1	7	14	9	160
S_3	5	8	9	6	15	300
Demand	150	100	250	90	230	

12. For some transportation problems it may be impossible to ship the product P from some supply point to some demand point. In this case it suffices to enter a number in the body of the cost table that is large enough to ensure that at the minimal solution no units of P are shipped from this supply point to the demand point. In the following cost table ∞ indicates that no supplies can be shipped from the corresponding supply point to the corresponding demand point. Replace ∞ with an appropriately large number (say 100) and solve the problem.

Supply points	Demand points			
	D_1	D_2	D_3	Supply
S_1	4	∞	8	240
S_2	6	12	5	160
S_3	10	7	∞	200
S_4	8	10	16	300
Demand	350	400	150	

13. § For some transportation problems it may occur that the demands exceed the supplies. In this case there can be no solution. However, it still may be of interest to determine how the available supplies can be shipped to minimize costs and meet as many demands as possible. To do this set up a "dummy" supply point S^* whose supply consists of the difference between the total demand and the total supply. To the original cost table add a row corresponding to S^*. In this row you should enter large enough numbers so that no units are actually shipped from this fictitious supply point.

Apply this procedure to the following cost table where the total demand exceeds the total supplies.

Supply points	Demand points				
	D_1	D_2	D_3	D_4	Supply
S_1	6	10	8	14	80
S_2	7	13	5	9	120
Demand	55	85	40	60	

14. Solving the LP problem

$$\begin{array}{lrcrcl} \text{Maximize: } z = & 5x_1 & + & 27x_2 & & \\ \text{Subject to:} & 2x_1 & + & 11x_2 & \leq & 59 \\ & x_1 & - & x_2 & \leq & 7 \\ & & x_1, x_2 \geq 0 & & & \end{array}$$

with *LP SOLVER*, you will find that the optimal solution is

$$x_1 = 10\tfrac{6}{13}, x_2 = 3\tfrac{6}{13}; z = 145\tfrac{10}{13}$$

Suppose you want an optimal *integer* solution to the problem. Show that of the four "round-off" possibilities

$$\begin{array}{ll} x_1 = 10, x_2 = 3 & \qquad x_1 = 10, x_2 = 4 \\ x_1 = 11, x_2 = 3 & \qquad x_1 = 11, x_2 = 4 \end{array}$$

only $x_1 = 10, x_2 = 3$ is a feasible solution (satisfies all the constraints). Show, however, that $x_1 = 2, x_2 = 5$ is also a feasible solution with a z value larger than that obtained with the solution $x_1 = 10, x_2 = 3$.

15. This exercise introduces the inequality generation features of *LP SOLVER*. We illustrate the ideas with the small transportation problem below. As you will see, you can type a small transportation problem more easily than you can type the commands to generate it; however, you can use these same commands, with different data, to solve very large transportation problems.

Minimize:

$$z = 4x_{11} + 2x_{12} + 3x_{13} + 2x_{21} + 4x_{22} + 3x_{23}$$

Subject to:

$$\begin{array}{lllllllll} x_{11} + x_{12} + x_{13} & & & & \leq 30 \\ & x_{21} + x_{22} + x_{23} & & & \leq 20 \\ x_{11} & + x_{21} & & & \geq 15 \\ x_{12} & & + x_{22} & & \geq 10 \\ x_{13} & & & + x_{23} & \geq 25 \end{array}$$

$$\text{All } x_{ij} \geq 0$$

You can enter this LP problem in *LP SOLVER* as follows.

```
&m := 2; {The number of supply points}
&n := 3; {The number of demand points}
data 30, 20; {The supply amounts}
data 15, 10, 25; {The demand amounts}
data 4, 2, 3, 2, 4, 3; {The transportation costs}
for i = 1 to &m do read &si; {Read the supplies}
for j = 1 to &n do read &dj; {Read the demands}
for i = 1 to &m do for j = 1 to &n do read &cij; {Read the costs}
for i = 1 to &m do sum(j = 1 to &n, -&cij*xij); {Objective function}
for i = 1 to &m do sum(j = 1 to &n, xij) <= &si; {Supply Constraints}
for j = 1 to &n do sum(i = 1 to &m, xij) >= &dj; {Demand constraints}
```

We give only a short explanation of this program here because you can read more detailed instructions with the *Help* facilities of *LP SOLVER*.

- The symbols that begin with & are called *symbolic variables*. The statement &m := 2; assigns the value 2 to the symbol &m. Wherever this symbol occurs later in the program, it will be replaced by 2.
- Symbolic variables are also assigned values with *read* statements. To use *read* you first prepare a list of data values. These lists begin with the word *data* in the above example. Each time a read statement is encountered the next value from the list is assigned to the symbol being read into. There are three *data* statements in

the example. We used three for clarity. Actually we could have listed all of the data in one statement because the data statements are all concatenated into one list.

- The *for* statements are similar to the *for loops* of the BASIC programming language. Here is how the statement

  for i = 1 to &m do read &si;

  is interpreted. First, 2 is substituted for &m. Next, i is set to 1 and substituted for the i that occurs in &si. Then the next value from the data list is assigned to &s1 (this value is 30 since this is the first read statement). Next, i is set to 2 and substituted for the i that occurs in &si. Then the next value from the data list (20) is assigned to &s2. Since i only ranges from 1 to 2, the loop stops. The other *for loops* execute in a similar way.

- The statement "sum(i = 1 to 5, yi) <= 69" generates the sum

$$y1 + y2 + y3 + y4 + y5 <= 69;$$

  The *sum* statements in the example are nested in a *for loop* and so generate a sequence of constraints.

Enter the above example in *LP SOLVER*. Before you solve the problem, press [F4] to choose the option of listing the generated inequalities. (You will see $F4$: *Options* in the menu at the top of the screen.) The generated inequalities will not appear on the screen until you solve the problem by pressing [F1].

You should also press [F9] and read the instructions for using *symbolic variables* and *for loops*.

When you have this small transportation problem entered correctly, change the values of &m, &n, and all the data to those appropriate to Example 4. Solve the problem and check the answer we gave.

16. Given the information

Month	Monthly demand	Normal production capacity	Overtime production capacity	Unit storage costs	Normal unit costs	Overtime additional unit costs
Jan.	160	120	60	6	30	10
Feb.	200	160	50	3	35	15
March	120	200	30	5	25	10
April	120	130	30	6	30	10
May	200	150	40	6	30	15
June	250	200	60	4	25	20

find a production schedule that meets the monthly demands at minimal cost.

6 Advanced Linear Programming*

Thus far we have been concerned with finding an optimal solution to an LP problem. In this chapter we turn our attention to other conclusions that can be drawn from the final simplex tableau. For instance, in an LP problem whose objective function gives the profit of an enterprise, it is natural to ask how much one can increase the profit by reducing costs or by providing additional resources. The process of deriving such additional information is called *postoptimality analysis.* In broad terms this analysis involves study of the dual problem, which we do in Section 6.2, and sensitivity analysis, which we take up in Section 6.3. Both of these analyses involve manipulation of the coefficients that occur in the objective function and the constant term that occurs in each constraint. To ease the problem of referring to these terms we make the following definitions which we will use throughout this chapter.

Definition. In an LP problem the coefficients of the objective function are called *cost coefficients.* This name is commonly applied in the literature of LP problems even when the coefficients represent something other than a cost, even a profit. When a constraint inequality in an LP problem is written with all variable terms to the left and the constant term to the right of the inequality sign, the constant term is called a *resource value.* This name is commonly applied even when the right-hand term is not identifiable as a resource.

For example, in the LP problem (1) on page 172, the coefficients $3, 2, 4$ in the objective function are cost coefficients, and the numbers $5, 6, 4$ on the right-hand side of the constraints are resource values.

6.1 DUAL SIMPLEX METHOD

The *dual simplex method*, a modification of the simplex method, can be used to solve LP problems whose initial tableaus have the properties

(i) Some of the resource values are negative.

(ii) All entries in the objective row are positive or zero.

We illustrate this method with the LP problem

$$\begin{array}{lrcrcrcl} \text{Minimize}: z = & 3x_1 & + & 2x_2 & + & 4x_3 & & \\ \text{Subject to}: & 2x_1 & + & x_2 & - & x_3 & \leq & 5 \\ & x_1 & - & 3x_2 & + & 2x_3 & \geq & 6 \\ & 2x_1 & + & 2x_2 & + & x_3 & \geq & 4 \\ & \multicolumn{7}{l}{x_i \geq 0, 1 \leq i \leq 3} \end{array} \tag{1}$$

To set up the initial tableau for this problem introduce (as before) a non-negative slack variable to the first inequality. Now multiply the second and third inequalities through by -1 and then introduce nonnegative slack variables to these inequalities as well. You must also convert the minimization problem to a maximization problem. Thus (1) becomes

$$\begin{array}{lrcrcrcrcr} \text{Maximize}: z = & -3x_1 & - & 2x_2 & - & 4x_3 & & & & \\ \text{Subject to}: & 2x_1 & + & x_2 & - & x_3 & + & s_1 & = & 5 \\ & -x_1 & + & 3x_2 & - & 2x_3 & + & s_2 & = & -6 \\ & -2x_1 & - & 2x_2 & - & x_3 & + & s_3 & = & -4 \\ & \multicolumn{9}{l}{x_i \geq 0, 1 \leq i \leq 3; s_i \geq 0, 1 \leq i \leq 3} \end{array} \tag{2}$$

The corresponding initial simplex tableau

	x_1	x_2	x_3	s_1	s_2	s_3	RHS
s_1	2	1	−1	1	0	0	5
s_2	−1	3	−2	0	1	0	−6
s_3	−2	−2	−1	0	0	1	−4
OBJ	3	2	4	0	0	0	0

satisfies conditions (i) and (ii) above.

Notice that although s_1, s_2, s_3 are basic variables, they do not give us a feasible solution since s_2 and s_3 have negative values. To implement the *dual simplex method* we first select the *pivot row*, which corresponds to the *departing basic variable.* The *pivot row* is the row containing the largest negative resource value (actually any row containing a negative resource value will do). The *departing basic variable* is the basic variable associated with this row. In our example this row is the second row and the departing basic variable is s_2 since -6 is the most negative resource value.

To find the *pivot column* and the *entering basic variable* we form the ratios of entries in the objective row to *negative* entries in the pivot row.

The pivot column is the column associated with the largest of these ratios and the entering basic variable is the variable heading this column. In our example the pivot column is the column headed by the entering basic variable x_3 since the ratio $\frac{4}{-2}$ is larger than the ratio $\frac{3}{-1}$. The *pivot entry* is the entry found in the pivot row and pivot column, which in our case is -2. Using *PIVOTER* to pivot on this entry we obtain the next tableau

	x_1	x_2	x_3	s_1	s_2	s_3	RHS
s_1	$\frac{5}{2}$	$-\frac{1}{2}$	0	1	$-\frac{1}{2}$	0	8
x_3	$\frac{1}{2}$	$-\frac{3}{2}$	1	0	$-\frac{1}{2}$	0	3
s_3	$-\frac{3}{2}$	$-\frac{7}{2}$	0	0	$-\frac{1}{2}$	1	-1
OBJ	1	8	0	0	2	0	-12

Note that all entries in the objective row remain nonnegative; this will always be the case for this method. However, there remains a negative entry in the right-hand side. This entry corresponds to the pivot row with the departing basic variable s_3. To find the pivot column form again the ratios of entries in the objective row to negative entries in the pivot row. Since $\frac{1}{\frac{-3}{2}}$ is the largest of the ratios $\{\frac{1}{-\frac{3}{2}}, \frac{8}{-\frac{7}{2}}, \frac{2}{-\frac{1}{2}}\}$, the first column is the pivot column and x_1 is the entering basic variable. The pivot entry is $-\frac{3}{2}$. Pivoting on this entry you obtain the tableau

	x_1	x_2	x_3	s_1	s_2	s_3	RHS
s_1	0	$-\frac{19}{3}$	0	1	$-\frac{4}{3}$	$\frac{5}{3}$	$\frac{19}{3}$
x_3	0	$-\frac{8}{3}$	1	0	$-\frac{2}{3}$	$\frac{1}{3}$	$\frac{8}{3}$
x_1	1	$\frac{7}{3}$	0	0	$\frac{1}{3}$	$-\frac{2}{3}$	$\frac{2}{3}$
OBJ	0	$\frac{17}{3}$	0	0	$\frac{5}{3}$	$\frac{2}{3}$	$-\frac{38}{3}$

Since all resource values are nonnegative as are all objective row entries, we are done. The solution to (2) is

$$x_1 = \tfrac{2}{3}, x_2 = 0, x_3 = \tfrac{8}{3}, s_1 = \tfrac{19}{3}, s_2 = 0, s_3 = 0$$

The value of z is $\frac{38}{3}$. Notice that in the last tableau we obtained $z = -\frac{38}{3}$. However, this value corresponds to the maximization problem (2); the original problem (1) was a minimization problem so we must change the sign of $-\frac{38}{3}$ — if you have any doubts about this, just substitute the solution values into the original objective function and you will obtain $z = \frac{38}{3}$.

In general, the steps of the dual simplex method are:

1. Determine the pivot row (and the departing basic variable) by selecting the row corresponding to the most negative entry found in the right-hand side.

2. Determine the pivot column (and the entering basic variable) by selecting the column corresponding to the largest ratio of entries in the objective row to negative entries in the pivot row. If there are no negative entries in the pivot row, then the problem is infeasible; that is, there is no feasible solution to the problem. (See Problem 8 of Exercises 6.1A.)

3. Pivot on the pivot element (which lies in both the pivot row and pivot column).

4. Repeat steps 1 – 3 until no negative entry appears in the right-hand side.

It is important to re-emphasize that you can use the dual simplex method only when there are negative resource values and all entries in the objective row of the tableau are nonnegative. Finally we mention that in the first step the pivot row need not correspond to the most negative entry in the right-hand side; any negative entry will do.

6.1A Exercises: Dual Simplex Method

Use the dual simplex method to solve Problems 1–4 by hand and 5–7 with *PIVOTER*.

1. §

$$\begin{array}{llcl}
\text{Minimize: } z = & 4x_1 + 12x_2 & & \\
\text{Subject to:} & x_1 + 5x_2 & \geq & 120 \\
 & x_1 + 3x_2 & \geq & 100 \\
 & x_1, x_2 \geq 0 & &
\end{array}$$

2.

$$\begin{array}{llcl}
\text{Minimize: } z = & 3x_1 + 2x_2 & & \\
\text{Subject to:} & -x_1 + x_2 & \geq & 3 \\
 & x_1 + x_2 & \leq & 6 \\
 & 3x_1 + x_2 & \geq & 5 \\
 & x_1, x_2 \geq 0 & &
\end{array}$$

3. §

$$\begin{array}{lrcrcrcl} \text{Minimize: } z = & 2x_1 & + & x_2 & + & 2x_3 & & \\ \text{Subject to:} & x_1 & + & x_2 & - & x_3 & \geq & 5 \\ & -2x_1 & + & x_2 & + & 4x_3 & \geq & 8 \\ & \multicolumn{7}{l}{x_1, x_2, x_3 \geq 0} \end{array}$$

4.

$$\begin{array}{lrcrcrcr} \text{Minimize: } z = & x_1 & + & 3x_2 & + & 4x_3 & & \\ \text{Subject to:} & 3x_1 & - & 2x_2 & + & 6x_3 & \geq & 1 \\ & x_1 & + & x_2 & - & 3x_3 & \leq & 10 \\ & -2x_1 & + & x_2 & - & 4x_3 & \geq & 2 \\ & \multicolumn{7}{l}{x_1, x_2, x_3 \geq 0} \end{array}$$

5.

$$\begin{array}{lrcrcrcr} \text{Minimize: } z = & 150x_1 & + & 120x_2 & + & 75x_3 & & \\ \text{Subject to:} & 3x_1 & + & 4x_2 & + & x_3 & \geq & 15 \\ & 12x_1 & - & 15x_2 & + & x_3 & \geq & 30 \\ & 4x_1 & + & 2x_2 & + & x_3 & \geq & 15 \\ & 3x_1 & + & x_2 & + & 2x_3 & \geq & 10 \\ & \multicolumn{7}{l}{x_i \geq 0, 1 \leq i \leq 3} \end{array}$$

6. §

$$\begin{array}{lrcrcrcr} \text{Minimize: } z = & x_1 & + & x_2 & + & x_3 & & \\ \text{Subject to:} & -x_1 & + & 2x_2 & + & x_3 & \leq & 1 \\ & x_1 & & & - & 2x_3 & \leq & -4 \\ & x_1 & - & x_2 & + & 2x_3 & \leq & 4 \\ & \multicolumn{7}{l}{x_i \geq 0, 1 \leq i \leq 3} \end{array}$$

7.

$$\begin{array}{lrcrcrcrcl} \text{Minimize: } z = & 10x_1 & + & 6x_2 & + & x_3 & + & x_4 & & \\ \text{Subject to:} & x_1 & + & 3x_2 & - & 2x_3 & - & x_4 & \geq & 2 \\ & -x_1 & - & x_2 & - & x_3 & - & x_4 & \leq & 3 \\ & -x_1 & - & x_2 & - & x_3 & + & x_4 & \geq & 1 \\ & \multicolumn{9}{l}{x_i \geq 0, 1 \leq i \leq 4} \end{array}$$

8. § Explain why a problem is infeasible when there is a negative resource value (that is, there is a pivot row), but all the other entries in the pivot row are nonnegative.

6.2 DUALITY

In this section we define the *dual* of an LP problem and indicate various relationships between an LP problem and its dual. We also look at some economic interpretations of duality.

Definition. The *dual* of the LP problem

$$\begin{array}{llcllclcll} \text{Maximize}: z = & c_1x_1 & + & c_2x_2 & + & \cdots & + & c_nx_n & & \\ \text{Subject to}: & a_{11}x_1 & + & a_{12}x_2 & + & \cdots & + & a_{1n}x_n & \leq & b_1 \\ & a_{21}x_1 & + & a_{22}x_2 & + & \cdots & + & a_{2n}x_n & \leq & b_2 \\ & \vdots & & & & & & & & \\ & a_{m1}x_1 & + & a_{m2}x_2 & + & \cdots & + & a_{mn}x_n & \leq & b_m \\ & \multicolumn{9}{l}{x_i \geq 0, 1 \leq i \leq n} \end{array} \tag{1}$$

is the LP problem

$$\begin{array}{llcllclcll} \text{Minimize}: w = & b_1y_1 & + & b_2y_2 & + & \cdots & + & b_my_m & & \\ \text{Subject to}: & a_{11}y_1 & + & a_{21}y_2 & + & \cdots & + & a_{m1}y_m & \geq & c_1 \\ & a_{12}y_1 & + & a_{22}y_2 & + & \cdots & + & a_{m2}y_m & \geq & c_2 \\ & \vdots & & & & & & & & \\ & a_{1n}y_1 & + & a_{2n}y_2 & + & \cdots & + & a_{mn}y_m & \geq & c_n \\ & \multicolumn{9}{l}{y_i \geq 0, 1 \leq i \leq m} \end{array} \tag{2}$$

In this context, the original LP problem (1) is called the *primal problem* and (2) is called its *dual problem.* It is also possible to refer to (2) as the primal problem and consider (1) to be its dual problem.

Example 1: The Dual of an LP Problem

The dual of the LP problem

$$\begin{array}{llclclcl} \text{Maximize}: z = & 7x_1 & - & 3x_2 & + & 3x_3 & & \\ \text{Subject to}: & 2x_1 & - & 4x_2 & - & 3x_3 & \leq & 7 \\ & 3x_1 & + & 2x_2 & + & x_3 & \leq & 16 \\ & 2x_1 & - & x_2 & + & 2x_3 & \leq & 20 \\ & -6x_1 & + & 4x_2 & + & x_3 & \leq & 2 \\ & \multicolumn{7}{l}{x_i \geq 0, 1 \leq i \leq 3} \end{array} \tag{3}$$

is the LP problem

$$\begin{array}{lrcrcrcrcr}
\text{Minimize}: w = & 7y_1 & + & 16y_2 & + & 20y_3 & + & 2y_4 & & \\
\text{Subject to}: & 2y_1 & + & 3y_2 & + & 2y_3 & - & 6y_4 & \geq & 7 \\
 & -4y_1 & + & 2y_2 & - & y_3 & + & 4y_4 & \geq & -3 \\
 & -3y_1 & + & y_2 & + & 2y_3 & + & y_4 & \geq & 3 \\
y_i \geq 0, 1 \leq i \leq 4 & & & & & & & & &
\end{array} \quad (4)$$

The dual of the LP problem

$$\begin{array}{lrcrcrcr}
\text{Minimize: } z = & 3x_1 & + & 7x_2 & - & 4x_3 & & \\
\text{Subject to:} & x_1 & - & 2x_2 & + & 3x_3 & \geq & 4 \\
 & 2x_1 & + & 5x_2 & - & x_3 & \geq & -7 \\
 & x_i \geq 0, 1 \leq i \leq 3 & & & & & &
\end{array}$$

is the LP problem

$$\begin{array}{lrcrcr}
\text{Maximize: } w = & 4y_1 & - & 7y_2 & & \\
\text{Subject to:} & y_1 & + & 2y_2 & \leq & 3 \\
 & -2y_1 & + & 5y_2 & \leq & 7 \\
 & 3y_1 & - & y_2 & \leq & -4 \\
 & y_1, y_2 \geq 0 & & & &
\end{array}$$

The following observations are in order:

1. The dual of a maximization problem is a minimization problem and vice-versa.

2. The entries of the right-hand side of the primal (original) problem are the coefficients of the objective function of the dual problem, and vice-versa.

3. The coefficient matrix of the dual problem is the transpose of the coefficient matrix of the primal problem, and vice-versa.

6.2A Exercises: Forming Dual Problems

Form the dual of each LP problem.

1. §

$$\begin{array}{lrcrcrcl}
\text{Maximize: } z = & 9x_1 & + & 3x_2 & - & 3x_3 & & \\
\text{Subject to:} & 4x_1 & + & 2x_2 & - & 2x_3 & \leq & 8 \\
& 4x_1 & + & x_2 & - & 2x_3 & \leq & 12 \\
& 3x_1 & + & 2x_2 & - & 2x_3 & \leq & 18 \\
& 6x_1 & + & x_2 & + & 2x_3 & \leq & 17 \\
& \multicolumn{7}{l}{x_i \geq 0, 1 \leq i \leq 3}
\end{array}$$

2.

$$\begin{array}{lrcrcrcrcl}
\text{Maximize: } z = & -x_1 & + & x_2 & - & 4x_3 & + & 2x_4 & & \\
\text{Subject to:} & 4x_1 & + & x_2 & + & 2x_3 & + & 3x_4 & \leq & 6 \\
& x_1 & - & 3x_2 & + & 4x_3 & + & x_4 & \leq & 1 \\
& 2x_1 & - & x_2 & - & 2x_3 & & & \leq & 32 \\
& -6x_1 & + & 3x_2 & - & 2x_3 & - & 3x_4 & \leq & 12 \\
& \multicolumn{9}{l}{x_i \geq 0, 1 \leq i \leq 4}
\end{array}$$

3. §

$$\begin{array}{lrcrcl}
\text{Minimize: } z = & x_1 & + & x_2 & & \\
\text{Subject to:} & 4x_1 & + & 2x_2 & \geq & 3 \\
& 2x_1 & + & 3x_2 & \geq & 2 \\
& x_1 & + & x_2 & \geq & 3 \\
& \multicolumn{5}{l}{x_1, x_2 \geq 0}
\end{array}$$

4.

$$\begin{array}{lrcrcrcrcl}
\text{Minimize: } z = & x_1 & + & x_2 & + & 3x_3 & + & 2x_4 & & \\
\text{Subject to:} & -4x_1 & + & 2x_2 & + & x_3 & + & 3x_4 & \geq & 6 \\
& 4x_1 & & & - & x_3 & - & 2x_4 & \geq & 1 \\
& 2x_1 & + & 3x_2 & - & x_3 & & & \geq & 2 \\
& 3x_1 & + & x_2 & + & 5x_3 & - & x_4 & \geq & 2 \\
& \multicolumn{9}{l}{x_i \geq 0, 1 \leq i \leq 4}
\end{array}$$

Next we look at other important relationships between the primal and dual problems. We also see how to use the solution to the primal problem to solve the dual problem.

The initial and final tableaus of the first primal problem (3) given in Example 1 are

Initial Tableau of the Primal Problem

	x_1	x_2	x_3	s_1	s_2	s_3	s_4	RHS
s_1	2	−4	−3	1	0	0	0	7
s_2	3	2	1	0	1	0	0	16
s_3	2	−1	2	0	0	1	0	20
s_4	−6	4	1	0	0	0	1	2
OBJ	−7	3	−3	0	0	0	0	0

Final Tableau of the Primal Problem

	x_1	x_2	x_3	s_1	s_2	s_3	s_4	RHS
x_1	1	1.25	0	0	0.5	−0.25	0	3
x_3	0	−1.75	1	0	−0.5	0.75	0	7
s_1	0	−11.75	0	1	−2.5	2.75	0	22
s_4	0	13.25	0	0	3.5	−2.25	1	13
OBJ	0	6.5	0	0	2	0.5	0	42

Thus, the solution of the primal problem is $x_1 = 3, x_2 = 0, x_3 = 7, s_1 = 22, s_2 = 0, s_3 = 0, s_4 = 13; z = 42$.

We use the dual simplex method of Section 6.1 to solve the dual (4) of this problem. First we convert all of the $\geq$ inequalities to $\leq$ inequalities by multiplying each constraint by -1 to obtain the equivalent problem

$$
\begin{array}{llcrcrcrcrcr}
\text{Minimize: } w = & 7y_1 & + & 16y_2 & + & 20y_3 & + & 2y_4 & & \\
\text{Subject to:} & -2y_1 & - & 3y_2 & - & 2y_3 & + & 6y_4 & \leq & -7 \\
& 4y_1 & - & 2y_2 & + & y_3 & - & 4y_4 & \leq & 3 \\
& 3y_1 & - & y_2 & - & 2y_3 & - & y_4 & \leq & -3 \\
& \multicolumn{9}{l}{y_i \geq 0, 1 \leq i \leq 3}
\end{array}
$$

The initial tableau for this dual problem is

Initial Tableau of the Dual Problem

	y_1	y_2	y_3	y_4	t_1	t_2	t_3	RHS
t_1	−2	−3	−2	6	1	0	0	−7
t_2	4	−2	1	−4	0	1	0	3
t_3	3	−1	−2	−1	0	0	1	−3
OBJ	7	16	20	2	0	0	0	0

and the final tableau (obtained by *PIVOTER*) is

Final Tableau of the Dual Problem

	y_1	y_2	y_3	y_4	t_1	t_2	t_3	RHS
y_3	−2.75	0	1	2.25	0.25	0	−0.75	0.5
t_2	11.75	0	0	−13.25	−1.25	1	1.75	6.5
y_2	2.5	1	0	−3.5	−0.5	0	0.5	2
OBJ	22	0	0	13	3	0	7	−42

The solution of the dual problem is $y_1 = 0, y_2 = 2, y_3 = 0.5, y_4 = 0, t_1 = 0, t_2 = 6.5, t_3 = 0; w = 42$ (−42 appears in the tableau, but remember, we change the sign because we are solving a minimum problem).

We make the following observations regarding the dual problem and the final tableau of the primal problem.

1. The optimal value of the primal problem is found in the lower right corner of the final tableau; it is $z = 42$ and is equal to the optimal value of the dual problem $w = 42$.

2. The optimal values of the dual variables $y_1 = 0, y_2, = 2, y_3 = .5, y_4 = 0$ coincide with the values found in the objective row of the final tableau of the primal problem in the columns headed by the slack variables s_1, s_2, s_3, s_4.

3. The optimal values of the primal variables $x_1 = 3, x_2 = 0, x_3 = 7$ are found in the objective row of the final tableau of the dual problem in the columns headed by the slack variables t_1, t_2, t_3.

Observation 1 illustrates one of the most important and useful results in linear programming, the Fundamental Principle of Duality, due to the great German mathematician John von Neumann.

Fundamental Principle of Duality: If an optimal solution $x_1^*, x_2^*, \ldots, x_n^*$, of the primal problem (1) exists and z^* is the optimal value of the objective function, then an optimal solution $y_1^*, y_2^*, \ldots, y_m^*$ of the dual problem (2) exists with optimal objective function value w^* and

$$z^* = c_1x_1^* + c_2x_2^* + \cdots + c_nx_n^* = b_1y_1^* + b_2y_2^* + \cdots + b_my_m^* = w^* \qquad (5)$$

Moreover, if an optimal solution $y_1^*, y_2^*, \ldots, y_m^*$ of the dual problem exists, then an optimal solution $x_1^*, x_2^*, \ldots, x_n^*$ of the primal problem exists and (5) also holds.

It is important to observe that von Neumann's result assures us that the maximal value for z in the primal problem is always equal to the minimal value for w in the dual problem.

Observations 2 and 3 are also valid in general and give us a way of using the solution of the primal problem to solve the dual problem and vice-versa. The optimal values of the dual variables are found in the objective row in the slack variable columns of the final tableau of the primal problem. The optimal values of the primal problem are found in the objective row in the slack variable columns of the dual problem.

CALIPSO

LP SOLVER gives the dual prices (the optimal values for the dual problem) as part of the sensitivity analysis. After solving the problem, press [*Alt*]+[*F3*] to see the dual prices. (If you forget what keys to press, you can recall them by choosing the menu of postoptimality analyses advertised at the top of the screen.)

Note: We did not define the dual of a problem with mixed constraints ($\leq, =,$ or $\geq$) because doing so is beyond the scope of this book; however, CALIPSO can find the dual of a problem with mixed constraints. Since you may use CALIPSO to find dual prices for such a problem we warn you that the report you get from CALIPSO may contain some negative dual prices, whereas until now, we have insisted that feasible solutions are always nonnegative. In order to deal with a mixture of $\leq$ and $\geq$ constraints, it is necessary to assign nonpositive (≤ 0) values to a dual variable associated with a $\geq$ constraint. For a more thorough discussion of this topic see [CV].

6.2B Exercises: Using the Final Primal Tableau to Solve the Dual

Solve Problems 1 and 2 in Exercises 6.2A with *PIVOTER*. From each final tableau you obtain, locate the solution of the dual problem in the objective row under the columns headed by slack variables. Next use *LP SOLVER* to solve the dual problems and check that you get the same answer.

To illustrate why duals of LP problems are of practical significance we consider the following typical LP problem. Suppose a small firm, Mixed Nuts, produces four different mixtures of nuts: A, B, C, and D. Each mixture is packaged in 10-pound bags and consists of some combination of peanuts, almonds, and pecans. Table 1 gives the contents of each mixture, the profits (in dollars) on the sale of each package, and the amounts (in pounds) of nuts available on a daily basis for forming the mixtures.

Table 1

	Peanuts (lb/package)	Almonds (lb/package)	Pecans (lb/package)	Profit ($/package)
A	6	2	2	6
B	4	3	3	8
C	3	5	2	12
D	3	7	0	9
lbs Avail.	120	80	90	

To determine how many packages of each mixture should be produced to maximize profits, Mixed Nuts sets up the LP problem

$$\begin{array}{llllllllll}\text{Maximize}: z = & 6x_1 & + & 8x_2 & + & 12x_3 & + & 9x_4 & & \\ \text{Subject to}: & 6x_1 & + & 4x_2 & + & 3x_3 & + & 3x_4 & \leq & 120 \\ & 2x_1 & + & 3x_2 & + & 5x_3 & + & 7x_4 & \leq & 80 \\ & 2x_1 & + & 3x_2 & + & 2x_3 & & & \leq & 90 \\ & x_i \geq 0, 1 \leq i \leq 4 & & & & & & & & \end{array} \tag{6}$$

With the aid of CALIPSO you can verify that the initial and final tableaus for this problem are

Initial Tableau

	x_1	x_2	x_3	x_4	x_5	x_6	x_7	RHS
x_5	6	4	3	3	1	0	0	120
x_6	2	3	5	7	0	1	0	80
x_7	2	3	2	0	0	0	1	90
OBJ	−6	−8	−12	−9	0	0	0	0

Final Tableau

	x_1	x_2	x_3	x_4	x_5	x_6	x_7	RHS
x_1	1	0	−1.1	−1.9	0.3	−0.4	0	4
x_2	0	1	2.4	3.6	−0.2	0.6	0	24
x_7	0	0	−3	−7	0	−1	1	10
OBJ	0	0	0.6	8.4	0.2	2.4	0	216

From the final tableau we find that to maximize profits Mixed Nuts should produce 4 packages of mixture A per day, 24 packages of mixture B, and should not produce either mixture C or D.

Now let us consider the problem from a different perspective. Suppose another firm, Nuts of America, offers to buy Mixed Nuts' resources (the nuts) rather than the finished products (the various mixtures). Naturally it would like to pay as little as possible for these nuts. In order to consider this proposal Mixed Nuts needs to determine a per pound price it should assign to the peanuts, almonds, and pecans to make it worthwhile to sell the nuts rather than the mixtures.

To resolve this problem, Mixed Nuts lets y_1 be the price for a pound of peanuts, y_2 the price for a pound of almonds, and y_3 the price for a pound of pecans. Then the cost to Nuts of America for all of the nuts is

$$w = 120y_1 + 80y_2 + 90y_3$$

This firm wants to minimize w.

From the standpoint of Mixed Nuts there is no reason to sell the nuts rather than the mixtures unless the firm gets at least the same amount of return. Since the profit on a 10-pound bag of Mixture A is \$6 and since Mixture A contains 6 pounds of peanuts, 2 pounds of almonds and 2 pounds of pecans, Mixed Nuts will not sell the nuts unless

$$6y_1 + 2y_2 + 2y_3 \geq 6$$

Similarly since the profit on a 10-pound bag of Mixture B is \$8 and since mixture B contains 4 pounds of peanuts, 3 pounds of almonds, and 3 pounds of pecans, Mix Nuts will not sell the nuts unless

$$4y_1 + 3y_2 + 3y_3 \geq 8$$

Applying the same reasoning to mixtures C and D leads to two additional inequalities

$$\begin{aligned} 3y_1 + 5y_2 + 2y_3 &\geq 12 \\ 3y_1 + 7y_2 \qquad\quad &\geq 9 \end{aligned}$$

From this discussion we get the LP problem

$$\begin{array}{llllllllll}
\text{Minimize: } w = & 120y_1 & + & 80y_2 & + & 90y_3 & & \\
\text{Subject to:} & 6y_1 & + & 2y_2 & + & 2y_3 & \geq & 6 \\
& 4y_1 & + & 3y_2 & + & 3y_3 & \geq & 8 \\
& 3y_1 & + & 5y_2 & + & 2y_3 & \geq & 12 \\
& 3y_1 & + & 7y_2 & & & \geq & 9 \\
& y_i \geq 0, 1 \leq i \leq 3 & & & & & &
\end{array}$$

Note that this problem is precisely the dual of the original problem (6). It follows from the final tableau for (6) and Observation 2 above that the optimal values for the dual variables are $y_1 = 0.2, y_2 = 2.4, y_3 = 0; w = 216$.

Let us examine two interpretations of this result. You can view the optimal values of y_1, y_2, y_3 as the minimal prices set by Mixed Nuts for selling the nuts (rather than the mixtures). If Mixed Nuts were to charge less than these amounts for its resources (the nuts), then it will earn less money than if it had sold the mixtures. Note that by the Fundamental Principle of Duality the total amount Mixed Nuts receives for its resources (or, equivalently, the amount Nuts of America pays for these resources) is the same as the maximal profit Mixed Nuts can earn from the mixtures.

The values $y_1 = 0.2, y_2 = 2.4, y_3 = 0$ have another useful interpretation. The value $y_1 = 0.2$ is the amount of *additional* profit Mixed Nuts can earn by selling the mixtures if it had available one more pound of peanuts. In essence, it is the value to Mixed Nuts of an additional unit (pound) of peanuts. To see why this is so we again use the Fundamental Principle of Duality which tells us that the maximal profit is

$$z^* = b_1y_1^* + b_2y_2^* + b_3y_3^* = 120 \cdot 0.2 + 90 \cdot 2.4 + 80 \cdot 0 = 216 \qquad (7)$$

Now note that if we increase the original peanut supply from 120 pounds to 121 pounds, then the maximal profit increases by 0.2, which is the value of y_1^*.

Similarly, $y_2 = 2.4$ is the additional profit gained if Mixed Nuts had one more pound of almonds available. In other words, $y_2 = \$2.40$ is the value to Mixed Nuts of one additional unit (pound) of almonds. This is also clear from (7) since if we increase 90 to 91, then there is a \$2.40 increase in the optimal profit z^*. Thus, you can also view the optimal dual values y_1 and y_2 as the maximum amounts Mixed Nuts is willing to pay for additional pounds of these nuts — willing to pay in the sense that Mixed Nuts' profit increases will balance the additional expenditures for these nuts.

Finally notice that $y_3 = 0$. This means that an additional pound of pecans will not lead to greater profits. This is clear if we look at the optimal solution to the original problem. There we see in the final tableau that the slack variable associated with the pecan constraint (the third row of the tableau) is positive ($x_7 = 10$). This tells us that at the optimal solution, 10 pounds of pecans were left over. Therefore, Mixed Nuts is not interested in acquiring more pecans since it already has a surplus of this resource. The value of an additional unit of pecans is 0.

The optimal values of the dual variables are often referred to as the *shadow prices*, the *marginal prices*, the *marginal values*, or the *fictitious prices* of the resources. They represent the increase in profit if an additional unit of a resource is made available.

6.2C Exercises: Economic Applications of Duality

1. § A company produces two blended drinks, A and B. Each gallon of blend A contains 3 quarts of apple juice and 1 quart of cherry juice, while each gallon of blend B contains 2 quarts of apple juice and 2 quarts of cherry juice. The profit on each gallon sold of blend A is 15 cents and the profit on each gallon of blend B is 20 cents. The company has 200 quarts of apple juice and 100 quarts of cherry juice available each day.

   (a) How many gallons of each of the blends should the company produce to maximize its profit?

   (b) Set up and give an interpretation of the dual of this problem.

   (c) Use the final tableau obtained in (a) to solve the dual problem.

2. Cats at a kennel are fed goat meat and squid meat. Each ounce of goat contains 4 units of fat, 6 units of carbohydrate, and 2 units of protein. Each ounce of squid contains 3 units of fat, 2 units of carbohydrate, and 4 units of protein. Healthy cats require at least 10 units of fat, 16 units of carbohydrate, and 12 units of protein each day. The cost per ounce of goat meat is 10 cents and the cost per ounce of squid meat is 15 cents.

   (a) How many ounces of each meat should be served to ensure healthy cats at minimum cost?

   (b) Set up and give an interpretation of the dual of this problem.

(c) Solve the dual problem.

3. § A small company manufactures four types of toy cars, Models A, B, C, and D. Model A requires 6 units of raw materials and 4 units of time; Model B, 5 units of raw materials and 6 units of time; Model C, 8 units of raw materials and 5 units of time; Model D, 4 units of raw materials and 8 units of time. Each day the company has available 450 units of raw materials and 300 units of time. The profits on Models A, B, C, and D are \$4.50, \$6.50, \$7.00, and \$6.00, respectively.

   (a) Use *LP SOLVER* to find how many of each model should be produced to maximize profits.

   (b) Set up and give an economic interpretation of the dual of this problem.

   (c) Use the final tableau obtained in (a) to determine the optimal values of the dual variables. Interpret your results. (Note that after solving the problem you get the final tableau from the postoptimality menu by pressing [*Alt*]+[*F2*]).

4. A lumber mill produces construction-grade and finish-grade boards from raw logs. Suppose that it takes 3 hours to rough-saw each 1000 board feet of the construction-grade boards and 3 hours to plane 1000 board feet of these boards. Suppose also it takes 3.5 hours to rough-saw each 1000 board feet of the finish-grade boards and 6 hours to plane each 1000 board feet of these boards. There are 120 saw-hours and 150 plane-hours available each day. The profit for each 1000 board feet sold of construction-grade boards is \$80 and the profit for each 1000 board feet of finish-grade boards is \$110.

   (a) Use *LP SOLVER* to find how many board feet of each type of lumber should be produced to maximize profits.

   (b) Set up and interpret the dual of this problem.

   (c) Use the final tableau obtained in (a) to determine the optimal values of the dual problem. Interpret your results. (Note that after solving the problem you get the final tableau from the postoptimality menu by pressing [*Alt*]+[*F2*]).

5. § A small manufacturer produces two steel grates, Grate A and Grate B. Table 2 gives data about the production of these grates.

Table 2

Resources	Grate *A*	Grate *B*
Steel per grate (lb)	80	130
Labor per grate (h)	9	8
Machine time per grate (h)	5	7
Profit per grate ($)	75	115

The company has available on a daily basis 1000 pounds of steel, 100 hours of labor, and 120 hours of machine time.

(a) Use *LP SOLVER* to determine how many grates of each should be produced daily to maximize profits.

(b) Use the final tableau obtained in (a) to determine the optimal values of the dual problem. Interpret your results.

(c) Set up and solve the dual problem to determine the marginal prices (shadow prices) of the steel, the labor, and the machine time.

6. A metallurgical company manufactures three alloys, A, B, and C. The company can use four processes, each of which can produce the alloys. Table 3 gives the tons of alloys produced per hour by these processes and the cost per hour of each process.

Table 3

Processes	Tons per hour			Cost per hour ($)
	A	*B*	*C*	
Process 1	3	8	5	400
Process 2	5	7	4	350
Process 3	2	9	5	400
Process 4	4	6	7	300

The company must meet demands of at least 6000 tons of alloy A, 5000 tons of alloy B, and 7000 tons of alloy C.

(a) Use *LP SOLVER* to determine how best to meet the demand at minimal cost.

(b) Use the dual problem to determine the marginal cost (shadow price) of each kind of alloy.

7. § Set up and solve the dual of Problem 7 in the Exercises 5.4A. Interpret your results.

8. Dogs at the local kennel are given three types of meals, A, B, and C. Table 4 gives the units of fats, carbohydrates, proteins, and multivitamins found in each ounce of these meals

Table 4

Nutrients	A	B	C
Fat	4	3	7
Carbohydrate	6	2	5
Protein	3	5	4
Multivitamins	8	6	10

To maintain healthy dogs, the kennel's policy is to provide at least 20 units of fat, 18 units of carbohydrate, 15 units of protein, and 25 units of multivitamins with each meal. The per ounce costs of meals A, B, C are 20 cents, 15 cents, and 12 cents, respectively.

(a) Use *LP SOLVER* to determine the number of ounces of each of the meals that should be served to maintain healthy dogs at a minimum cost.

(b) Set up and solve the dual problem with *LP SOLVER*. Interpret your results in terms of a competing firm that sells artificial foods F, C, P and V where each unit of F provides 1 unit of fat, each unit of C provides 1 unit of carbohydrates, each unit of P provides 1 unit of protein, and each unit of V provides one unit of multivitamins.

9. § Set up, solve, and interpret your solution to the dual problem of Problem 4 in Exercises 5.4A.

6.3 SENSITIVITY ANALYSIS

Thus far we have assumed that the data given for an LP problem is precise and not subject to change. In practice, however, this is rarely the case. For instance, you may not know exactly your profit on the sale of a particular item. It may also occur that after solving an LP problem you find that profits (or costs) have changed. We will see that there are ranges for profit (cost) changes which do not cause changes in the optimal solution to the original problem.

Similarly, there may be changes in the amounts of available resources. We want to determine how these changes affect the optimal results already obtained.

Investigating the effects of changes in the original data is the subject of *sensitivity analysis*. This study is also often referred to as *postoptimality analysis.*

In this section we show you how to use CALIPSO to draw conclusions about changes in the cost coefficients or in changes in the resource values of an LP problem. For a more detailed discussion of this topic we encourage you to consult [CV].

Change in a Single Cost Coefficient

Using CALIPSO to solve the LP problem

$$\begin{array}{llllllll} \text{Maximize}: z = & 9x_1 & + & 6x_2 & + & x_3 & & \\ \text{Subject to}: & 2x_1 & + & \frac{1}{3}x_2 & + & \frac{1}{3}x_3 & \leq & 8 \\ & 2x_1 & + & x_2 & & & \leq & 12 \\ & 23x_1 & + & \frac{7}{3}x_2 & + & \frac{4}{3}x_3 & \leq & 78 \\ & x_1, x_2, x_3 \geq 0 & & & & & & \end{array} \quad (1)$$

you will find that the optimal solution is

$$x_1 = 0, x_2 = 12, x_3 = 12, s_1 = 0, s_2 = 0, s_3 = 34; z = 84 \quad (2)$$

We are interested first in finding to what extent we can change a single cost coefficient in the original problem (1) without causing a change in the optimal solution (2). Later in this section we deal with simultaneous changes of two or more cost coefficients. Of course, a change in any cost coefficient may lead to a change in the value of the objective function; here, however, by changes in the *optimal solution* we mean changes in the *optimal values* for the decision and slack variables – the value of z may change even though the optimal solution does not. CALIPSO provides the information we need.

CALIPSO

The first step in performing a cost coefficient sensitivity analysis for the LP problem (1) is to use *LP SOLVER* to solve this problem. As usual, enter this problem in *LP SOLVER* and press [*F1*] to obtain the optimal values given in (2). Next press [*F4*] to see the menu of postoptimality functions. You see that you are presented four alternatives. We are interested in sensitivity analysis so press [*Alt*]+[*F3*]. On the screen you see the heading Cost

Coefficient Ranges. Note that for each basic and nonbasic variable various allowable increases and decreases in the cost coefficients are given.

For example, for the variable x_3 you find that for the original cost coefficient $c_3 = 1$ the allowable increase is 5 and the allowable decrease is 1. This means that if you replace $c_3 = 1$ with any value from 0 to 6, then the optimal solution (2) of the original problem (1) is also the optimal solution of the new problem with this changed cost coefficient. On the other hand, if you replace the cost coefficient $c_3 = 1$ with $c_3 = 7$, then the optimal solution of the new LP problem will no longer be (2).

The allowable decrease in the cost coefficient $c_1 = 9$ is ∞. This shows that you can decrease the cost coefficient c_1 any amount you want and the original optimal solution (2) will also be the optimal solution to the problem with the changed cost coefficient. Similarly, the allowable increase in the cost coefficient $c_2 = 6$ is ∞ ; thus you can increase this cost coefficient by any amount without changing the original optimal solution (2).

One note of caution: The ranges given for each cost coefficient are not necessarily valid if you change two or more cost coefficients simultaneously. For instance, if in (1) you replace the cost coefficient $c_3 = 1$ with $c_3 = 5$ and the cost coefficient $c_2 = 6$ with $c_2 = 3$, then both of these changes fall in the allowable ranges. However, you will find that the optimal solution of the problem with these changed coefficients is

$$x_1 = 0, x_2 = 0, x_3 = 24, s_1 = 0, s_2 = 12, s_3 = 46$$

which differs from the optimal solution (2) of (1).

6.3A Exercises: Cost Coefficient Ranges

In Problems 1 - 3 use *LP SOLVER* to determine the allowable ranges of the cost coefficients that will not affect the optimal solution.

1.

$$\begin{array}{llcccccll}
\text{Maximize: } z = & 3x_1 & + & 2x_2 & + & 4x_3 & & \\
\text{Subject to:} & 3x_1 & + & 2x_2 & + & 6x_3 & \leq & 6 \\
& 5x_1 & + & x_2 & + & 4x_3 & \leq & 8 \\
& 2x_1 & - & 2x_2 & + & 3x_3 & \leq & 9 \\
& x_1, x_2, x_3 \geq 0 & & & & & &
\end{array}$$

2. §

$$\begin{array}{llllllllll} \text{Minimize: } z = & 2x_1 & - & 3x_2 & + & 4x_3 & + & 5x_4 & & \\ \text{Subject to:} & 3x_1 & - & x_2 & + & 4x_3 & - & x_4 & \leq & 120 \\ & 2x_1 & + & x_2 & - & 5x_3 & + & 3x_4 & \geq & 20 \\ & 4x_1 & + & 5x_2 & + & x_3 & & & \leq & 80 \\ & \multicolumn{9}{l}{x_i \geq 0, \ 1 \leq i \leq 4} \end{array}$$

3.

$$\begin{array}{llllllllll} \text{Maximize: } z = & 2x_1 & - & x_2 & - & x_3 & + & 6x_4 & & \\ \text{Subject to:} & 3x_1 & + & 2x_2 & + & 4x_3 & + & 6x_4 & \leq & 24 \\ & 5x_1 & + & x_2 & + & 2x_3 & + & 3x_4 & \geq & 12 \\ & \multicolumn{9}{l}{x_i \geq 0, \ 1 \leq i \leq 4} \end{array}$$

4. § This problem refers to Problem 7, Exercises 5.4A.
   (a) Will a change in the profit of Drink 1 from \$.60 to \$.80 affect the original optimal solution?
   (b) Will a change in the profit of Drink 3 from \$.80 to \$.50 affect the original optimal solution?
   (c) Verify that an increase of 20 cents in the profit of Drink 2 will not affect the original optimal solution, nor will a decrease of 25 cents in the profit of Drink 3. Will changing these profits simultaneously by these amounts affect the optimal solution?

5. This problem refers to Problem 11, Exercises 5.4A.
   (a) Will a change from 13 to 11 in the cost table cause a change in the original optimal solution?
   (b) Over what range can the cost coefficient of x_{23} vary without changing the original optimal solution?
   (c) Can the coefficient of x_{15} be increased to 10 without changing the original optimal solution?

6. In Problem 5 of Exercises 6.2C how much must the profit level on Grates A be increased so that more small grates and fewer large grates are produced?

Single Changes in the Resource Values

We consider what happens if just one of the resource values is changed. Unlike the case of cost coefficients, any change in the resource values may affect the values of the optimal solution of the original problem. What interests us now is to what extent we can change a resource value without changing the *set* of basic variables comprising the optimal solution. For instance, for the LP problem (1) the basic variables in the optimal solution are x_2, x_3, and s_3. To what extent can we change the resource value of, say, $b_1 = 8$ so that the basic variables in the optimal solution to the problem with this change are still x_2, x_3, s_3? Again CALIPSO provides the answer.

CALIPSO

As before, your first step is to use *LP SOLVER* to solve (1). Next press [*F4*] (or more directly, simply press [*Alt*]+[*F3*]) to obtain the sensitivity analysis option. You should [*PgDn*] if necessary until you see the heading, Resource Value Ranges. Under this heading you see the initial (original) resource values and then the allowable decreases and increases in these values. Thus, for example, you see that an increase of 7 in the resource value $b_1 = 8$ will result in an LP problem whose optimal solution still consists of the set of basic variables x_2, x_3, s_3. However, the optimal values of these variables may be different than the optimal values (2) for the LP problem (1). In fact, the optimal solution of (1) when $b_1 = 8$ has been changed to $b_1 = 15$ is

$$x_1 = 0, x_2 = 12, x_3 = 33, s_1 = 0, s_2 = 0, s_3 = 6$$

As was the case for cost coefficients, the ranges given for changes in the resource values do not necessarily apply to multiple changes in the RHS. For example, the changes of $b_1 = 8$ to $b_1 = 5$ and $b_2 = 12$ to $b_2 = 23$ in the resource values of (1) are both within the ranges given by CALIPSO. However, the optimal solution of (1) with these changes consists of the basic variables x_2, s_2, s_3, which differs from the set of basic variables x_2, x_3, s_3 in the optimal solution of (1).

6.3B Exercises: Single Changes in the Resource Values

In Problems 1–3 use *LP SOLVER* to determine the allowable range of each resource entry so that there will be no change in the set of basic variables in the optimal solution to the original problem.

1.

$$\begin{array}{lrcrcrcrcr}
\text{Maximize: } z = & 2x_1 & + & 3x_2 & + & x_3 & - & x_4 & & \\
\text{Subject to:} & 2x_1 & + & x_2 & - & 3x_3 & - & x_4 & \leq & 16 \\
 & x_1 & + & 5x_2 & + & 2x_3 & & & \leq & 9 \\
 & 4x_1 & & & + & 3x_3 & + & 2x_4 & \leq & 22 \\
 & \multicolumn{9}{l}{x_i \geq 0,\ 1 \leq i \leq 4}
\end{array}$$

2. §

$$\begin{array}{lrcrcrcr}
\text{Minimize: } z = & 2x_1 & + & 3x_2 & - & 4x_3 & & \\
\text{Subject to:} & x_1 & + & 4x_2 & + & 5x_3 & \geq & 9 \\
 & 2x_1 & + & 3x_2 & - & 4x_3 & \leq & 8 \\
 & 3x_1 & - & 2x_2 & + & 6x_3 & \leq & 10 \\
 & \multicolumn{7}{l}{x_i \geq 0,\ 1 \leq i \leq 3}
\end{array}$$

3.

$$\begin{array}{lrcrcr}
\text{Maximize: } z = & 3x_1 & + & 2x_2 & & \\
\text{Subject to:} & 4x_1 & + & 2x_2 & \leq & 16 \\
 & 3x_1 & - & 5x_2 & \leq & 34 \\
 & 2x_1 & + & 6x_2 & \leq & 20 \\
x_1, x_2 \geq 0 & & & & &
\end{array}$$

In Problems 4–7 answer the questions by studying the sensitivity analysis; do not solve the problems again unless instructed to do so.

4. § Suppose in Problem 7, Exercises 5.4A, Blended Fruits has 6000 rather than 5000 quarts of orange juice available. Will this change the kinds of drinks that Blended Fruits planned to produce to maximize profits as a result of solving the original problem?

5. Suppose in Problem 4, Exercises 5.4A, the company received orders for 625 rather than 1000 dishwashers. Will the factories used to obtain an optimal solution to the original problem be the same ones to meet the new demands? If not, what ones will be used? (In more advanced courses you will learn how to answer the latter question from the sensitivity analysis. Now you will need to solve the new problem to answer it.)

6. § Suppose in Problem 11 of Exercises 5.4A, supply point S_3 has 325 rather than 300 units available. What effect does this have on the original optimal solution?

7. Note that in Problem 1 above you can increase the resource value 9 to 30.5 without changing the optimal set of basic solutions; you can also increase the entry 22 to 31.556 without changing the optimal set of basic solutions. What happens if you try to simultaneously increase 9 to 30 and 22 to 31?

Multiple Changes in the Resource Values

In [CV] there is a detailed explanation of the following procedure that can be used to determine if multiple changes in the resource values produce a different set of basic variables in the optimal solution.

Step 1. Use *LP SOLVER* to solve the original problem.

Step 2. Press Alt+F4 to obtain the *inverse basis matrix* and the final resource column (for the original problem). These values will appear in an array of the form

$$\begin{bmatrix} a_{11} & a_{12} & \cdots & a_{1n} & d_1 \\ a_{21} & a_{22} & \cdots & a_{2n} & d_2 \\ \vdots & & & & \\ a_{m1} & a_{m2} & \cdots & a_{mn} & d_m \end{bmatrix}$$

where a_{ij} are entries in the final tableau (and are the entries of the inverse basis matrix) and d_i are the final resource values. Use these values to form the system of linear inequalities

$$\begin{array}{ccccccccccc} a_{11}\Delta b_1 & + & a_{12}\Delta b_2 & + & \cdots & + & a_{1n}\Delta b_n & + & d_1 & \geq & 0 \\ a_{21}\Delta b_1 & + & a_{22}\Delta b_2 & + & \cdots & + & a_{2n}\Delta b_n & + & d_2 & \geq & 0 \\ \vdots & & & & & & & & & & \\ a_{m1}\Delta b_1 & + & a_{m2}\Delta b_2 & + & \cdots & + & a_{mn}\Delta b_n & + & d_m & \geq & 0 \end{array}$$

Step 3. Substitute your proposed changes Δb_i in these inequalities. If all of the inequalities are satisfied, then your changes will not produce a change in the set of basic variables in the optimal solution. If one or more of these inequalities is not satisfied, then there will be a change in the set of these basic variables.

Example 1: Multiple Changes in the Resource Values

Consider the problem

$$\begin{array}{llrcrcrcr} \text{Maximize}: z = & 2x_1 & + & 3x_2 & - & 4x_3 & & \\ \text{Subject to}: & 2x_1 & - & x_2 & + & x_3 & \leq & 30 \\ & x_1 & + & 3x_2 & + & 2x_3 & \leq & 10 \\ & 3x_1 & - & x_2 & + & 2x_3 & \geq & 8 \\ & x_1, x_2, x_3 \geq 0 & & & & & & \end{array} \qquad (3)$$

Solving this problem you will find that the optimal basic variables are s_1, s_3, and x_1 with values $s_1 = 10, s_3 = 22, x_1 = 10$. From the sensitivity analysis of CALIPSO ([Alt]+[F4]) we find that the inverse basis matrix and final resource values are

$$\begin{bmatrix} 1 & -2 & 0 & 10 \\ 0 & 3 & -1 & 22 \\ 0 & 1 & 0 & 10 \end{bmatrix}$$

Thus the corresponding system of inequalities that must be satisfied if changes in the resource values do not produce a change in the optimal set of basic variables $\{s_1, s_3, x_1\}$ is

$$\begin{array}{rcrcrcrcl} \Delta b_1 & - & 2\Delta b_2 & & & + & 10 & \geq & 0 \\ & & 3\Delta b_2 & - & \Delta b_3 & + & 22 & \geq & 0 \\ & & \Delta b_2 & & & + & 10 & \geq & 0 \end{array} \qquad (4)$$

Suppose in the original problem (3) we want to change the resource value 30 to 25 and the resource value 8 to 10. Then $\Delta b_1 = -5, \Delta b_2 = 0$, and $\Delta b_3 = 2$. Substituting these values in (4) we find that all of the inequalities are satisfied. Therefore, in the optimal solution of the new problem (with these changes in the resources) the optimal *set* of basis variables will still be $\{s_1, s_3, x_1\}$ although as you can verify, the *values* of these variables change.

Suppose in (3) you change the resource value 30 to 23, the resource value 10 to 14, and the resource value 8 to 9. Then $\Delta b_1 = -7, \Delta b_2 = 4$, and $\Delta b_3 = 1$. Substituting these values in (4) you will find that the first inequality is not satisfied. Therefore, these changes will lead to a change in the set of optimal basis variables. In fact the new optimal basis set is $\{s_3, x_2, x_1\}$ with values $s_3 = 25.86, x_2 = .71, x_1 = 11.86$. You might note, however, that each of these changes lies in the allowable range for individual changes. So making any one of these changes will not cause a change in the set of basis variables.

6.3C Exercises: Multiple Changes in the Resource Values

In Problems 1–3 form the system of inequalities obtained from the inverse basis matrix and the final right-hand side. Use these inequalities to determine if the proposed multiple changes in the resource values will lead to a change in the original optimal set of basis variables.

1. Refer to Problem 1, Exercises 6.3B. In the resource values change 16 to 20, 9 to 6, and 22 to 28.

2. § Refer to Problem 2, Exercises 6.3B. In the resource values change 9 to 10, 8 to 11, and 10 to 11.

3. Refer to Problem 3, Exercises 6.3B. In the resource values change 16 to 25 and 20 to 23.

4. Refer to Problem 7, Exercises 5.4A. Will the mix of the kinds (not the amounts) of the drinks produced by Blended Fruits change from your original answer if the amounts available of the juices are: apple juice, 8000 quarts; orange juice, 5000 quarts; lime juice, 900 quarts; cranberry juice, 4200 quarts; peach juice, 1000 quarts?

5. § Refer to Problem 11, Exercises 5.4A. Analyse the changes resulting from the following changes to the supplies and demands for $S_1 = 390; S_2 = 160; S_3 = 340; D_1 = 160; D_2 = 80; D_3 = 250; D_4 = 95; D_5 = 210$.

6. Refer to Problem 4, Exercises 5.4A. Analyze the changes resulting from the following changes in the number of received orders: refrigerators, 1700; freezers, 900; dishwashers, 1100; ovens, 800.

Multiple Changes of Cost Coefficients

In [CV] there is a detailed explanation of the following procedure for determining whether specific multiple changes in the cost coefficients of an LP problem will result in a change in the optimal solution. The idea is to form the dual problem so that the cost coefficients of the primal problem become resource values in the dual problem. We then apply the above analysis of multiple changes in the resource values to the dual problem, thus obtaining a system of inequalities that must be satisfied by changes to the cost coefficients of the primal problem.

WARNING: It is possible for a problem to have alternative optimal solutions (see Section 5.2). That occurs when the optimal value of the objective function is attained for more than one feasible solution to the problem. When that happens, a solution to the dual problem may not correspond to the particular solution to the primal problem. Consequently, the procedure we describe here can give erroneous results if the optimal solution is not unique. Note that *LP SOLVER* informs you (at the beginning of the solution listing) when a problem has alternative optimal solutions.

We illustrate this procedure with (1).

Step 1. Formulate and solve the dual problem. The dual of (1) is

$$\begin{array}{llllllll}\text{Minimize: } w = & 8y_1 & + & 12y_2 & + & 78y_3 & & \\ \text{Subject to:} & 2y_1 & + & 2y_2 & + & 23y_3 & \geq & 9 \\ & \frac{1}{3}y_1 & + & y_2 & + & \frac{7}{3}y_3 & \geq & 6 \\ & \frac{1}{3}y_1 & & & + & \frac{4}{3}y_3 & \geq & 1 \\ & y_1, y_2, y_3 \geq 0 & & & & & & \end{array}$$

Step 2. Press [Alt]+[F4] to get the inverse basis matrix (see [CV]) and the final resource values for the dual problem. Use this matrix and the final resource values to write the system of linear inequalities

$$\begin{array}{rcrcrcrcl} 0\Delta c_1 & + & 1\Delta c_2 & - & 1\Delta c_3 & + & 5 & \geq & 0 \\ -\Delta c_1 & + & 2\Delta c_2 & + & 4\Delta c_3 & + & 7 & \geq & 0 \\ 0\Delta c_1 & + & 0\Delta c_2 & + & 3\Delta c_3 & + & 3 & \geq & 0 \end{array} \tag{5}$$

Note: You use $-\Delta c_i$ whenever the cost coefficient in the original problem is negative (see Example 2).

Step 3. Substitute your proposed changes Δc_i in (5). If all of these inequalities are satisfied, then your changes will maintain the original optimal solution. If one or more of these inequalities is not satisfied, then your changes will lead to a new optimal solution.

Example 2: Multiple Changes in the Cost Coefficients

To form the dual of the LP problem

$$\begin{array}{lrcrcrcl}
\text{Maximize: } w = & 2x_1 & - & 3x_2 & + & x_3 & & \\
\text{Subject to:} & x_1 & + & x_2 & - & x_3 & \le & 8 \\
& 2x_1 & + & x_2 & - & x_3 & \le & 6 \\
& x_1 & & & + & x_3 & \le & 5 \\
& 2x_1 & - & x_2 & + & x_3 & \ge & 5 \\
& -3x_1 & - & x_2 & + & 2x_3 & \ge & 3 \\
& x_1, x_2, x_3 \ge 0 & & & & & &
\end{array}$$

first multiply the last two constraints by -1 and then obtain

$$\begin{array}{lrcrcrcrcrcr}
\text{Minimize: } w = & 8y_1 & + & 6y_2 & + & 5y_3 & - & 5y_4 & - & 3y_5 & & \\
\text{Subject to:} & y_1 & + & 2y_2 & + & y_3 & - & 2y_4 & + & 3y_5 & \ge & 2 \\
& y_1 & + & y_2 & & & + & y_4 & + & y_5 & \ge & -3 \\
& -y_1 & - & y_2 & + & y_3 & - & y_4 & - & 2y_5 & \ge & 1 \\
& y_i \ge 0,\ 1 \le i \le 5 & & & & & & & & & &
\end{array}$$

Solving the dual with *LP SOLVER*, you will find from the sensitivity analysis ([Alt]+[F4]) that the inverse basis matrix and final resource values are

$$\begin{bmatrix} \frac{1}{5} & 0 & -\frac{1}{5} & \frac{1}{5} \\ \frac{1}{5} & 1 & -\frac{1}{5} & \frac{16}{5} \\ \frac{2}{5} & 0 & \frac{3}{5} & \frac{7}{5} \end{bmatrix} \tag{6}$$

We must take the negative of the entries in the second column of (6) because the second cost coefficient (-3) in the original problem is negative. (See [CV] for an explanation of the reason for this.) Thus the corresponding system of inequalities that must be satisfied if changes in the cost coefficients do not produce a change in the solution is

$$\begin{array}{rcrcrcrcl}
\frac{1}{5}\Delta c_1 & - & 0\Delta c_2 & - & \frac{1}{5}\Delta c_3 & + & \frac{1}{5} & \ge & 0 \\
\frac{1}{5}\Delta c_1 & - & 1\Delta c_2 & - & \frac{1}{5}\Delta c_3 & + & \frac{16}{5} & \ge & 0 \\
\frac{2}{5}\Delta c_1 & - & \Delta c_2 & + & \frac{3}{5}\Delta c_3 & + & \frac{7}{5} & \ge & 0
\end{array} \tag{7}$$

6.3D Exercises: Multiple Changes in the Cost Coefficients

In Problems 1–3 of Exercises 6.3A find a system of inequalities that must be satisfied by changes to the cost coefficients in order that the solution will not be changed. (The answer to Problem 2 is given.)

6.4 INTERPRETING LP REPORTS

Few corporate or business managers count on their own expertise to solve LP problems. They know it is better to hire someone with the depth of mathematical training to analyze a problem and produce a reliable report. Nevertheless, the allocation of resources and scheduling of operations is so basic to the success of any financial enterprise that it is important for a manager to be able to foresee the need for using LP techniques and to be able to understand a report based on them.

In this section we apply duality theory and sensitivity analysis to a case study. We have reduced the scale of our study from the truly practical so that it is more easily comprehensible; however, in all basic aspects, it demonstrates the techniques you should use and the foresight you must have in applying the principles of linear programming to evaluate the practices of an enterprise.

In our book [CV] we reported on the activities of Moscow Operations Management, Inc. (MOM), a new company at the time of that report. MOM offers production management services to businesses. Since our earlier report MOM has continued to expand and diversify its services to manufacturing enterprises. A typical client supplies MOM with data describing its manufacturing processes and the various supplies, demands, and costs that affect the production process and profits therefrom. MOM uses LP methods to find an optimal solution to the client's problem and advises the client on how to best manage its operation to maximize profit. After the initial report, clients can call on MOM for advice on the best response to changing conditions. We will describe one of MOM's cases and discuss the responses that MOM gives to several queries.

Pea and Lentil Soup, Inc. (PEALS) operates farms at Genesee and Potlatch. The Genesee field produces peas and the Potlatch field produces lentils. PEALS also operates a factory that makes soup from peas and lentils and small amounts of other ingredients that we will ignore in this study. The soup factory can use any of four recipes to produce two products: *Spicy Peal Soup* and *Mellow Peal Soup.* The Genesee field produces peas at a cost of \$2.40 per bushel, and the Potlatch field produces lentils at a cost of \$3.60 per bushel. Table 1 summarizes the amounts of labor, electric power, peas, and lentils that are used in each recipe, and the number of cans of each product produced by each recipe. The Genesee field can produce up to 11,000 bushels of peas per season and the Potlatch field can produce up to 26,000 bushels of lentils. Because of the size of the plant, the soup factory can employ no more than 6000 person-hours of labor per season. The average salary paid to these workers is \$4.75 per hour. Electricity costs \$.05 per kwh, and the supply is virtually unlimited.

PEALS wants to determine the most profitable way of operating its soup factory and two farms in order to produce 9,000 cases of Spicy soup and 12,000 cases of Mellow soup each season.

Table 1

	Resources				Output	
	Labor	Power	Peas	Lentils	Spicy	Mellow
Recipe	(worker-h)	(kwh)	(bushels)	(bushels)	(cases)	(cases)
1	24	800	32	155	40	50
2	32	1200	35	100	45	40
3	12	800	50	200	75	22
4	27	750	69	84	10	100

MOM's Solution

Let x_1, x_2, x_3, x_4 denote the number of hours per week that PEALS uses recipes 1,2,3,4, respectively. To find the cost of labor, first note that the number of hours required is

$$24x_1 + 32x_2 + 12x_3 + 27x_4$$

Since the average cost of an hour of labor is \$4.75, the total cost is

$$\begin{aligned} &4.75(24x_1 + 32x_2 + 12x_3 + 27x_4) \\ &= 114x_1 + 152x_2 + 57x_3 + 128.25x_4 \end{aligned} \tag{1}$$

To find the cost of peas from the Genesee field, first note that the number of bushels of peas needed is

$$32x_1 + 35x_2 + 50x_3 + 69x_4$$

Since the cost of a bushel of Genesee peas is \$2.40, the total cost of peas is

$$\begin{aligned} &2.4(32x_1 + 35x_2 + 50x_3 + 69x_4) \\ &= 76.8x_1 + 84x_2 + 120x_3 + 165.6x_4 \end{aligned} \tag{2}$$

Similarly the cost of Potlatch lentils is

$$\begin{aligned} &3.6(155x_1 + 100x_2 + 200x_3 + 84x_4) \\ &= 558x_1 + 360x_2 + 720x_3 + 302.4x_4 \end{aligned} \tag{3}$$

The cost of electricity is

$$\begin{aligned} &0.05(800x_1 + 1200x_2 + 800x_3 + 750x_4) \\ &= 40.0x_1 + 60.0x_2 + 40.0x_3 + 37.5x_4 \end{aligned} \tag{4}$$

The objective function is the sum of the expressions (1), (2), (3), and (4). The LP problem is

$$\begin{array}{lllllllllll} \multicolumn{11}{l}{\text{Minimize}: z = 788.8x_1 + 656x_2 + 937x_3 + 633.75x_4} \\ \multicolumn{11}{l}{\text{Subject to}:} \\ 24x_1 & + & 32x_2 & + & 12x_3 & + & 27x_4 & \le & 6000 & \text{(Labor)} \\ 32x_1 & + & 35x_2 & + & 50x_3 & + & 69x_4 & \le & 11,000 & \text{(Peas)} \\ 155x_1 & + & 100x_2 & + & 200x_3 & + & 84x_4 & \le & 26,000 & \text{(Lentils)} \\ 40x_1 & + & 45x_2 & + & 75x_3 & + & 10x_4 & \ge & 9,000 & \text{(Spicy)} \\ 50x_1 & + & 40x_2 & + & 22x_3 & + & 100x_4 & \ge & 12,000 & \text{(Mellow)} \\ \multicolumn{11}{l}{x_i \ge 0, 1 \le i \le 4} \end{array} \tag{5}$$

MOM expects that PEALS will be calling throughout the year with questions about the effect of various changes in the labor force, the supplies of legumes, the demand for soup, and the costs of operation. MOM plans to gather all data from the solution to the LP problem that might prove useful in responding to such queries. Clearly, the sensitivity ranges on resource values will be useful; however, the sensitivity ranges on the individual cost coefficients will not be very useful because any change in the cost of extraction or electricity will affect all of the cost coefficients. For instance, changing the processing cost of peas will change all of the cost coefficients in (2) and hence in (5). Therefore, MOM also plans to solve the dual of (5) to obtain a system of equations that can be used to test whether a certain change in the cost of one aspect of the operation (resulting in multiple cost coefficient changes) will change the solution. MOM uses CALIPSO to find the following solution to the primal problem (5).

The minimum value of the objective function is

$$z = \$155,964.72 \tag{6}$$

The solution is

$$\begin{aligned} x_1 &= 0 \\ x_2 &= 118.666 \\ x_3 &= 40.3120 \\ x_4 &= 63.6651 \end{aligned} \tag{7}$$

SENSITIVITY ANALYSIS

Dual Prices

Row	Dual Price
1	2.62458
2	0
3	0
4	−11.1742
5	−5.92872

(8)

Ranges for Single Changes to Cost and Resource Values

Cost Coefficient Ranges

Variable	Initial Cost Coeff.	Range				
Basic Variables						
x_4	−633.75	−1025.79	$\le$	c_4	$\le$	−464.667
s_2	0	−3.82335	$\le$	c_6	$\le$	7.72399
s_3	0	−1.30293	$\le$	c_7	$<$	∞
x_2	−656	−702.684	$\le$	c_2	$\le$	∞
x_3	−937	−1387.56	$\le$	c_3	$\le$	−854.108
Nonbasic Variables						
x_1	−788.8	$-\infty$	$<$	c_1	$\le$	−680.413
s_1	0	$-\infty$	$<$	c_5	$\le$	2.62458
s_4	0	$-\infty$	$<$	c_8	$\le$	11.1742
s_5	0	$-\infty$	$<$	c_9	$\le$	5.92872

(9)

Resource Value Ranges

Row Number	Initial Resource Value	Range				
1	6000	5640.99	$\le$	b_1	$\le$	7273.17
2	11000	10561.8	$\le$	b_2	$<$	∞
3	26000	25276.8	$\le$	b_3	$<$	∞
4	9000	6537.74	$\le$	b_4	$\le$	9271.76
5	12000	7567.74	$\le$	b_5	$\le$	12533.2

The inverse basis matrix and final resource values are

$$\left[\begin{array}{lllll|l} -.0155224 & 0 & 0 & -.00172986 & 0.014364 & 63.6651 \\ 0.686462 & 1 & 0 & -0.535439 & -0.821801 & 438.211 \\ 2.01436 & 0 & 1 & -2.66109 & -1.117777 & 723.176 \\ 0.0562206 & 0 & 0 & -0.0046799 & -0.0147116 & 118.666 \\ -0.0316627 & 0 & 0 & 0.0163719 & 0.00691173 & 40.312 \end{array}\right] \tag{10}$$

Using (10) MOM calculates that multiple changes to the resource values must satisfy the system of inequalities

$$\begin{array}{rcrcrcr} -.0156\Delta b_1 & + & 0\Delta b_2 & + & 0\Delta b_3 & - & 0.00173\Delta b_4 \\ & & & + & 0.0144\Delta b_5 & + & 63.665 \geq 0 \\ 0.686\Delta b_1 & + & 1\Delta b_2 & + & 0\Delta b_3 & - & 0.535\Delta b_4 \\ & & & - & 0.822\Delta b_5 & + & 438.211 \geq 0 \\ 2.014\Delta b_1 & + & 0\Delta b_2 & + & 1\Delta b_3 & - & 2.661\Delta b_4 \\ & & & - & 1.1178\Delta b_5 & + & 723.176 \geq 0 \\ 0.0562\Delta b_1 & + & 0\Delta b_2 & + & 0\Delta b_3 & - & 0.00468\Delta b_4 \\ & & & - & 0.0147\Delta b_5 & + & 118.666 \geq 0 \\ -.0317\Delta b_1 & + & 0\Delta b_2 & + & 0\Delta b_3 & + & 0.01637\Delta b_4 \\ & & & + & 0.00691\Delta b_5 & + & 40.312 \geq 0 \end{array} \tag{11}$$

We rounded off the numbers in (11) to about three significant digits. The variables Δb_i are the amounts that the resource values b_i of row i are changed. Rearranging (11) gives the simpler system of inequalities

$$\begin{array}{rcrcrcrcr} 0.0156\Delta b_1 & & & + & 0.0017\Delta b_4 & - & 0.0144\Delta b_5 & \leq & 63.665 \\ -0.686\Delta b_1 & - & 1\Delta b_2 & + & 0.535\Delta b_4 & + & 0.822\Delta b_5 & \leq & 438.211 \\ -2.014\Delta b_1 & - & 1\Delta b_3 & + & 2.661\Delta b_4 & + & 1.1177\Delta b_5 & \leq & 723.176 \\ -0.0562\Delta b_1 & & & + & 0.00468\Delta b_4 & + & 0.0147\Delta b_5 & \leq & 118.666 \\ 0.0317\Delta b_1 & & & - & 0.01637\Delta b_4 & - & 0.00691\Delta b_5 & \leq & 40.312 \end{array} \tag{12}$$

Next MOM forms the dual of (5). You will notice that the dual constraints we use in (13) are different from the ones you would obtain by the method we described in Section 6.3. In Section 6.3 we formed the dual by multiplying $\geq$ constraints by -1 to convert them to $\leq$ constraints. We did this to simplify the solution of the dual problem by the simplex method. However, the multiplication by -1 introduces a problem in applying the Fundamental Principle of Duality because you must remember to use the negative of the resource value in the computation. Since it is difficult for the reader of a computer report to know which constraints were multiplied by -1, it is customary to assign a negative value to a dual price that corresponds to a $\geq$ constraint. That is why there are some negative values in (8) and why (13) appears unfamiliar. This difficulty is more thoroughly discussed in more advanced books such as [CV]. The dual problem is

$$\begin{array}{l}
\text{Minimize}: w = 6{,}000y_1 + 11{,}000y_2 + 26{,}000y_3 + 9{,}000y_4 + 12{,}000y_5 \\
\text{Subject to}: \\
\begin{array}{rcrcrcrcrcr}
24y_1 & + & 32y_2 & + & 155y_3 & + & 40y_4 & + & 50y_5 & \geq & 788.8 \\
32y_1 & + & 35y_2 & + & 100y_3 & + & 45y_4 & + & 40y_5 & \geq & 656 \\
12y_1 & + & 50y_2 & + & 200y_3 & + & 75y_4 & + & 22y_5 & \geq & 937 \\
27y_1 & + & 69y_2 & + & 84y_3 & + & 10y_4 & + & 100y_5 & \geq & 633.75
\end{array} \\
y_1, y_2, y_3 \geq 0;\ y_4, y_5 \leq 0
\end{array} \tag{13}$$

MOM uses CALIPSO to find the solution of (13) and to print the inverse basis matrix and final resource column. The maximum value of the objective function of (13) is

$$w = \$155{,}964.72$$

which, of course, is the same as (6). The maximal solution is

$$y_1 = 2.6249, y_2 = 0, y_3 = 0, y_4 = -11.1741, y_5 = -5.9287$$

which is the same as (8).

The inverse basis matrix and final resource values are

$$\left[\begin{array}{rrrr|r}
1 & -.426519 & -0.240559 & -0.276469 & 108.387 \\
0 & -0.0147116 & 0.00691173 & 0.014364 & 5.92872 \\
0 & -0.0046799 & 0.0163719 & -0.00172986 & 11.1742 \\
0 & -0.0562206 & 0.0316627 & 0.0155224 & 2.62458
\end{array}\right] \tag{14}$$

MOM uses (14) to find the system of inequalities that must be satisfied by multiple changes to the cost coefficients of (5). Recall that we must change the sign of each Δc_i because the cost coefficients in (5) are all negative. The system of inequalities is

$$\begin{array}{rcrcrcrcrl}
-1\Delta c_1 & + & .427\Delta c_2 & + & 0.241\Delta c_3 & + & .276\Delta c_4 & + & 108.387 & \geq 0 \\
0\Delta c_1 & + & .0147\Delta c_2 & - & .0069\Delta c_3 & - & .0144\Delta c_4 & + & 5.929 & \geq 0 \\
0\Delta c_1 & + & .0047\Delta c_2 & - & .0163\Delta c_3 & + & .0017\Delta c_4 & + & 11.174 & \geq 0 \\
0\Delta c_1 & + & .0562\Delta c_2 & - & .0317\Delta c_3 & - & .0155\Delta c_4 & + & 2.625 & \geq 0
\end{array} \tag{15}$$

We rounded the numbers in (15) to about 3 significant digits. The variables Δc_i are the amounts that the values of the cost coefficients c_i are changed. Rearranging (15) gives the simpler system of inequalities

$$\begin{array}{lrrrrrrrl} \Delta c_1 & - & 0.427\Delta c_2 & - & 0.241\Delta c_3 & - & 0.276\Delta c_4 & \leq & 108.387 \\ & & -.0147\Delta c_2 & + & 0.0069\Delta c_3 & + & 0.0144\Delta c_4 & \leq & 5.929 \\ & & -.0047\Delta c_2 & + & 0.0163\Delta c_3 & - & 0.0017\Delta c_4 & \leq & 11.174 \\ & & -.0562\Delta c_2 & + & 0.0317\Delta c_3 & + & 0.0155\Delta c_4 & \leq & 2.625 \end{array} \qquad (16)$$

Report to PEALS

MOM reports the solution to (5) to PEALS as

Production Schedule

Recipe	Hours per season
1	0
2	118.67
3	40.31
4	63.67

Resources Required

	Peas	Lentils	Electricity	Labor
Quantity	10,562 bu	25,277 bu	47,752.5 kwh	6,000.25 h
Cost	$25,348.80	$90,997.20	$2,387.63	$28,501.19

In Problem 12 you are asked to check the values in the table of Resources Required.

PEALS sets up its operation according to MOM's advice and at various times asks MOM the following questions.

Question 1 Why did you advise us to use Recipe 3? Its labor costs are low, but it uses more peas and lentils than the other recipes.

MOM sees from (8) that the dual price of labor is about 2.62. That means that an additional hour of labor will increase the production cost by $2.62. Since the dual price of peas and lentils are both 0, it is the lack of enough labor that constrains the solution.

From the resource value ranges in (9), MOM sees that the labor force needs to be increased to 7,273.17 hours to change the mix of recipes used. Adding 1,273.17 person-hours of labor to the existing 6,000 hours would increase the total cost by $1273.17 \times 2.62458 = \$3,341.54$. To find out what the new mix of recipes would be, MOM changes the labor constraint in (5) to 7,274 and solves the resulting problem to discover that using Recipe 2 for 190.24 hours and Recipe 4 for 43.9 hours and not using the other recipes will reduce the total operating cost to $152,623.17. (Note: In a more advanced linear programming course you will learn how to calculate the new solution from the sensitivity analysis.)

MOM advises PEALS of this option; however, PEALS determines that the lack of enough trained workers in the job market precludes taking advantage of the savings.

Question 2 We are starting production next week and have only been able to hire 5700 person-hours of labor. Should we change the mix of recipes? How much is this reduced labor force going to cost us?

MOM sees from the resource value ranges in (9) that a reduction to 5700 is at the upper limit of the range that would cause a change in the set of basic variables in the solution, and, therefore, no change in the mix of recipes is advantageous. Since the dual price of labor is 2.62458, the cost of reducing the force by 300 person-hours is $2.62458 \times 300 =$ $787.37.

By the start of production PEALS manages to hire another 300 person-hours of labor and did not need to incur this additional cost.

Question 3 The projected cost of peas has gone up to $3.00 per bushel and of lentils to $3.80 per bushel. Do we need to change our production schedule? How much is this increase going to cost us?

Since these changes will affect all of the cost coefficients, MOM must calculate the change in each cost coefficient and see if any constraint in the system of inequalities (16) is violated.

Substituting the new costs of legumes in (2) and (3) gives

$$\begin{aligned}&3.0(32x_1 + 35x_2 + 50x_3 + 69x_4)\\&= 96x_1 + 105x_2 + 150x_3 + 207x_4\end{aligned}$$

Similarly the cost of Potlatch lentils is

$$\begin{aligned}&3.8(155x_1 + 100x_2 + 200x_3 + 84x_4)\\&= 589x_1 + 380x_2 + 760x_3 + 319.2x_4\end{aligned}$$

Therefore the changes are the differences between the new costs and the old costs of legumes. These are

$$\begin{aligned}
\Delta c_1 &= (96+589)-(76.8+558) &&= 50.2 \\
\Delta c_2 &= (105+380)-(84+360) &&= 41 \\
\Delta c_3 &= (150+760)-(120+720) &&= 70 \\
\Delta c_4 &= (207+319.2)-(165.6+302.4) &&= 58.2
\end{aligned} \qquad (17)$$

Substituting the values in (17) into (16) gives

$$\begin{array}{rrrrrrrrrr}
1\cdot 50.2 & - & .427\cdot 41 & - & .241\cdot 70 & - & .276\cdot 58.2 & = -.24 & \le & 108.387 \\
 & & -.0147\cdot 41 & + & .0069\cdot 70 & + & .0144\cdot 58.2 & = .718 & \le & 5.929 \\
 & & -.0047\cdot 41 & + & .0163\cdot 70 & - & .0017\cdot 58.2 & = .849 & \le & 11.174 \\
 & & -.0562\cdot 41 & + & .0317\cdot 70 & + & .0155\cdot 58.2 & = .817 & \le & 2.625
\end{array}$$

Since all of the inequalities are satisfied, MOM concludes that the solution does not change. MOM calculates the additional cost resulting from the higher legume prices by using the values of Δc_i and the solution (7) to be

$$50.2\cdot 0 + 41\cdot 118.666 + 70\cdot 40.3120 + 58.2\cdot 63.6651 = \$11,392.21$$

CALIPSO

We had to make several hand calculations [(1)–(4)] to formulate the objective function for the PEALS problem. If we need to change the per-unit cost of any of the commodities, we must repeat some of these hand calculations. This is undesirable because of the added work and because errors are easily made in hand calculations. Fortunately, LP SOLVER has a feature that simplifies this process. Here is a good way to enter LP problem (5) in *LP SOLVER*.

$$\begin{array}{lrcrcrcrcr}
\text{Min} & 4.75(24x1 & + & 32x2 & + & 12x3 & + & 27x4); & & \\
\text{Min} & 2.4(32x1 & + & 35x2 & + & 50x3 & + & 69x4); & & \\
\text{Min} & 3.6(155x1 & + & 100x2 & + & 200x3 & + & 84x4); & & \\
\text{Min} & 0.05(800x1 & + & 1200x2 & + & 800x3 & + & 750x4); & & \\
 & 24x1 & + & 32x2 & + & 12x3 & + & 27x4 & <= & 6000; \\
 & 32x1 & + & 35x2 & + & 50x3 & + & 69x4 & <= & 11000; \\
 & 155x1 & + & 100x2 & + & 200x3 & + & 84x4 & <= & 26000; \\
 & 40x1 & + & 45x2 & + & 75x3 & + & 10x4 & >= & 9000; \\
 & 50x1 & + & 40x2 & + & 22x3 & + & 100x4 & >= & 12000;
\end{array}$$

LP SOLVER gathers the four objective terms into the single term

$$\text{Min} \quad 788.8x1 + 656x2 + 937x3 + 633.75x4;$$

When the problem is entered this way, the task of changing costs is easy.

6.4A Exercises: Getting Help from MOM

Answer the questions in Problems 1–6 that PEALS poses to MOM. When it is practical, use the sensitivity analysis in preference to solving new LP problems. Assume that each question refers to the original solution and not to some change made in response to an earlier question.

1. A fire in the Potlatch field destroyed about 10% of the lentil crop. We start production next week. Do we need to change the set of recipes we are using?

2. § We have an order for an additional 1500 cases of *Mellow Peal Soup*. Do we have enough resources to meet this demand? If so, how shall we change our plan and how much will production cost? (You will need to solve the original problem with these changes to answer the last question. In a more advanced LP course you will learn how to calculate the answer from the current solution.)

3. We have an order for an additional 100 cases of *Spicy Peal Soup* and 400 cases of *Mellow Peal Soup*. Do we have enough resources to meet this demand? If so, how shall we change our plan and how much will production cost?

4. § In Question 3, the cost of peas increased to \$3.00 per bushel and lentils increased to \$3.80 per bushel. Suppose that labor costs also increase to \$5.75 per hour. What is the new value of the objective function?

5. Due to a worldwide change in preference, the price of peas has shot up to \$4.50 per bushel and that of lentils has fallen to \$1.50 per bushel. Will the solution change? If so, what is the new solution?

6. § Suppose the cost of electricity increases to \$.07 per kwh, labor increases to \$6.00 per hour, and lentils increase to \$4.00 per bushel. What is the new value of the objective function?

7. From the resource value ranges given in (9) compute the allowable increase and decrease in b_1 (with other resource values held fixed) that does not change the set of basic variables.

8. § From the resource value ranges given in (9) compute the allowable increase and decrease in b_4 (with other resource values held fixed) that does not change the set of basic variables.

9. You can calculate from the ranges in (9) that the allowable changes (Δc_i) in c_1, c_2, c_3, and c_4 are

	Allowable Increase	Allowable Decrease
c_1	$-\infty$	108.387
c_2	46.6837	∞
c_3	450.563	82.8921
c_4	329.04	169.083

By setting $c_2 = 0, c_3 = 0, c_4 = 0$, check that the system of equations (16) gives the same allowable decrease and increase of c_1. Make a similar calculation for c_2, c_3, and c_4.

10. PEALS is changing Recipe 3 by adding 30 grams of saffron. It is also introducing a new Recipe 5 whose per hour requirements are 20 person-hours of labor, 750 kwh of power, 85 bushels of peas, 70 bushels of lentils, and 120 grams of saffron. Recipe 5 produces 60 cases of *Spicy Peal Soup* per hour and no *Mellow Peal Soup*. Saffron is a very expensive spice, costing \$1.50 per gram and PEALS can only obtain 2100 grams this season.

    Solve this problem and prepare a report to PEALS.

11. MOM has another client Potlatch Forest Products (PFP) that operates a clay mine and processing plant in Bovill to supply clay slurry for use in its paper mill in Lewiston. PFP makes two kinds of slurry: grade A to make magazine stock and grade B to make newsprint. The Bovill mine produces a gray clay at an extraction cost of \$30 per ton from one mining site and a white clay at a second site at an extraction cost of \$70 per ton. The slurry plant can be set up to use any of four processes to manufacture grade A and B slurry. Table 2 summarizes the amounts of labor, electrical power, and clay that each process utilizes. The mines can produce up to 200 tons of white clay and 600

tons of gray clay each week. The Bovill plant employs 30 workers who work 8-hour days for a total labor supply of 1200 worker-hours per week. PFP's labor contract requires that workers be paid for a full week even if they cannot all be used to make slurry. Electricity costs $0.04 per kwh and the supply is virtually unlimited.

PFP wants to determine the most economical way of operating the Bovill plant to produce 2000 tons of grade A and 1800 tons of grade B slurry each week.

Solve an appropriate LP problem and prepare MOM's report to PFP. Since PFP gave no data on wages and has no intention of varying the size of its work force, labor should be regarded as a fixed expense that need not be included in the objective function.

Table 2

	Resources				Output	
Process	Labor (worker-h)	Power (kwh)	White clay (tons)	Gray clay (tons)	Grade A (tons)	Grade B (tons)
1	24	900	3.0	15.0	40	50
2	32	1200	3.2	10.0	45	40
3	12	800	5.0	20.0	75	0
4	28	750	7.0	8.5	0	100

12. Use equations (1)–(4) to check the values in the table labelled Resources Required in MOM's report to PEALS.

7 Counting

Counting is a more subtle task than you might think. In this chapter we look at a variety of counting techniques ranging from the simple to the surprisingly difficult.

7.1 THE INCLUSION–EXCLUSION PRINCIPLE AND VENN DIAGRAMS

Before beginning this section you may want to review Section 0.1.

If a set A has 4 members and a set B has 6 members, can we conclude that the set $A \cup B$ has 10 members? In general, the answer is no since in $A \cup B$ we do not want to count twice those elements that belong to both A and B. For instance, if $A = \{0, 9, 7, 15\}$ and $B = \{2, 9, 0, 13, 22, 6\}$, then the set $A \cup B = \{0, 9, 7, 15, 2, 13, 22, 6\}$ contains 8 elements, not 10.

The *Inclusion–Exclusion Principle* for counting the number of elements in the union of two sets ensures that no elements in the union are counted twice. In what follows we let $n(X)$ denote the number of elements in a set X.

Inclusion-Exclusion Principle: If A and B are sets, then

$$n(A \cup B) = n(A) + n(B) - n(A \cap B)$$

Applying this principle to the example just discussed we have $n(A) = 4$, $n(B) = 6$, $n(A \cap B) = 2$ and so

$$n(A \cup B) = n(A) + n(B) - n(A \cap B) = 4 + 6 - 2 = 8$$

We can also use the Inclusion–Exclusion Principle to find $n(A \cap B)$ as the next example illustrates.

Example 1: Inclusion-Exclusion Principle

Suppose that 50 students are taking either French or Spanish, 27 students are taking French, and 38 students are taking Spanish. How many students are taking both French and Spanish? Let F be the set of students enrolled in French and S the set of students taking Spanish. Then we are given $n(F \cup S) = 50$, $n(F) = 27$, and $n(S) = 38$. From the Inclusion–Exclusion Principle we have

$$50 = 27 + 38 - n(F \cap S)$$

and so the number of students taking both, $n(F \cap S)$, is 15.

Venn diagrams are useful in dealing with counting problems involving two or three sets.

Example 2: Counting with Venn Diagrams

Of 60 people signed up for an exercise class, 32 are male, 40 are overweight, and 23 are overweight males. To represent this information in a Venn diagram we proceed as follows. Let X be the set of people in the class, M the set of males in the class and O the set of overweight exercisers. In Figure 1 the rectangle represents X and the circles marked M and O represent the two subsets of X.

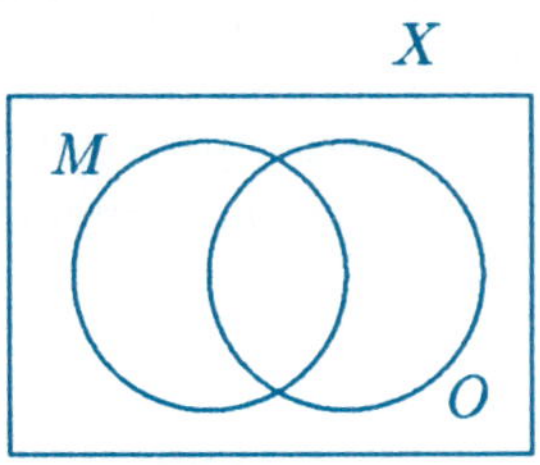

Figure 1

To each region in Figure 1 we now assign the number of people in the set corresponding to the region. In Figure 2 we begin with the number of male overweight persons: $n(M \cap O) = 23$. Working outwards we next find the number of males who are not overweight. Since there are 32 males in the class and 23 are overweight, it follows that there are $32 - 23 = 9$ males who are not overweight and this is the number found in the corresponding region in Figure 3. Similarly, since there are 40 overweight persons and 23 of these are male, it follows that the number $40 - 23 = 17$ appears in the region in Figure 3 corresponding to overweight people who are not male. Finally, since we have now accounted for $9 + 23 + 17 = 49$ people and since there are 60 people in the class, it follows that there are $60 - 49 = 11$ exercisers who are neither male nor overweight.

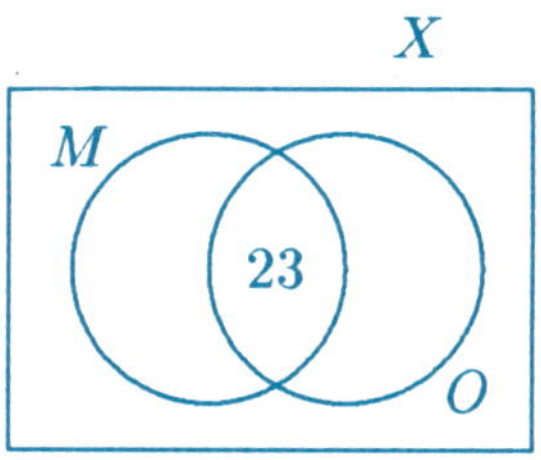

Figure 2

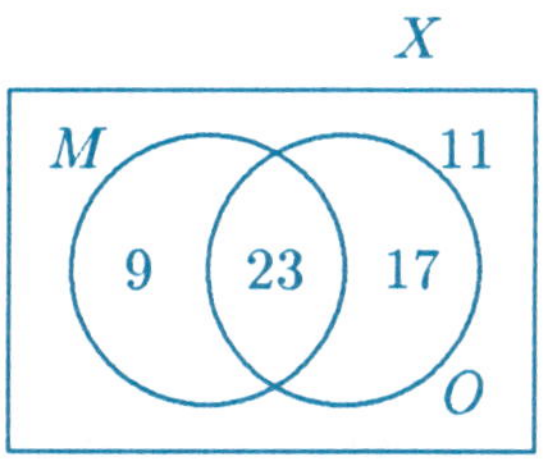

Figure 3

From Figure 3, we can now answer questions such as: How many overweight females are there in the class? The answer is 17, the number appearing in the region corresponding to overweight exercisers who are not male. How many class members are female or are not overweight? The answer is $17 + 11 + 9 = 37$ since there are $17 + 11$ females and there are 9 males who are not overweight.

Example 3: More Counting with Venn Diagrams

We represent the following information with a Venn diagram. Suppose that on a certain day 200 people enter a clothing store. Suppose further that 95 people buy sweaters, 120 buy pants, 50 buy coats, 40 buy sweaters and pants, 30 buy sweaters and coats, 25 buy pants and coats, and 10 people buy all three items. To include this information in a Venn diagram, let X be the set of people entering the store, let S be the set of people buying sweaters, P the set of people buying pants, and C the set of people buying coats. Figure 4 gives the corresponding Venn diagram.

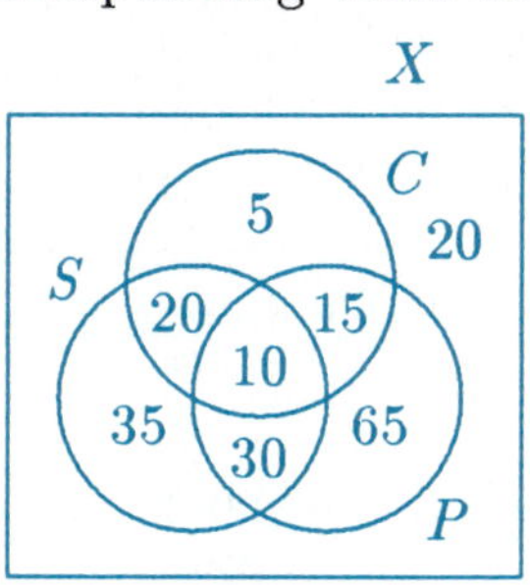

Figure 4

To obtain the numbers in the various regions of Figure 4 begin with the region corresponding to the intersection $S \cap C \cap P$ of all three sets.

This region represents the set of people who bought all three items; this set has 10 members. Next, working outwards, consider the intersection $S \cap P$, which represents the set of people who bought both sweaters and pants. The number of these people is 40. We have already counted 10 of these people and so the number $40 - 10 = 30$ appears in the region indicated in Figure 4 corresponding to those people who bought sweaters and pants but did not buy all three items. Similarly, since 30 people bought sweaters and coats and 10 bought all three items, the region corresponding to the set of people buying both sweaters and coats (but not pants) is assigned the number $30 - 10 = 20$. For the same reason, the region corresponding to those who bought pants and coats but not sweaters is assigned the number $25 - 10 = 15$.

To see how to calculate the number 35 appearing in Figure 4 note that this region corresponds to the set of people who bought sweaters, but did not buy coats or pants. Since 95 people bought sweaters and since $10+20+30 = 60$ of these people did not buy either coats or pants, it follows that we should assign the number $95 - 60 = 35$ to this region. In a similar way we find that we should assign the number $120 - 55 = 65$ to the region corresponding to the set of people who bought pants but did not buy either sweaters or coats. Likewise, there are $50 - 45 = 5$ people who bought coats but not pants or sweaters.

Finally, note that $n(S \cup P \cup C)$ is equal to the sum of the numbers calculated thus far. Since this sum is 180 and since 200 persons entered the store we know that $200 - 180 = 20$ did not buy any of the items.

From Figure 4 we can now make other calculations. For instance, the number of people who bought only pants or who bought both sweaters and coats is $65 + 20 + 10 = 95$. The number of people who bought nothing or who bought coats only is $20 + 5 = 25$. The number of people who bought sweaters or who bought pants only is $35 + 30 + 10 + 20 + 65 = 160$.

7.1A Exercises: The Inclusion–Exclusion Principle and Venn Diagrams

In Problems 1 – 4, use the Inclusion-Exclusion Principle to find the required values.

1. Find $n(A \cup B)$ if $n(A) = 16, n(B) = 41, n(A \cap B) = 9$.

2. § Find $n(A \cup B)$ if $n(A) = 6, n(B) = 12, n(A \cap B) = 4$.

3. Find $n(A)$ if $n(A \cup B) = 36, n(B) = 19, n(A \cap B) = 8$.

4. § Find $n(A \cap B)$ if $n(A) = 126, n(B) = 83, n(A \cup B) = 155$.

Use Venn diagrams to answer the questions in Problems 5 and 6.

5. A survey of 115 students showed that 40 are taking Finite Math and 83 are taking Statistics. Suppose that 22 students take both of these courses.

   (a) How many of the students surveyed are not taking either Finite Math or Statistics?

   (b) How many students are taking Finite Math but not Statistics?

   (c) How many students are taking only Statistics or neither of the two courses?

6. § A basketball team consisting of 18 players has 10 members who can play power–forward and 6 who can play center. Suppose that 4 of these players can play both positions.

   (a) How many team members can play neither power–forward nor center positions?

   (b) How many team members can play either power–forward or center?

   (c) How many team members can play center only or can play both center and power–forward?

In Problems 7 – 14 find the number of elements in the indicated regions of the following Venn diagram.

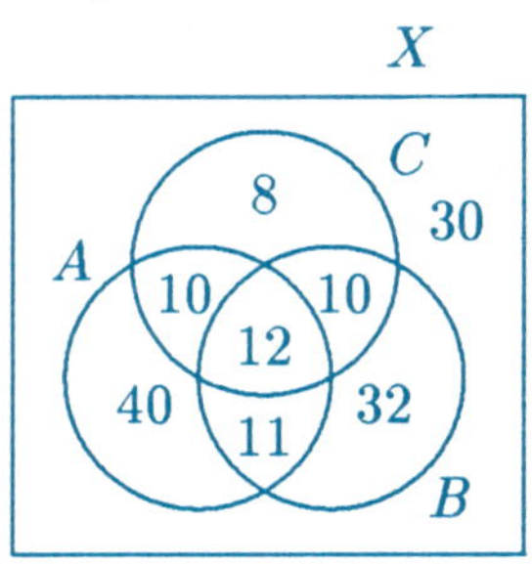

Figure 5

7. $n(A \cap B \cap C)$ 8.§ $n((A \cup B) \cap C)$ 9. $n(X - ((A \cap C) \cup B))$

10. § $n((A \cap B) - C)$ 11. $n((A \cap C) \cup (B - C))$

12. § $n((X - B) \cap (X - C))$ 13. $n(B \cap (C - A))$ 14.§ $n(B \cup C)$

15. A survey of 250 people showed that 125 owned a television set, 160 owned a dishwasher, and 135 owned one dog. The survey also found that 90 people owned a television set and a dishwasher, 60 owned a television set and at least one dog, 99 owned at least one dog and a dishwasher, and 32 people owned a television set, dishwasher, and at least one dog.

    (a) Draw a Venn diagram that incorporates all of this information.
    (b) How many people own neither a dishwasher nor a television set?
    (c) How many people own a television set but do not own either a dishwasher or at least one dog?
    (d) How many people own at least one dog but no television set?
    (e) How many people do not own any of the items surveyed?

16. § A survey of 200 students showed that during one week 150 students ate at least two pizzas, 95 had at least three dates, and 140 slept through at least one class. The survey also showed that 65 ate at least two pizzas and had at least three dates, 45 had at least three dates and slept through at least one class, 90 ate at least two pizzas and slept through at least one class, and 15 ate at least two pizzas, had at least three dates and slept through at least one class.

    (a) Draw a Venn diagram that incorporates all of this information.
    (b) How many students had fewer than three dates but slept through at least one class?
    (c) How many students ate at least two pizzas and had at least three dates?
    (d) How many students had fewer than three dates and did not sleep through any classes?
    (e) How many students ate at most one pizza and slept through at least one class or had at least three dates?

17. Use a Venn diagram to show that the Inclusion–Exclusion principle for three sets is

$$\begin{aligned} n(A \cup B \cup C) &= n(A) + n(B) + n(C) - n(A \cap B) \\ &\quad - n(A \cap C) - n(B \cap C) + n(A \cap B \cap C) \end{aligned}$$

7.2 THE FUNDAMENTAL COUNTING PRINCIPLE

Suppose you can take 4 different routes from Moscow to Coeur d'Alene and 6 routes from Coeur d'Alene to Spokane. How many different routes are there from Moscow to Spokane (via Coeur d'Alene)? To solve this problem observe that for each of the four routes from Moscow to Coeur d'Alene there are 6 routes you can take from Coeur d'Alene to Spokane. Thus there are $4 \cdot 6$ possible routes between Moscow and Spokane.

This illustrates what is often called the *Fundamental Counting Principle* or the *Multiplication Principle.*

Fundamental Counting Principle

If you can perform one task in q_1 ways and a second task in q_2 ways, then you can do both tasks in $q_1 \cdot q_2$ ways. More generally, if there are q_1 ways of doing task 1, q_2 ways of doing task 2, $\ldots, q_k$ ways of doing task k, then there are $q_1 \cdot q_2 \cdot \cdots \cdot q_k$ ways of doing all of the tasks.

Example 1: Using the Fundamental Counting Principle

If a car comes in 5 models and 7 colors, how many possible selections can you make? Since there are 5 ways to choose the model and 7 ways to choose the color, it follows from the Fundamental Counting Principle that there are $5 \cdot 7$ different selections.

Example 2: Code Words

We use the Fundamental Counting Principle to determine the number of code words consisting of 2 digits followed by 5 letters that can be formed if no digits or letters are repeated. (A typical code word is $07DEKGH$.) You can select the first digit in 10 ways and since you cannot repeat digits you can choose the second digit in 9 ways. Similarly, you can select the first letter in 26 ways, and since letters cannot be repeated there are 25 ways to select the second letter, 24 ways to select the third letter, 23 ways to select the fourth letter, and 22 ways to select the fifth letter. By the Fundamental Counting Principle you can form

$$10 \cdot 9 \cdot 26 \cdot 25 \cdot 24 \cdot 23 \cdot 22 = 710{,}424{,}000$$

different code words.

Example 3: Heads and Tails

If you toss a coin 10 times, how many different sequences of heads (H) and tails (T) can you obtain? (A typical sequence is HHTHTHHHTH.) Note that there are 2 possibilities, H or T, for the first toss, 2 for the second toss, and in fact 2 for each of the 10 tosses. Thus, by the Fundamental Counting Principle there are $2 \cdot 2 \cdot 2 \cdot 2 \cdot 2 \cdot 2 \cdot 2 \cdot 2 \cdot 2 \cdot 2 = 2^{10} = 1024$ possible sequences.

CALIPSO

You can use CALIPSO's *Calculator* to make arithmetical calculations such as finding $10 \cdot 9 \cdot 26 \cdot 25 \cdot 24 \cdot 23 \cdot 22$ or 2^{10}. To do this select *STATISTICS* from CALIPSO's menu. Next select *Calculator* from the *STATISTICS* options. *Calculator* will evaluate arithmetic expressions that involve addition (+), subtraction (−), multiplication (*), division (/), and power (e.g., 5^3 for 5^3). You can use parentheses to group terms. *Calculator* also recognizes the functions $P(m,n), C(m,n)$, and $n!$ that are introduced in the next two sections.

The expression you enter, some intermediate calculations, and the answer are displayed in a worksheet. You can look at parts of the worksheet that have moved off the screen by scrolling with one of the keys [↑], [↓], [*PgUp*], [*PgDn*], [*Home*], [*End*]. You can print the worksheet by leaving *Calculator* with the [*Esc*] key and choosing *Print* from the menu.

Example 4: Using CALIPSO's *Calculator*

We give examples of expressions that can be evaluated by *Calculator*. For later reference we include the functions $P(m,n), C(m,n), n!$ that will be introduced in Sections 7.3 and 7.4.

Expression	Answer	Comment
10 * 9 * 26 * 25 * 24 * 23 * 22	7.1042E+08	Answer is $7.1042 \cdot 10^8$
2^10	1024	
$12(2/3 + 5/7)$	16.571	Answers are rounded to 5 digits. Internal accuracy is 15 digits.
$(2+3)/(3+7)$	0.5	The denominator is $3+7$
$C(12,5) + 5C(10,3)$	1392	You can omit * between 5 and C
$P(12,2)P(5,3) * P(6,3)$	950,400	* is optional here
$6! * 3!$	4320	* is not optional here
$6!/(3! * 2!)$	60	
$C(12,5)/C(5,3) + C(12,2)$	145.2	The denominator is $C(5,3)$
$C(12,5)/(C(5,3) + C(12,2))$	10.421	The denominator is $C(5,3) + C(12,2)$

7.2A Exercises: Counting Problems

Solve the following problems either by hand or with CALIPSO. If you do the problems by hand, you do not need to simplify your answer.

1. License plates in some states consist of a sequence of 2 letters followed by 4 digits. How many plates can be issued before some other system must be used? Repetition of the letters and the digits is allowed.

2. § How many sorority names can be formed from the Greek alphabet (24 letters) if each name consists of three letters and repetitions of the letters are not allowed? How many if repetitions are allowed?

3. If you have 7 pairs of socks, 4 pairs of pants, and 14 shirts, in how many different ways can you present yourself to the world?

4. § In how many ways is it possible that the birthdays of 6 people fall in different months?

5. A slot machine has 4 pictures of fruits on one wheel, 6 pictures of animals on a second wheel, and 5 pictures of plants on a third wheel. If each picture is distinguishable from the others, how many different patterns can the machine create?

6. § A restaurant offers 5 salads, 12 entrees, 8 desserts, and 16 drinks. How many different meals are there?

7. How many code words can you form if each code word consists of 3 digits followed by 6 letters? You may repeat the digits, but not the letters.

8. § A lock's dial has 19 numbers. You first turn the dial to the right, then to the left, and then again to the right. If the first two numbers you select must be different, how many lock combinations are there?

9. Next semester you plan to take 1 science course, 1 language course, 1 math course, 1 literature course, and 1 health and recreation course. If there are 15 science courses, 6 language courses, 4 math courses, 4 literature courses, and 8 health and recreation courses offered, how many different schedules can you form?

10. § The Fortran programming language recognizes variables (code words) formed from the letters A–Z, the underscore character "_", and the digits 0 – 9. A variable cannot be longer than 7 characters and the first character must be a letter (A–Z). The language does not distinguish between upper- and lower-case letters. (a) How many variables of length 1 can be formed? (b) How many variables of length 1 or 2 can be formed? (c) How many variables of length less than or equal to 7 can be formed?

7.3 PERMUTATIONS

Suppose 5000 people run a marathon. Prizes of \$2000, \$1000, \$500, \$250, and \$100 are awarded to the first, second, third, fourth, and fifth place finishers, respectively. In how many ways can these prizes be awarded? In essence you are asked to find in how many ways you can select and arrange (permute) 5 objects taken from a pool of 5000 objects. The number of such arrangements (or *permutations*) is denoted by $P(5000, 5)$ and is determined as follows. There are 5000 possible first place winners. Once first place is decided, there are 4999 possible second place winners. Similarly, once first and second place are decided, there are 4998 possible third place winners. There are then 4997 possible fourth place winners and 4996 possible fifth place winners. Thus by the Fundamental Counting Principle there are

$$P(5000, 5) = 5000 \cdot 4999 \cdot 4998 \cdot 4997 \cdot 4996 \doteq 3.1 \cdot 10^{18}$$

possible ways of awarding the prizes. (The symbol $\doteq$ means approximately equal to.) In general we have the following rule.

Permutations. The number of permutations of r objects selected from n distinct objects is

$$P(n, r) = n(n-1)(n-2)\cdots(n-r+1)$$

Example 1: Calculating a Permutation

The number of ways of selecting 4 objects (in some order) from a pool of 11 distinct objects is $P(11, 4) = 11 \cdot 10 \cdot 9 \cdot 8$.

Notation: The symbol $n!$ (read n *factorial*) is defined as

$$n! = n(n-1)(n-2)\cdots 2 \cdot 1$$

It is customary (and useful) to set $0! = 1$.

Note that $5! = 5 \cdot 4 \cdot 3 \cdot 2 \cdot 1$ and $P(13, 5) = 13 \cdot 12 \cdot 11 \cdot 10 \cdot 9 = \frac{13!}{8!}$. In general, it is easy to verify that in factorial notation,

$$P(n, r) = \frac{n!}{(n-r)!}$$

Example 2: Factorial Notation and Permutations

You can verify that

$$P(14, 4) = \tfrac{14!}{4!} = 14 \cdot 13 \cdot 12 \cdot 11$$
$$P(23, 8) = \tfrac{23!}{15!}$$
$$P(n, n-1) = n!$$
$$P(n, 1) = n$$
$$P(n, 0) = 1$$
$$P(n, n) = n!$$

Example 3: Batting Orders

If a baseball club has 20 players, then the number of 9-person batting orders is $P(20,9)$ since we are selecting 9 members of the team and arranging them in some order.

Example 4: Queueing Up

Eight associate professors and 5 full professors are standing in line at the university cafeteria. We find the number of ways they can form this line if the associate professors stand together and so do the full professors. Suppose the full professors are first in line. They can line up in 5! ways. The associate professors can line up in 8! ways. By the Fundamental Counting Principle there are $5! \cdot 8!$ possibilities if the full professors are first in line. Similarly there are $8! \cdot 5!$ different lines if the associate professors head the line. Thus, altogether there are $2 \cdot 8! \cdot 5!$ possible lines. If you calculate this value with CALIPSO or a hand calculator you will see that there are 9,676,800 lines.

7.3A Exercises: Permutations and Factorials

If you do these problems by hand, you may express your answers in factorial or permutation notation.

1. Without using CALIPSO make the following calculations. Then check your answers with CALIPSO.

   (a) $P(8,3)$ (c) $P(7,0)$
   (b) $P(100,2)$ (d) $P(6,6)$

2. § In how many different orders can you draw 8 cards one at a time from a 52-card deck (without replacing the cards drawn)?

3. Suppose the call letters of a radio station are 4-letter words from the 26-letter alphabet and must begin with either K or W. If no letters can be repeated, how many possible call letters can you form?

4. § In how many ways can 6 members of a large family of 14 line up for a picture?

5. In how many ways can 10 people sit around a table? (Two ways are considered the same if one is a rotation of the other.)

6. § In how many ways can 9 women and 9 men stand in a line for a picture? In how many ways can these people stand in line if the order must be woman, man, woman, man, ... ?

7. Suppose a traveling salesman must start and end in Pocatello and visit 7 other Idaho cities exactly once. How many routes can he form?

8. § Twelve people are to be assigned to twelve different tasks. In how many ways can this be done?

9. (a) How many batting orders are possible in a nine-member baseball team? (b) How many if the pitcher bats last? (c) How many if the pitcher bats last and the catcher bats first? (d) How many if the pitcher bats last and the catcher bats second?

7.4 COMBINATIONS

Suppose that 5 equal prizes are given for the first 5 finishers in the marathon race described in Section 7.3. If the finishing order of the leading 5 runners is not considered, in how many ways can the prizes be awarded?

In essence, you are asked to determine in how many ways you can select 5 objects from a pool of 5000 objects if the order (arrangement) of the 5 objects selected is not considered. The number of ways that this can be done is denoted by $C(5000, 5)$. ($C(5000, 5)$ is called the number of *combinations* of 5000 objects taken 5 at a time without regard to order.) To calculate $C(5000, 5)$ we first relate it to $P(5000, 5)$ as follows. We can find $P(5000, 5)$ in two steps.

Step 1. Select 5 runners from 5000 without considering the order (arrangement) of the 5 runners. This is done in $C(5000, 5)$ ways.

Step 2. Arrange the selected 5 runners in some order. This is done in $5!$ ways.

By the Fundamental Counting Principle we can do both *Step 1* and *Step 2* to calculate $P(5000,5)$ in $C(5000,5)\cdot 5!$ ways. That is,

$$P(5000,5) = C(5000,5)\cdot 5!$$

and so

$$C(5000,5) = \frac{P(5000,5)}{5!}$$

Note that since $P(5000,5) = \frac{5000!}{4995!}$ it follows that

$$C(5000,5) = \frac{5000!}{4995!\cdot 5!}$$

In general, we have the following rule.

Combinations. The number of ways of selecting r objects from a pool of n distinct objects without regard to order (the number of *combinations* of n distinct objects taken r at a time) is

$$C(n,r) = \frac{n!}{(n-r)!\cdot r!}$$

Note: The notation C_n^r, $\binom{n}{r}$, or ${}_nC_r$ is often used in place of $C(n,r)$.

Example 1: Calculating Combinations

The number of ways of selecting 7 objects from a pool of 17 distinct objects (where the 7 objects are not ordered or arranged) is

$$C(17,7) = \frac{17!}{10!7!}$$

Example 2: Calculating Numbers of Card Hands

(a) To determine the number of 7-card hands in a 52-card deck you can use CALIPSO to find that

$$C(52,7) = \frac{52!}{45!7!} = 133,784,560$$

(b) We find the number of 7-card hands (from a 52-card deck) consisting of 2 spades, 4 hearts, and 1 club. In a 52-card deck there are 13 spades. We want to select 2 of these. This can be done in $C(13,2)$ ways. Similarly, we can select 4 hearts in $C(13,4)$ ways and 1 club in $C(13,1)$ ways. By the Fundamental Counting Principle we can do all this in

$$C(13,4) \cdot C(13,2) \cdot C(13,1) = 725,010$$

ways. Note that you can enter the above expression in CALIPSO's *Calculator* as $C(13,4)C(13,2)C(13,1)$ or as $C(13,4) * C(13,2) * C(13,1)$.

Example 3: Defective Light Bulbs

In a shipment of 500 light bulbs it is known that there are 20 defective bulbs. In how many ways can you select 7 bulbs so that no more than 2 are defective? We must find in how many ways we can choose 7 bulbs so there are no defective bulbs, exactly 1 defective bulb, or exactly 2 defective bulbs. In order that there are no defective bulbs you must choose the 7 bulbs from the 480 nondefective bulbs. This can be done in $C(480,7)$ ways. In order to have chosen exactly 1 defective bulb (and 6 nondefective bulbs) you must choose 1 of the bulbs from the 20 defective bulbs (done in $C(20,1)$ ways) and 6 from the 480 nondefective bulbs (done in $C(480,6)$ ways). By the Fundamental Counting Principle you can do both of these tasks in $C(20,1) \cdot C(480,6)$ ways. Finally, to obtain exactly 2 defective (and, consequently, 5 nondefective) bulbs you need to choose two of the 20 defective bulbs (done in $C(20,2)$ ways) and 5 of the 480 nondefective bulbs (done in $C(480,5)$ ways). Again by the Fundamental Counting Principle you can do both of these tasks in $C(20,2) \cdot C(480,5)$ ways. Thus, the total number of ways to have no more than 2 defective bulbs is

$$C(480,7) + C(20,1) \cdot C(480,6) + C(20,2) \cdot C(480,5) \qquad (1)$$

(Why don't you multiply all of these numbers together?)

Some of the combinations in (1) are so large that (even) CALIPSO will not give you a numerical value for them.

Example 4: Calculating Numbers of Poker Hands

First we find how many 5-card hands (from a 52-card deck) contain exactly 2 pairs (a pair consists of 2 cards of the same denomination – for instance, two Jacks). A typical hand with two pairs is 2 Jacks, 2 eights, and 1 four. You can choose the denomination (for instance, Jacks and eights) of the pairs in $C(13,2)$ ways and once the denominations are chosen there are $C(4,2)$ ways of selecting each of the pairs. You can select one of the remaining 44 cards in 44 ways, so by the Fundamental Counting Principle there are

$$C(13,2) \cdot C(4,2) \cdot C(4,2) \cdot 44 = 123,552$$

possible hands consisting of 2 pairs.

Next we calculate the number of hands having three of a kind; that is, exactly 3 cards of one denomination and 2 other cards that do not form a pair. Since there are 13 denominations and we want 3 cards from one of these, there are $13 \cdot C(4,3)$ ways to form 3 cards of the same denomination. In how many ways can we select the remaining 2 cards? There are 48 cards from which to choose the two cards (this can be done in $C(48,2)$ ways), but we must be careful not to choose a pair from these 48 cards. The number of possible pairs in these 48 cards is $12 \cdot C(4,2)$ since there are now 12 denominations. Thus the number of ways of selecting 2 cards from 48 without selecting a pair is $C(48,2) - 12 \cdot C(4,2)$. By the Fundamental Counting Principle the number of hands with three of a kind is

$$13 \cdot C(4,3) \cdot [C(48,2) - 12 \cdot C(4,2)] = 54,912$$

Note that you could enter this in *Calculator* as $13C(4,3)(C(48,2)-12C(4,2))$.

Example 5: Counting Routes on a Grid

Figure 1 shows part of a map of a city whose streets are arranged in a grid of squares. Maria lives at point H and must ride her bicycle to work at point W every morning. She observes that there are many routes she can take that all require traveling the shortest distance from H to W along the streets. She travels the same distance in cycling from H to W along any route that involves moving only from left to right or top to bottom in the grid. How many different routes can she take?

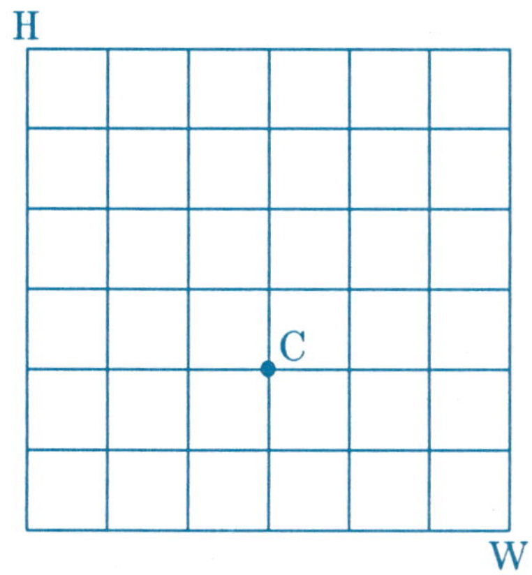

To analyze this problem we make a code word to describe each route. Let R denote a trip from left to right along an edge of one square, and let D denote a trip from top to bottom along an edge of one square. We can specify a route by a code word of length 12 that contains 6 $R's$ and 6 $D's$. For instance $RDRDRDRDRDRD$ specifies a trip in which Maria turns at each corner.

We can count the number of such code words by counting the number of positions (1–12) in the word in which we can put the 6 $R's$. That number is the number of ways of choosing 6 numbers from the set $\{1, 2, \ldots, 12\}$ which is $C(12, 6) = 924$.

7.4A Exercises: Combination Problems

If you do these problems by hand, you may leave your answer in factorial or combination notation.

1. Without using CALIPSO make the following calculations; then check your answers with CALIPSO.

   (a) $C(10, 7)$ (d) $C(123, 121)$
   (b) $C(6, 6)$ (e) $C(7, 0)$
   (c) $C(100, 1)$ (f) $C(9, 3)$

2. § How many soccer teams consisting of 11 players can be formed from a pool of 34 players?

3. How many 7-card hands (selected from a 52-card deck) contain exactly 4 diamonds, 2 hearts, and 1 club.

4. § In how many ways can you choose 4 different pizza toppings from a list of 11 toppings?

5. A shipment of 60 television sets is known to contain 6 defective sets. In how many ways can you select 10 of these sets so that you have at most three defective sets?

6. § How many different 13-card bridge hands are there?

7. A multiple choice test has three incorrect and one correct answer for each of 20 questions. In how many ways can you get exactly 5 wrong answers? Can you solve this problem in two different ways?

8. § An experimenter wishes to test the effect of three variables, temperature, humidity, and sex on a new formula for body deodorant. If she wants to use 6 temperatures, 8 levels of humidity, and 2 sexes, how many experimental runs will have to be conducted to study all possible combinations of the three variables?

9. The personnel director at a multinational corporation has hired 15 new college graduates. Three jobs are open at its Paris office: sales manager, consumer relations director, and secretary. Five jobs are open at the London office. In how many ways can these positions be filled by the 15 new employess?

10. § A 6-member discussion group is to be chosen from a group of 14 people. Two of the group form a significant pair who prefer to work together; the rest do not have a significant other with whom they prefer to work. If the significant pair stays together, how many different 6-member discussion groups can be formed? (Suggestion: Consider the following questions. How many discussion groups of 4 can you make from the people who have no significant other, and how is that number related to the number of 6-member discussion groups that include the significant pair? How many discussion groups of 6 can you make that do not include the significant pair?)

11. Using the streets shown on the grid below, in how many ways can you drive from location A to location B without backtracking, that is, along a shortest route?

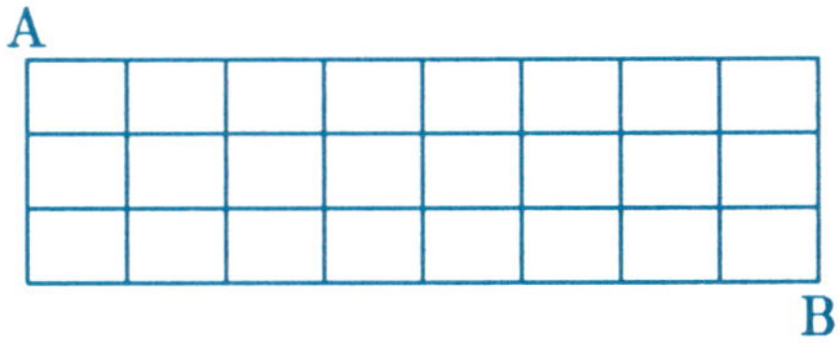

12. § Suppose that Maria stops for coffee at the corner labeled C in the street grid of Example 3. How many routes (without backtracking) can she take from home to work with a stop at C?

13. Ten lines are drawn in the plane. What is the maximum number of points of intersection of the ten lines? What is the minimum number?

7.5 ADDITIONAL COUNTING TECHNIQUES

Permutations with Repeated Objects

The counting techniques looked at so far have depended on being able to distinguish all of the objects. For instance, if you are given 10 indistinguishable marbles, then it makes no sense to ask in how many ways 4 of these marbles can be selected: one selection of 4 marbles is indistinguishable from any other such selection. Nevertheless, problems involving permutations and combinations of objects that may be alike are of interest.

Consider, for example, the problem of finding the number of distinct words (possibly nonsense words) that can be formed using all the letters in the word MISSISSIPPI. We assume that in this word it is impossible to distinguish among the 4 S's or among the 4 I's or between the 2 P's.

To solve this problem consider 11 boxes into which we can put the letters of MISSISSIPPI to form the new words

Since there are 4 I's we can fill 4 of these boxes with I's. There are $C(11,4)$ ways of choosing these boxes to be filled by the I's. There now remain 7 boxes to be filled. Since there are 2 P's we select two of these boxes for the P's. This you can do in $C(7,2)$ ways. There now remain 5

boxes. We need 4 of these for the 4 S's. We can select these 4 boxes in $C(5,4)$ ways. There is just 1 box left for the 1 M. This can be chosen in $C(1,1) = 1$ way. By the Fundamental Counting Principle we can make all of these selections in

$$C(11,4) \cdot C(7,2) \cdot C(5,4) \cdot C(1,1) = \frac{11!}{7!4!} \frac{7!}{5!2!} \frac{5!}{4!1!} \frac{1!}{1!0!} = \frac{11!}{4!2!4!1!}$$

ways. This example illustrates the following general result.

Permutations With Repeated Objects: The number of permutations of n objects, r_1 of which are of type 1, r_2 of which are of type 2, $\ldots, r_k$ of which are of type k (where $r_1 + r_2 + \cdots + r_k = n$) is

$$\frac{n!}{r_1!r_2!\cdots r_k!} \tag{1}$$

Example 1: Counting Permutations with Repeated Objects

Prizes for a marathon race with 5000 contestants are distributed as follows: The first 5 finishers receive \$1000; the next 10 runners receive \$500; the next 15 runners receive \$250; and the next 70 runners receive \$100. We want to find the number of ways the prizes can be distributed. In this problem we have five "types": 5 runners receiving \$1000, 10 runners receiving \$500, 15 runners receiving \$250, 70 runners receiving \$100, and 4900 runners receiving no money. Therefore the number of ways to distribute the prize money is

$$\frac{5000!}{5!10!15!70!4900!}$$

7.5A Exercises: Counting Permutations with Repeated Objects

1. How many distinct words can you form by rearranging the letters in ASSASSIN?

2. § How many distinct words can you form by rearranging the letters in ARKANSAS?

3. Suppose an urn contains 4 red balls, 5 blue balls, and 3 purple balls. In how many ways can you arrange the balls in a line without distinguishing the balls of the same color?

4. § To determine the winning number in a lottery, 20 cards each marked with a single digit are placed in an urn. The winning number is formed by drawing the cards one at a time and recording the digit on the card as the next digit in the winning number. If the cards in the urn contain six 6's, five 5's, four 4's, three 3's, and two 2's, how many possible numbers can be drawn?

5. A company with 100 employees decides to distribute Christmas bonuses as follows. The top 10 employees get \$500, the next 10 employees get \$250, the next 30 employees get \$100, and the rest get nothing. How many distinct ways are there to distribute the bonus money?

6. § How many distinct license plates can be made using 4 A's, 3 B's, 2 C's, and 1 D?

Partitions

Closely related to the ideas just discussed is the notion of a *partition.* Suppose a group of 37 people is to be divided into four committees, A, B, C, and D, where committee A has 5 members, committee B has 10 members, committee C has 10 members, and committee D has 12 members. If no one can be assigned to more than one committee, in how many ways can these committees be formed? In essence, we have partitioned the set of 37 people into four smaller subsets consisting of 5, 10, 10, and 12 people. We need to find how many such partitions are possible.

To solve this problem note that there are $C(37,5)$ ways to assign five people to committee A. Then there are $C(32,10)$ ways to assign people to committee B (since 5 of the 37 people have already been assigned). There are now $C(22,10)$ ways to assign people to committee C, and finally $C(12,12) = 1$ way to assign the remaining 12 people to committee D. By the Fundamental Counting Principle we can do all this in

$$C(37,5) \cdot C(32,10) \cdot C(22,10) \cdot C(12,12) = 1.8184 \times 10^{19}$$

ways. Observe also that

$$\begin{aligned} C(37,5) \cdot C(32,10) \cdot C(22,10) \cdot C(12,12) &= \frac{37!}{32!5!}\,\frac{32!}{22!10!}\,\frac{22!}{12!10!}\,\frac{12!}{0!12!} \\ &= \frac{37!}{5!10!10!12!} \end{aligned} \qquad (2)$$

The kind of partition just discussed is called an *ordered partition* of type $(5, 10, 10, 12)$. The word "ordered" is used in the sense that assigning, say, persons 1 - 10 to committee B and persons 11 - 20 to committee C is regarded as different from assigning persons 1 - 10 to committee C and persons 11 - 20 to committee B. In other words, the committees are viewed as being distinct.

Definition. An *ordered partition* of a set S of n objects of type $(r_1, r_2, \ldots, r_k)$ is a decomposition of S into k disjoint subsets $A_1, A_2, \ldots, A_k$ of S where the number of elements in A_1 is r_1, the number of elements in A_2 is r_2, ... , and the number of elements in A_k is r_k. It is assumed that $r_1 + r_2 + \cdots + r_k = n$.

Example 2: An Ordered Partition

One ordered partition of type (3,2,5,2) of the 12 letters $a, b, c, d, e, f, g, h, i, j, k, l$ is $A_1 = \{a, d, g\}, A_2 = \{h, e\}, A_3 = \{k, b, j, b, i\}, A_4 = \{c, l\}$.

The result obtained in (2) illustrates the following general rule.

Number of Ordered Partitions: The number of ordered partitions of type $(r_1, r_2, \ldots, r_k)$ of a set with n elements is

$$\frac{n!}{r_1! r_2! \cdots r_k!}$$

Example 3: Number of Ordered Partitions

Thirty people are selected to test the effectiveness of a drug. The 30 people are divided into 3 groups; 12 of the people are given the drug, 8 are not given the drug, and 10 are given a placebo in place of the drug. The number of ways to divide the people into these 3 groups is

$$\frac{30!}{12!8!10!}$$

7.5B Exercises: Partitions

1. One hundred people are to be divided into 3 groups to test a new vaccine. The first group of 50 people receives a full-strength dose, the second group of 35 people receives a half-strength dose, and the third group of 15 people receives a placebo. In how many ways can the 100 people be divided into 3 such groups?

2. § A gym class of 30 students is to be divided into a soccer team of 11 players, a football team of 11 players, a basketball team of 5 players, and a running team of 3 players. In how many ways can the class be divided into such teams?

3. Ten people want ice cream cones. In how many ways can you serve them 5 vanilla cones, 3 chocolate cones, and 2 licorice cones?

4. § Twenty stores are selected to test market a new recipe for Coke. Eight of the stores receive only the new recipe, seven receive some bottles of the new and some of the old recipe, and five get only the old recipe. In how many ways can we divide the stores into three such groups?

5. In how many ways can you divide 18 people into committees of sizes 3, 5, and 10 if no person can serve on more than one committee?

6. § In how many ways can you divide 100 people into committees of size 40, 25, and 35 if no person can serve on more than one committee?

Binomial Theorem

Determining the coefficients of the expansion of the binomial

$$(x + y)^n$$

provides another application of combinations. Recall that to expand the expression $(x+y)^2 = (x+y)(x+y)$ you add all the products with one term from the first factor and one from the second; that is,

$$(x + y)^2 = xx + xy + yx + yy = x^2 + 2xy + y^2$$

Similarly, you expand $(x + y)^4 = (x + y)(x + y)(x + y)(x + y)$ by adding all of the products obtained by choosing one term from each factor to get

$$\begin{aligned}(x + y)^4 &= xxxx + xxxy + xxyx + xyxx + yxxx + xxyy + xyxy \\ &\quad + xyyx + yxyx + yyxx + xyyy + yxyy + yyxy + yyyx + yyyy \\ &= xxxx + 4xxxy + 6xxyy + 4xyyy + yyyy \\ &= x^4 + 4x^3y + 6x^2y^2 + 4xy^3 + y^4\end{aligned}$$

There is an easier way to find the coefficient of, say, x^2y^2 in $(x + y)^4$, than multiplying everything out and combining like terms. Note that the number of terms in the expansion of $(x+y)^4$ having 2 x's and 2 y's is $C(4,2)$, the number of ways placing 2 x's (or 2 y's) in 4 positions. Thus, the coefficient of x^2y^2 is $C(4,2) = 6$.

In a similar way you should convince yourself that the coefficient of x^5y^2 in the expansion of $(x+y)^7$ is $C(7,5)$, the number of ways you can place 5 x's in 7 positions. In general, we have the following result.

Combinatorial Form of the Binomial Theorem

$$\begin{aligned}(x+y)^n &= C(n,n)x^n + C(n,n-1)x^{n-1}y + C(n,n-2)x^{n-2}y^2 + \cdots \\ &\quad + C(n,2)x^2y^{n-2} + C(n,1)xy^{n-1} + C(n,0)y^n\end{aligned}$$

Example 4: Binomial Expansion of $(x+y)^6$

$$\begin{aligned}(x+y)^6 = &\ C(6,6)x^6 + C(6,5)x^5y + C(6,4)x^4y^2 \\ &+ C(6,3)x^3y^3 + C(6,2)x^2y^4 + C(6,1)xy^5 + C(6,0)y^6\end{aligned}$$

We can use the Binomial Theorem to count the number of subsets of a set of given size n. If we set $x = y = 1$ in the expansion of $(x+y)^6$ given in Example 4, we obtain

$$(1+1)^6 = 2^6 = C(6,6)+C(6,5)+C(6,4)+C(6,3)+C(6,2)+C(6,1)+C(6,0)$$

$C(6,0) = 1$ is the number of subsets with 0 elements from a set of 6 elements, $C(6,1) = 6$ is the number of subsets with 1 element, $C(6,2) = 3$ is the number of subsets with 3 elements, and so on up to $C(6,6) = 1$, the number of subsets with 6 elements. Thus, $(1+1)^6 = 2^6 = 64$ is the number of subsets that can be formed from a set of 6 elements. In general, there are 2^n subsets of a set with n elements.

Example 5: Using the Binomial Theorem to Count

An ice cream parlor offers 10 flavors of ice cream and 5 toppings. How many different sundaes are possible if each sundae uses one kind of ice cream and any number of toppings (a sundae with 0 toppings is a plain dish of ice cream)? Solution: You can select the kind of ice cream in 10 ways and top your sundae with any subset of the 5 toppings, that is, in any of $2^5 = 32$ ways. Thus, $10 \cdot 2^5 = 320$ different sundaes can be served.

7.5C Exercises: Counting Problems Involving the Binomial Theorem

1. Use CALIPSO's *Calculator* to verify that

$$2^6 = C(6,0) + C(6,1) + C(6,2) + C(6,3) + C(6,4) + C(6,5) + C(6,6)$$

2. § Use the Binomial Theorem to evaluate

$$\left(\frac{1}{3}\right)^4 + C(4,3)\left(\frac{1}{3}\right)^3\left(\frac{2}{3}\right) + C(4,2)\left(\frac{1}{3}\right)^2\left(\frac{2}{3}\right)^2 + C(4,1)\left(\frac{1}{3}\right)\left(\frac{2}{3}\right)^3 + \left(\frac{2}{3}\right)^4$$

3. Use the Binomial Theorem to evaluate

$$C(8,1) + C(8,2) + C(8,3) + C(8,4) + C(8,5) + C(8,6) + C(8,7)$$

4. § Determine the number of subsets of a set with 10 elements. How many of these subsets have two or more elements?

5. What is the coefficient of x^7y^3 in the expansion of $(x+y)^{10}$?

6. § How many different tips can you leave a waiter if you have a penny, a nickel, a dime, a quarter, and a half-dollar?

7. A pizza parlor sells small, medium, and large pizzas. Each pizza has a base of sauce and cheese and can be topped with any combination of pepperoni, sausage, mushrooms, anchovies, onions, olives, and garbanzos. How many different pizzas can you order?

8. § In the ice cream parlor of Example 5, how many sundaes can you order consisting of any number of flavors of ice cream and any number of toppings?

9. In how many ways can you select sets of finite–mathematics books for the library from the 254 books on the market?

10. § Show that for any positive integer n

$$C(n,n) - C(n,n-1) + C(n,n-2) - C(n,n-3) + \cdots \pm C(n,0) = 0$$

8 Probability

You are accustomed to using the word probability. We refer easily to the probability of rain tomorrow or the probability that the local football team will emerge victorious in its traditional game with Moo U. A moment's thought though should convince you that common usage of the idea of probability is vague at best. What does it really mean that the probability of rain tomorrow is 65%? (65% of what?) In this chapter we review some basic probability notions and use the counting techniques developed in the last chapter to calculate (often with the aid of CALIPSO) various probabilities.

8.1 BASIC CONCEPTS OF PROBABILITY

For our purposes an *experiment* is an activity or happening that has a number of observable *outcomes* (results). For instance, the experiment of tossing a coin three times yields the outcomes HHH, HHT, HTH, THH, HTT, THT, TTH, TTT.

The set of all possible outcomes of an experiment is called the *sample space* of the experiment. An *event* is any subset of the sample space. Events consisting of just a single outcome are called *simple events*. We say that an event *occurs* if one of the outcomes comprising the event occurs.

Example 1: Sample Spaces and Events

- The experiment is to select 2 of the 5 numbers $2, 3, 5, 8, 9$. The sample space S for this experiment is defined by the set of all possible outcomes

  $$S = \{\{2,3\}, \{2,5\}, \{2,8\}, \{2,9\}, \{3,5\}, \{3,8\}, \{3,9\}, \{5,8\}, \{5,9\}, \{8,9\}\}.$$

  One event is the subset $E = \{\{2,5\}, \{3,8\}, \{5,9\}, \{8,9\}\}$ of S. If you were to select, say, the numbers $\{3,8\}$, then the event E occurs. The event E also occurs if you select the pair $\{9,8\}$. The event E does not occur if you select the pair $\{2,3\}$.

- The experiment is to select 5 cards from a 52-card deck. The sample space is the set of all possible 5-card hands. One event is the set of all 5-card hands that contain only spades. Another event is the set of all 5-card hands that contain 4 of a kind. A simple event is the particular hand consisting of the Ace of hearts, Jack of clubs, 5 of diamonds, 7 of diamonds, and 3 of spades.

Assigning Probabilities. Suppose $S = \{s_1, s_2, s_3, \ldots, s_n\}$ is the sample space for some experiment. If we assign a number p_1 to outcome s_1; a number p_2 to outcome s_2; $\cdots$; and a number p_n to outcome s_n, where

(i) $0 \leq p_i \leq 1$ for each i, $1 \leq i \leq n$

(ii) $p_1 + p_2 + \ldots + p_n = 1$

then the numbers $p_1, p_2, \ldots, p_n$ form a *probability distribution* for the sample space S. For $1 \leq i \leq n$, the probability of the outcome or simple event s_i occurring is p_i.

Example 2: A Probability Distribution

A sack contains 5 copies of the letter A, 8 copies of the letter B, 10 copies of the letter C, and 2 copies of the letter D. The experiment is to draw one letter from the sack. We can view the sample space as the set of outcomes $\{A, B, C, D\}$. If we assign the probability of $\frac{5}{25}$ to A, $\frac{8}{25}$ to B, $\frac{10}{25}$ to C, and $\frac{2}{25}$ to D, then you can easily verify that these values give us a probability distribution.

Probability distributions are used to define the probability of an event. Recall that an event is simply a subset of the sample space (the set of all possible outcomes of an experiment). A probability distribution gives you a probability for each of these outcomes. The *probability of an event E* is the sum of the probabilities of the outcomes comprising the event, and is denoted by $P(E)$; P is called a *probability function.*

Example 3: Probabilities of Events

If in Example 2 we define E to be the event that you draw either A or C from the sack (that is, $E = \{A, C\}$), then $P(E) = \frac{5}{25} + \frac{10}{25} = \frac{15}{25}$. If E is the event that you draw A, B, C, or D (in other words, $E = \{A, B, C, D\}$), then $P(E) = \frac{8}{25} + \frac{5}{25} + \frac{10}{25} + \frac{2}{25} = 1$.

Equally Likely Outcomes. For many experiments all of the outcomes occur with equal likelihood; that is, if s_i and s_j are any two outcomes with probabilities p_i and p_j, respectively, then $p_i = p_j$. For instance, if you roll a fair die, then the likelihood of getting a face up with two pips is just as likely as getting one with five. In this instance it is reasonable to assign a probability of $\frac{1}{6}$ to each of the 6 possible outcomes. The sample space for this experiment is $S = \{1, 2, 3, 4, 5, 6\}$. The probability of the event E of rolling either a 2 or a 5 (that is, $E = \{2, 5\}$) is

$$P(E) = \frac{1}{6} + \frac{1}{6} = \frac{2}{6}$$

Note that this is equal to

$$\frac{\text{Number of members of } E}{\text{Number of members of } S}$$

In general, we have the following useful rule.

Equally Likely Outcome Rule. Let S be a sample space consisting of n outcomes that are equally likely to occur and let E be an event in S consisting of r outcomes. Then $P(E) = \frac{r}{n}$. In particular, if E is a simple event consisting of just one outcome s, then $P(E) = P(s) = \frac{1}{n}$.

Occasionally in this context if we are interested in the occurrence of an event E, we say that E consists of all the favorable outcomes. Thus the probability that E occurs is given by

$$P(E) = \frac{\text{Number of favorable outcomes}}{\text{Number of total outcomes}}$$

Example 4: Calculating Probabilities for Equally Likely Outcomes

(a) In Example 4 of Section 7.4 we found that there are 54,912 ways of obtaining 3 of a kind in a 5-card hand dealt from a 52-card deck. Since there $C(52, 5) = 2,598,960$ possible 5-card hands and since each hand is equally likely to occur, the probability of obtaining 3 of a kind is

$$\frac{54,912}{2,598,960} \doteq 0.02113$$

(b) Many games involve tossing 2 dice and taking action based on the sum of the number of pips on the 2 dice. A good way to analyze the probabilities involved is to suppose that the dice are distinguishable, say one is red and the other green, and to make the sample space consist of the ordered pairs (x, y), where x is the number showing on the red die and y is the number showing on the green die. Figure 1 gives the sample space listed in rows for which $x + y$ is constant.

SUM						
2	$(1,1)$					
3	$(1,2)$	$(2,1)$				
4	$(1,3)$	$(2,2)$	$(3,1)$			
5	$(1,4)$	$(2,3)$	$(3,2)$	$(4,1)$		
6	$(1,5)$	$(2,4)$	$(3,3)$	$(4,2)$	$(5,1)$	
7	$(1,6)$	$(2,5)$	$(3,4)$	$(4,3)$	$(5,2)$	$(6,1)$
8	$(2,6)$	$(3,5)$	$(4,4)$	$(5,3)$	$(6,2)$	
9	$(3,6)$	$(4,5)$	$(5,4)$	$(6,3)$		
10	$(4,6)$	$(5,5)$	$(6,4)$			
11	$(5,6)$	$(6,5)$				
12	$(6,6)$					

Figure 1

Since each outcome (x, y) is equally likely and there are 36 outcomes, each outcome has probability $\frac{1}{36}$. The probability of the event "the sum of the number of pips on a pair of dice is 4" is $3 \cdot \frac{1}{36} = \frac{1}{12}$ since there are 3 pairs that add to 4 (they are in the third row). Since six pairs add to 7, the probability that the "sum is 7" is $6 \cdot \frac{1}{36} = \frac{1}{6}$. Notice that 7 is the most likely sum.

(c) If you draw a marble from a jar containing 4 distinguishable black marbles, 12 distinguishable green marbles, and 7 distinguishable grey marbles, what is the probability that the marble drawn is either grey or green? The number of possible outcomes is $4 + 12 + 7 = 23$ and the number of favorable outcomes is $12 + 7 = 19$. Since all outcomes are equally likely, the probability that you draw a grey or green marble is $\frac{19}{23}$.

(d) Suppose you draw 7 marbles from the jar described in (c). Let E be the event that you draw exactly 2 black marbles, 1 green marble, and

4 grey marbles from the jar. The number of favorable outcomes (the number of ways of drawing 2 black, 1 green, and 4 grey marbles) is $C(4,2) \cdot C(12,1) \cdot C(7,4) = 2,520$. The total number ways of drawing 7 marbles from the jar is $C(23,7) = 245,157$. Since each outcome is equally likely, it follows that

$$P(E) = \frac{2,520}{245,157} \doteq 0.010279$$

Two Basic Probability Properties

If S is a sample space and P is a probability function defined on the events in S, then

(a) If E and F are events in S, then

$$P(E \cup F) = P(E) + P(F) - P(E \cap F)$$

If, in addition, P and F are *disjoint* (that is, they have no elements in common), then

$$P(E \cup F) = P(E) + P(F)$$

(b) If E is an event and E' is the *complement* of E in S (E' is the set of all outcomes in S that are not in E), then

$$P(E') = 1 - P(E)$$

Example 5: Using the Two Basic Probability Properties

One hundred students were required to take 2 courses, Finite Math and English: 80 students passed Finite Math, 75 students passed English, and 65 students passed both. We find the probability that a student passed at least of one of these courses. If E is the event that a student passed Finite Math, then $P(E) = \frac{80}{100} = .8$. If F is the event that a student passed English, then $P(F) = \frac{75}{100} = .75$. From the given information you also know that $P(E \cap F) = \frac{65}{100} = .65$. The event $E \cup F$ is the event that a student passes at least one of these courses. By property (a) we have

$$P(E \cup F) = P(E) + P(F) - P(E \cap F) = .80 + .75 - .65 = .90$$

To find the probability of the event G that a student fails both of these courses note that G is the complement of the event $E \cup F$. Therefore, by property (b)

$$P(G) = 1 - P(E \cup F) = 1 - .90 = .10$$

Example 6: Disjoint (Mutually Exclusive) Events

First observe that if two events E and F are disjoint, then by the basic probability property (a) above, $P(E \cup F) = P(E) + P(F)$. To illustrate this we calculate the probability that if we randomly select 4 people from a group of 8 men and 10 women, then we get either 4 men or 4 women. Let E be the event of selecting 4 men and F the event of choosing 4 women. Then E and F are disjoint. Therefore, the probability $P(E \cup F)$ of selecting either 4 men or 4 women is given by

$$P(E \cup F) = P(E) + P(F) = \frac{C(8,4)}{C(18,8)} + \frac{C(10,4)}{C(18,8)} \doteq .0063988$$

Example 7: Defective TV Sets

A shipment of 100 television sets contains 7 defective sets. If you decide to check at random 10 of the sets, what is the probability that you will find at least 1 defective set? Let E be the event that you find at least 1 defective set. Note that this event includes finding exactly 1 defective set, exactly 2 defective sets, ... , or exactly 7 defective sets. Though you could find the probability of each of these (mutually disjoint) possibilities and then add them together to obtain the answer, it is much easier to calculate the probability of the complement E' of E, where E' is the event that you do not find any defective sets in the 10 sets you selected. Since $P(E') = C(93,10)/C(100,10) \doteq .46674$, it follows that

$$P(E) = 1 - P(E') \doteq 1 - .46674 = .53326$$

CALIPSO

The next example involves numbers that are very large or very small. Recall that CALIPSO displays such numbers in the computer representation of scientific notation that uses the letter E to stand for multiplication by a power of 10. For instance, $1.2345 \cdot 10^{23}$ is displayed as 1.2345E+23 by CALIPSO and $3.456 \cdot 10^{-15}$ is displayed as 3.456E-15. However, when you enter such numbers into CALIPSO's *Calculator*, you use a notation similar to the ordinary scientific notation. For instance, you enter $1.2345 \cdot 10^{23}$ as 1.2345 * 10^(23) and $3.456 \cdot 10^{-15}$ as 3.456 * 10^(−15).

Example 8: Birthdays

We find the probability that at least 2 people in a room with 10 people have the same birthday. To simplify matters we assume that each year has 365 days. Let E be the event that at least 2 people have the same birthday; we want to find $P(E)$. It turns out that it is easier to find the probability of E', the event that no 2 people have the same birthday. Note that E' is the complement of E, so $P(E) = 1 - P(E')$. To find $P(E')$, we count the number of ways that no 2 people have the same birthday. The first person has 365 choices for her birthday; the second person then has only 364 choices since no two people have the same birthday; the third person has 363 choices, ... ; the tenth person has 356 choices. Thus, by the Fundamental Counting Principle the 10 people can have different birthdays in $365 \cdot 364 \cdot 363 \cdots 356 \doteq 3.70608 \cdot 10^{25}$ ways.

Next observe that the total number of birthday possibilities for 10 people is $365^{10} \doteq 4.1969 \cdot 10^{25}$. Therefore,

$$P(E') \doteq \frac{3.70608 \cdot 10^{25}}{4.1969 \cdot 10^{25}} \doteq .883$$

and $P(E) = 1 - P(E') \doteq 1 - .833 = .117$. As you might expect this is not a particularly high probability. As you might not expect, if there are 23 people in a room, and E is the event that at least 2 of the people have the same birthday, then $P(E) = .507$. So there is a better than 50-50 chance that among 23 people at least 2 have the same birthday. The probability that at least 2 people in a group of 50 have the same birthday is .970.

8.1A Exercises: Finding Probabilities

If in the following problems you do not use CALIPSO, then, where appropriate, you may leave your answers in permutation or combination notation.

1. § Select a 5-card hand from a 52-card deck.

   (a) Find the probability that the hand contains all hearts.

(b) Find the probability that the hand contains cards of just 1 suit.

(c) Find the probability that the hand is a full house (contains 3 cards of 1 denomination and 2 cards of another).

(d) Find the probability that the hand contains at least 1 heart.

2. Roll two distinct dice and record the total of the numbers appearing.

(a) Find the probability that the total is 10.

(b) Find the probability that the total is 10 or greater.

(c) Find the probability that the total is at least 3.

3. § Seven people are selected randomly from a pool of 12 men and 15 women.

(a) Find the probability that 3 men and 4 women are selected.

(b) Find the probability that either 7 men or 7 women are selected.

(c) Find the probability that at least 1 man is selected.

4. If the numbers 1, 2, 3, 4, 5, 6 are arranged randomly in a row, find the probability that the numbers 3 and 6 will be adjacent.

5. § Suppose that in a group of 140 people, 85 drink Pepsi, 64 drink Coke, and 32 drink both.

(a) Find the probability that a person chosen at random from the group drinks either Pepsi or Coke.

(b) Find the probability that a person chosen at random drinks Pepsi but not Coke.

(c) Find the probability that a person chosen at random drinks neither Pepsi nor Coke.

6. A loaded die has the property that the probability of throwing an odd number is three times the probability of throwing an even number. If the even numbers are equally likely to appear, find the probability of rolling either a 3 or a 6.

7. § In Example 7, find the probability that in the shipment there are at least 3 defective sets.

8. The table below gives probabilities that in a randomly selected group of r people, at least 2 have the same birthday. Calculate the two missing entries.

r	5	10	15	20	22	23	30	40	50
$\Pr(r)$		.117		.411	.476	.507	.706	.891	.970

9. § If you have 10 pairs of socks, each pair of a different color, and you select (without looking) two socks, find the probability that you selected a matching pair.

10. Certain code words can use the letters A, B, C, D, X, Y, Z and the numbers 1, 4, 7, 9. Each code word consists of 4 letters followed by 5 numbers. You can repeat the numbers but not the letters.

    (a) Find the probability that a code word does not contain the letters C and X.

    (b) Find the probability that a code word contains the number 9 exactly 3 times.

11. § Twelve jurors are to be selected at random from a panel of 30 women and 50 men. What is the probability that 6 of the jurors are women and 6 are men?

12. If you draw 12 marbles from a jar containing 9 distinguishable red and 15 distinguishable blue marbles, what is the probability you drew 6 red and 6 blue marbles?

13. § A manufacturer has 24 system boards in stock, 8 of which come from supplier A and the rest from supplier B. The manufacturer uses the boards in assembling 4 computers today. What is the probability that all 4 of the system boards come from the same supplier?

14. Three cards are drawn at random from a 52-card deck. What is the probability that 1 is an ace and 2 are face cards?

15. § The Civil Rights Committee of the City of Pocatello considered a complaint of sex discrimination from a city employee. The Committee was composed of 3 men and 4 women. The 3 men voted against the complainant and the 4 women voted in his favor. Of all possible votes with 3 against and 4 in favor, what is the probability of this split along gender lines?

16. The Civil Rights Committee of Pocatello considered a complaint of racial discrimination against a local labor contractor. Twenty unemployed workers sought day jobs from the labor contractor. Four were assigned to pick lettuce, 8 were assigned lifeguard duty at the beach, 6 were assigned to be ushers in a theatre, and 2 were assigned to be waiters at the race track. As it happened, the 4 lettuce pickers were black and the rest were white. The lawyer for the labor contractor argued that the jobs were randomly assigned and that there was no intent to discriminate against the black people. Find the probability of this event occurring if the workers were randomly assigned jobs. (Suggestion: Review ordered partitions in Section 7.5.)

8.2 CONDITIONAL PROBABILITY; INDEPENDENT EVENTS

We now consider problems of the following sort.

Suppose that on a typical day 500 people enter an appliance store and that 200 of these people buy a washer, 100 buy a dryer, and 35 buy both. We select one of the customers at random and find that he has bought a washer. What is the probability that the customer has also bought a dryer? In other words:

> Given that a person buys a washer, what is the probability that the person also buys a dryer?

To solve this problem we let W be the set of people who buy a washer and D be the set of people who buy a dryer. Since we are told that the customer bought a washer, we can restrict our sample space to W. Thus, there are $|W| = 200$ possible outcomes of this event. (Here and elsewhere, we let $|A|$ denote the number of members of the set A.) Next note that there are $|W \cap D| = 35$ favorable outcomes since $W \cap D$ is the set of people who bought both washers and dryers. It follows that the probability that a person buys a dryer, given that the person bought a washer is

$$\frac{\text{Favorable outcomes}}{\text{Total outcomes}} = \frac{|W \cap D|}{|W|} = \frac{35}{200} = \frac{7}{40}$$

Observe that this probability is equal to

$$\frac{P(W \cap D)}{P(W)} = \frac{35/500}{200/500} = \frac{7}{40} \tag{1}$$

In general, suppose that E and F are events in a sample space and suppose further that the event E occurs. Then the probability that F occurs given that E occurred is called the *conditional probability of E given F* and is denoted by $P(E|F)$. Equation (1) illustrates the following rule for finding (and defining) $P(E|F)$.

$$P(E|F) = \frac{P(E \cap F)}{P(F)} \tag{2}$$

Example 1: Finding a Conditional Probability

A large corporation employs 1000 men and 1500 women. A survey indicates 725 of the men are college graduates as are 1200 of the women. A person is selected at random. We find the probability that a person is a female given that the person is a college graduate. Let G be the event that a person is a college graduate, M the event that a person is male, and F the event that a person is female. We need to find $P(F|G)$. By (2) we have

$$P(F|G) = \frac{P(F \cap G)}{P(G)} = \frac{1200/2500}{1925/2500} = \frac{1200}{1925} \doteq .6234$$

Next we find $P(G|M)$, the probability that the person selected is a college graduate given that the person is a male. We have

$$P(G|M) = \frac{P(G \cap M)}{P(M)} = \frac{725/2500}{1000/2500} = \frac{725}{2500} = .29$$

Example 2: Finding Another Conditional Probability

Workers in a plant are given 2 types of security clearance, Type I and Type II; 160 of the 200 workers have Type I clearance, 90 have Type II clearance, and 75 have both. Given that a worker has Type II clearance, what is the probability that she has Type I clearance? Let the event E be the set of all persons with Type I clearance, and let the event F be the set of all persons with Type II clearance. We want to find $P(E|F)$. By (2)

$$P(E|F) = \frac{P(E \cap F)}{P(F)} = \frac{75/200}{90/200} = \frac{75}{90} = \frac{15}{18}$$

Examples 3 and 4 illustrate how you can use a tree diagram and conditional probabilities to compute various other probabilities.

Example 3: Using a Tree Diagram

Suppose that 40% of the patients at a weight clinic are female and 60% are male. Suppose further that 75% of the females are mildly overweight and the rest are severely overweight. Suppose, also, that 65% of the males are mildly overweight and the rest are severely overweight. We use the following tree diagram to calculate a number of probabilities.

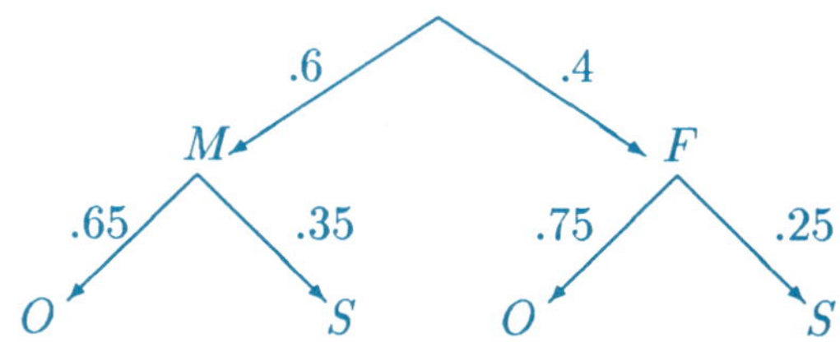

In this diagram M is the event that the patient is a male, F is the event that the patient is female, O is the event that the patient is mildly overweight, and S is the event that the patient is severely overweight. From the given data we have $P(M) = .6, P(F) = .4, P(O|M) = .65, P(S|M) = .35, P(O|F) = .75, P(S|F) = .25$.

Multiplying the two probabilities in the leftmost branch of the tree gives the probability that a patient is both male and mildly overweight. This follows from (2) since from that formula we see that $P(M \cap O) = P(M)P(O|M) = (.6)(.65) = .39$.

Similarly we see from the tree diagram and (2) that the probability that a patient is female and severely overweight is obtained by multiplying the two probabilities in the rightmost branch since $P(F \cap S) = P(F)P(S|F) = (.4)(.75) = .3$.

Observe also that the probability that a patient is severely overweight is $P(S) = P(M \cap S) + P(F \cap S) = (.6)(.35) + (.4)(.75) = .51$.

Example 4: Using a Tree Diagram to Analyze a More Complicated Problem

Suppose that one jar contains 12 red and 8 blue marbles, and a second jar contains 7 red and 10 blue marbles. If you (blindfolded) transfer 1 marble from the first jar to the second jar, mix up the marbles, and then remove a marble from the second jar, what is the probability that this marble is red?

We draw a tree whose first stage depicts drawing a red marble (R) with probability $\frac{12}{20}$ or a blue marble (B) with probability $\frac{8}{20}$. The left subtree at R depicts drawing marbles when the second jar contains 8 red and 10 blue marbles (a red marble from the first jar has been added to the second jar). The right subtree depicts drawing marbles when the second jar contains 7 red and 11 blue marbles (a blue marble from the first jar has been added to the second jar).

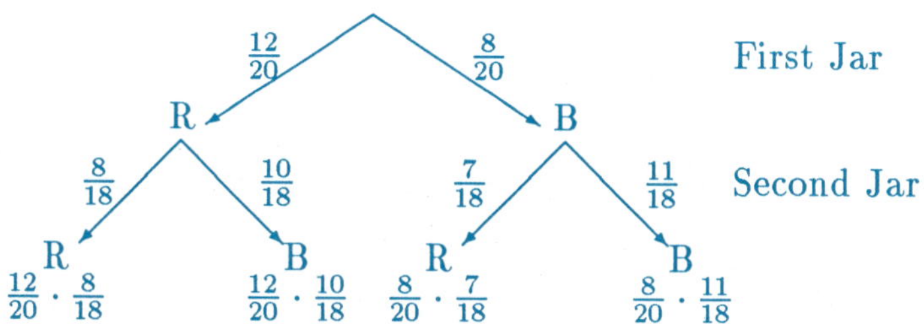

The leftmost R in the tree represents the event that a red marble is drawn from the first jar and a red marble is drawn from the second jar. The probability of this event is

$$P(\text{red at stage 1}) \cdot P(\text{red at stage 2}|\text{red at stage 1}) = \frac{12}{20} \cdot \frac{8}{18} = \frac{4}{15}$$

That is, the probability of this event is the product of the probabilities associated with each event on the path up to the root of the tree. The other probabilities at the bottom of the tree are computed in a similar way.

The probability of drawing a red from the second jar is the sum of the probability of getting a red after moving a red from jar 1 to jar 2 and the probability of getting a red after moving a blue. That is,

$$P(\text{drawing a red from jar 2}) = \frac{4}{15} + \frac{7}{45} = \frac{19}{45}$$

Independent Events

Intuitively, events E and F are independent if the occurrence of one does not depend in any way on the occurrence of the other. More precisely, events E and F are *independent* if

$$P(E|F) = P(E) \text{ and } P(F|E) = P(F) \tag{3}$$

Observe that since $P(E|F) = P(E \cap F)/P(F)$ or, equivalently, $P(E \cap F) = P(E|F)P(F)$, it follows from (3) that if E and F are independent, then

$$P(E \cap F) = P(E|F)P(F) = P(E)P(F)$$

Moreover, if $P(E \cap F) = P(E)P(F)$, then (3) holds. Thus, we can also say that events E and F are *independent* if

$$P(E \cap F) = P(E)P(F) \tag{4}$$

More generally, we have the following definition.

Definition. The events $E_1, E_2, \ldots, E_n$ are *independent* if

$$P(E_1 \cap E_2 \cap \cdots \cap E_n) = P(E_1)P(E_2) \ldots P(E_n) \tag{5}$$

Example 5: Independent Events

Suppose you draw 2 cards from a 52-card deck, toss a coin 3 times, and roll 2 dice. What is the probability of drawing 2 hearts (event E), getting exactly 2 heads (event F), and rolling a sum of 10 with the dice (event G)? Since it is clear that drawing cards from a deck, tossing a coin, and rolling a sum of 10 are independent events, we have

$$P(E \cap F \cap G) = P(E)P(F)P(G) = \frac{C(13,2)}{C(52,2)} \cdot \frac{3}{8} \cdot \frac{3}{36} \doteq .00184$$

Example 6: Determining if Events Are Independent

Sometimes it is not clear from the description of a problem whether two events are independent without checking that (4) holds. Suppose a jar contains 4 blue marbles, numbered 1,2,3,4; 5 red marbles, numbered 5,6,7,8,9; and 3 green marbles, numbered 10,11,12. The experiment is to select a marble and note its color and number. Let O be the event that an odd numbered marble is selected, G be the event that a green marble is selected, and B be the event that a blue marble is selected. Are these events independent? Observe that $P(O) = \frac{6}{12}, P(G) = \frac{3}{12}, P(B) = \frac{4}{12}, P(O \cap G) = \frac{1}{12}$, and $P(O \cap B) = \frac{2}{12}$. The events O and G are not independent since

$$P(O \cap G) = \frac{1}{12} \neq \frac{6}{12} \cdot \frac{3}{12} = P(O)P(G)$$

However, the events O and B are independent since

$$P(O \cap B) = \frac{2}{12} = \frac{6}{12}\frac{4}{12} = P(O)P(B)$$

8.2A Exercises: Conditional Probabilities and Independence

If in the following problems you do not use CALIPSO, then, where appropriate, you may leave your answers in permutation of combination notation.

1. § There are 8 men and 12 women on a committee. Three members of the committee are chosen randomly. Find the probability that the 3 chosen are women given that at least 1 is a woman.

2. Suppose a product P is advertised on a television program. The probability that a person saw the advertisement is .4; the probability that a person buys the product is .15; and the probability that a person saw the advertisement and buys the product is .1. Find the probability that a person saw the advertisement, given that he bought the product.

3. § Let S be the sample space of all families with exactly 2 children. Assume (somewhat incorrectly) that the probability of having a boy is the same as having a girl. Find the probability that a family has 2 girls, given that the family has at least 1 girl.

4. Suppose that 30% of the businesses in town support the college basketball program, 45% support the college tennis program, and 20% support both programs.

   (a) Find the probability that a business supports the tennis program, given that it supports the basketball program.

   (b) Find the probability that a business supports the basketball program, given that it supports the tennis program.

5. § Find the probability that a tossed coin will come up heads, given that it came up heads on the previous 5 tosses.

6. A medieval game was played with 3 cards, 1 red on both sides, 1 red on one side and blue on the other, and 1 blue on both sides. One of the cards is placed on a table. You win a dollar if the down side of the card is different than the up side; otherwise, you lose a dollar. If the up side is red, find the probability that the down side is also red to see if you should play this game.

7. § Find the probability that in a 7-card hand (dealt from a 52-card deck) there are 3 clubs, given that there are 4 hearts.

8. Roll 2 distinct dice. Find the probability that one of the dice comes up 2, given that the sum of the dice is 6 or 7.

9. § In 1927 a chapter of the Temperance League in Whyanip, New Jersey formed the opinion that 1/4 of their residents drink illicit alcohol. In order to scientifically verify their opinion, the chapter surveyed 100 residents of Whyanip asking the question, "Do you drink alcohol?" Since drinking alcohol at that time and place was an illegal act likely to result in a jail sentence, the Temperance League suspected that many of the respondents would not respond truthfully. To compensate for this tendency against self-incrimination the surveyors were instructed to have each respondent toss a die, shielding the outcome of the toss from the surveyor. If the die showed 1 or 2 pips, the respondent was to lie, and if the die showed 3, 4, 5, or 6 pips, the respondent was to tell the truth. The surveyors recorded each answer as a yes or a no. If, in fact, 1/4 of the people surveyed drink alcohol and the rest do not, what fraction of the people sampled would respond yes? (Suggestion: Draw a tree diagram as we did in Example 3.)

10. Gallop Polls, Inc. has been commissioned to ask 1000 fathers the question, "Did you ever sexually molest one of your children?" Gallop expects that sexual molesters will be unlikely to respond "yes" and thereby face the disapproval of the questioner. To lessen this possibility the questioner gives the respondent a spinner (a card on which is mounted an arrow that can be spun). With probability 2/5 the arrow stops on the area that instructs the respondent to lie and with probability 3/5 the arrow stops on the area that instructs him to tell the truth. The respondent spins the arrow without letting the questioner see the result and then answers the question: If 402 of the fathers say they are child molesters, how many are likely to be child molesters?

11. § If the Temperance League discovered that 3/5 of the respondents in the survey described in Problem 10 answered "yes," what fraction of the respondents drink alcohol?

12. Find the probability that in a 5-card hand (dealt from a 52-card deck) there are exactly 3 aces, given that the hand contains at least 2 aces.

13. § Two tests T_1 and T_2 are used to detect the presence of a disease. Test T_1 gives a positive reaction in 95% of the cases the patient has the disease; T_2 gives a positive reaction 88% of the time. The tests are assumed to be independent.

    (a) Find the probability that if a patient has the disease he will test positive on both tests.

    (b) Find the probability that both tests fail to detect the disease.

14. Two integers between 1 and 100 (inclusive) are selected by the random number generator of a computer.

    (a) Find the probability that both integers are divisible by 5.

    (b) Find the probability that one integer is less or equal to 25 and the other integer is greater than 90.

    (c) Find the probabilty that one integer is less than 20 and the other integer is less than 15.

15. § Toss a coin 3 times. Let E_1 be the event that tails is obtained on the first toss, let E_2 be the event that tails is obtained on the second toss, and let E_3 be the event that tails is obtained on the third toss. Use (5) to determine if these events are independent.

16. An electrical appliance uses 4 types, W, X, Y, Z, of transistors. It is known that 2% of Type W transistors are defective, as are 4% of type X, 1% of type Y, and 2% of type Z. The appliance contains 6 transistors of type W, 8 of type X, 3 of type Y, and 1 of type Z. Assume that all the defects are independent of one another. Find the probability of manufacturing a nondefective appliance.

17. § The probability that a pet goldfish will live at least 3 years is .2. The probability that a pet iguana will live at least 3 years is .6.

    (a) Find the probability both pets will live at least 3 years.

    (b) Find the probability that exactly 1 of the pets will live at least 3 years.

    (c) Find the probability that neither pet will live at least 3 years.

18. Fred and Jane decide independently to go to a movie. The probability that Fred chooses to see *The Terminator* is 0.62 and the probability that Jane chooses it is 0.35.

    (a) What is the probability that both see *The Terminator*?

    (b) What is the probabilty that one sees *The Terminator* and the other does not?

    (c) What is the probability that neither sees *The Terminator*?

8.3 BAYES' FORMULA

We now look at conditional probabilities in a slightly different light. Thus far we have considered $P(E|F)$, the probability that event E occurs, given that an earlier event F has already occurred. We are now interested in problems that use $P(E|F)$ to find $P(F|E)$. This type of problem often involves finding the probability that some outcome was due to a particular cause. The next example illustrates this idea.

Example 1: Testing for Diabetes

A test is used to detect the presence of diabetes. The test gives a positive result 90% of the time the person actually has diabetes. It also gives a positive result 15% of the time the person does not have diabetes. Studies have shown that 20% of a certain test group has diabetes. We want to find the probability that given a person tests positively, then the person actually has diabetes.

Before developing Bayes' Formula to solve this problem let us look at the following tree diagram

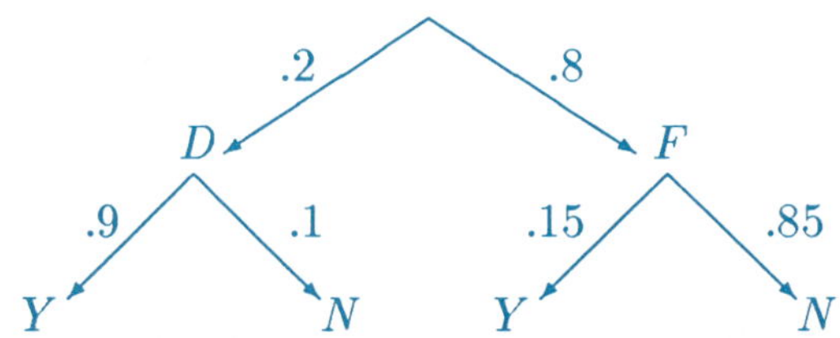

In this diagram D is the event "having diabetes," F is the event "not having diabetes," Y is the event "tests positively," and N is the event "tests negatively." We want to find

$$P(D|Y) = \frac{P(D \cap Y)}{P(Y)}$$

From the tree you can deduce that $P(Y) = (.2)(.9) + (.8)(.15)$ and $P(D \cap Y) = (.2)(.9)$. Therefore,

$$P(D|Y) = \frac{P(D \cap Y)}{P(Y)} = \frac{(.2)(.9)}{(.2)(.9) + (.8)(.15)}$$

Example 2: Defective Merchandise

A store sells three kinds of watches that, though identical in outward appearance, have different timing devices, A, B, and C. Fifty five percent of the watches have timing device A, twenty five percent utilize timing device

B, and twenty percent have timing device C. It is known that 2% of the watches having timing device A are defective, 3% of the watches having device B are defective, and 5% of the watches having device C are defective.

(a) You buy a watch and discover that it is defective. What is the probability that the watch has timing device B?

(b) You buy a watch that works. What is the probability that the watch has timing device C?

To find these probabilities we use the tree diagram

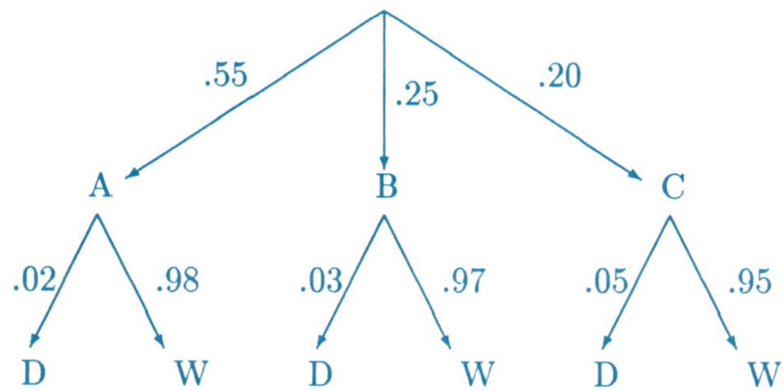

The letters in this diagram have the obvious interpretation. To solve the problem posed in (a) we need to find $P(B|D) = \frac{P(B\cap D)}{P(D)}$. From the tree diagram you can deduce that $P(B \cap D) = (.25)(.03) = .0075$ and $P(D) = (.55)(.02) + (.25)(.03) + (.20)(.05) = .0285$. It follows that

$$P(B|D) = \frac{P(B \cap D)}{P(D)} = \frac{.0075}{.0285} \doteq .2632$$

To solve the problem posed in (b) we need to find $P(C|W) = \frac{P(C\cap W)}{P(W)}$. From the tree diagram we see that $P(C \cap W) = (.2)(.95) = .19$ and $P(W) = (.55)(.98) + (.25)(.97) + (.2)(.95) = .9715$ and so

$$P(C|W) = \frac{P(C \cap W)}{P(D)} = \frac{.19}{.9715} \doteq .1956$$

We now proceed somewhat more formally to develop a formula that uses $P(Y|D)$ to find $P(D|Y)$. We use Example 1 to illustrate the formulas. We have from the definition of conditional probability that

$$\begin{aligned} P(Y|D) &= \frac{P(Y \cap D)}{P(D)} \quad \text{or, equivalently,} \\ P(D \cap Y) &= P(Y|D)P(D) = (.9)(.2) \end{aligned} \tag{1}$$

$$\begin{aligned} P(Y|F) &= \frac{P(Y \cap F)}{P(F)} \quad \text{or, equivalently,} \\ P(F \cap Y) &= P(Y|F)P(F) = (.15)(.8) \end{aligned} \tag{2}$$

Since $Y = (D \cap Y) \cup (F \cap Y)$, and since $D \cap Y$ and $F \cap Y$ are disjoint, it follows from the basic probability property (a) given in Section 8.1 that

$$P(Y) = P(D \cap Y) + P(F \cap Y) = (.90)(.2) + (.15)(.8) \tag{3}$$

From (1), (2), and (3) we obtain the result

$$P(D|Y) = \frac{P(D \cap Y)}{P(Y)} = \frac{P(Y|D)P(D)}{P(D \cap Y) + P(F \cap Y)} \tag{4}$$

This is one version of Bayes' Formula. Note that applying (4) to Example 1 we have

$$P(D|Y) = \frac{P(Y|D)P(D)}{P(D \cap Y) + P(F \cap Y)} = \frac{(.9)(.2)}{(.9)(.2) + (.15)(.8)} \tag{5}$$

Observation: Though we also obtained the result (5) directly from the tree diagram, tree diagrams can have many branches and are not particularly useful for large problems. Note that in (5) $P(D \cap Y) = P(D)P(Y|D)$ and $P(F \cap Y) = P(F)P(Y|F)$. Thus, (5) is a special case of the following general result.

In general, we have the following extension of (5).

Bayes' Formula: Let $E_1, E_2, \ldots, E_n$ be disjoint events whose union is the sample space S. Let E be an event in S. Then

$$P(E_i|E) = \frac{P(E_i)P(E|E_i)}{P(E_1)P(E|E_1) + P(E_2)P(E|E_2) + \cdots + P(E_n)P(E|E_n)} \tag{6}$$

Example 3: Using (6) to Calculate the Probabilities in Example 2

To find the probabilities in Example 2 we can use (6) to write

$$P(B|D) = \frac{P(B)P(D|B)}{P(A)P(D|A) + P(B)P(D|B) + P(C)P(D|C)}$$

$$= \frac{(.25)(.03)}{(.55)(.02) + (.25)(.03) + (.20)(.05)} \doteq .2632$$

Similarly, to find $P(C|W)$ in (b) of Example 2 we apply (6) to get

$$P(C|W) = \frac{P(C)P(W|C)}{P(A)P(W|A) + P(B)P(W|B) + P(C)P(W|C)}$$

$$= \frac{(.2)(.95)}{(.55)(.98) + (.25)(.97) + (.2)(.95)} \doteq .1956$$

Example 4: Using Bayes' Formula

For marketing purposes a state is divided into Regions I, II, III, IV, and V. It is known that 10% of the population is in Region I, 25% in Region II, 20% in Region III, 30% in Region IV, and 15% in Region V. A certain product is favored by 45% of the people in Region I, 80% of the people in Region II, 65% of the people in Region III, 40% of the people in Region IV, and 50% of the people in Region V. Given that a person favors the product, what is the probability that the person came from Region II? Let E be the event that a person favors the product, and let E_1, E_2, E_3, E_4, E_5 be the events that a person is from Region I, II, III, IV, V, respectively. Then from Bayes' Formula we have that P(From Region II | Favors the product) is

$$\frac{(.25)(.80)}{(.10)(.45) + (.25)(.80) + (.20)(.65) + (.30)(.40) + (.15)(.50)} \doteq .351$$

Next we use Bayes' Formula to find the probability that a person comes from Region I or Region IV, given that the person does not favor the product. We assume that no person lives in two regions simultaneously. Thus we have

$$P(\text{From Region I or IV}|\text{Does Not Favor the Product})$$

$$= P(\text{From Region I}|\text{Does Not Favor the Product})$$

$$+ P\ (\text{From Region IV}|\text{Does Not Favor the Product})$$

$$= \frac{(.10)(.55)}{(.10)(.55) + (.25)(.20) + (.20)(.35) + (.30)(.60) + (.15)(.50)}$$

$$+ \frac{(.30)(.60)}{(.10)(.55) + (.25)(.20) + (.20)(.35) + (.30)(.60) + (.15)(.50)} \doteq .5465$$

8.3A Exercises: Applying Bayes' Formula

Use either a tree diagram or (6) to calculate the following probabilities.

1. § Suppose Jar 1 contains 3 red and 5 green marbles, Jar 2 contains 6 red and 4 green marbles, Jar 3 contains 8 red marbles only, and Jar 4 contains 4 red and 7 green marbles. Someone selects a marble from one of the jars and hands it to you.

   (a) If the marble is green, find the probability that it came from Jar 2.

   (b) If the marble is red, find the probability that the marble did not come from Jar 1 or Jar 4.

2. An auto supply house receives a particular auto part X from 3 sources, A, B, and C. Experience has shown that 4% of the parts received from A are defective. 2% of the parts from B are defective, and 5% of the parts from C are defective. In recent shipments, the supply house received 100 units of X from A, 250 units from B, and 300 units from C. A customer is inadvertently sold a defective part. Find that the probability that the unit came from B.

3. § A new test for an arterial disease tests positive 85% of the time that a person has the disease. It also tests postive 5% of the time for persons without the disease. It is estimated that 20% of the population have

this disease. Find the probability that if a person tests positive, the person actually has the disease.

4. From a psychology class a researcher selected 20 students between the ages of 25 and 29 (inclusive), 25 students between the ages of 20 and 24, and 15 students between the ages of 15 and 19. Eight students in the first group, 15 in the second group, and 6 in the third group are female. If a student is selected at random from the 3 groups and the student is male, what is the probability that he was a member of the third group?

5. § In the St. Joe river, it is estimated that 45% of the fish are cutthroat trout, 35% are rainbow trout, 15% are squawfish, and 5% are "other." Moreover, 20% of the cutthroat, 25% of the rainbow, 10% of the squawfish, and 5% of the "other" are over 12 in long. If you land a fish that is 9 in long, what is the probability that it is a cutthroat?

6. Suppose four weighted coins, A, B, C, D, have probabilities, $.7, .3, .4, .8$, respectively, of tails occurring. One of the coins is selected randomly and tossed 3 times. If heads occurs each time, what is the probability that coin C was selected?

7. § A small firm use 4 different methods (I, II, III, IV) for quality control of its product P. It uses Method I for 30% of units, Method II for 20% of the units, Method III for 25% of the units, and Method IV for 25% of the units tested. Experience has shown that the rates of failure for Methods I, II, III, IV are $.03, .05, .02, .04$, respectively. If you buy a unit that turns out to be not defective, what is the probability that Method III was used to inspect that unit?

8. Of the voters in New Hampshire 42% are Republican and 58% are Democrat. It is reliably reported that 75% of the Republicans and 58% of the Democrats favor "NO NEW TAXES," A person is chosen from the population and discovered to favor "NO NEW TAXES." Find the probability that this person is a Democrat.

9. § A diagnostic test for the HIV virus is 90% dependable in the sense that of those tested who have HIV, 90% are diagnosed as having it; and of those tested who do not have HIV 90% are reported as free of it. Only 1% of the population has HIV. If a person is chosen at random from the population and the test shows that he has HIV, what is the

conditional probability that he does, in fact, have the disease. Do you think this test is reliable?

10. Two methods, A and B, are used to teach driving at a high school. Students fail the state driving test at the rate of 20% for A and 10% for B. Method B takes more equipment and instruction time than method A and hence A is used 30% of the time and B is used 70% of the time. A student failed the driving test. What is the probability that he was taught by method A?

8.4 BERNOULLI TRIALS

The idea of independence is important in probability problems involving the repetition of an experiment. The experiments we consider in this section have exactly two outcomes, often denoted *success* and *failure*.

Definition. An experiment is called a *Bernoulli trial* if the experiment has two outcomes (success and failure). A *binomial experiment* (also called a *Bernoulli process*) consists of a sequence of repetitions of a Bernoulli trial where

- The trials are independent
- The probability of success (or failure) is the same for each trial.

Example 1: Bernoulli Trials and Binomial Experiments

A consistent basketball player shoots free throws. From past experience we know that the probability she makes a free throw is $\frac{2}{3}$. If she shoots 10 free throws, what is the probability that she makes exactly 8 shots? In this example the Bernoulli trial is to shoot a free throw. There are exactly two outcomes: success (making the free throw) and failure (missing the free throw). The probability of success is $\frac{2}{3}$ and the probability of failure is $\frac{1}{3}$. The binomial experiment (or Bernoulli process) is to repeat this Bernoulli trial 10 times.

To find the probability of exactly 8 successes, we can proceed as follows. Since the trials are independent, the probability of each outcome of the binomial experiment is

$$\left(\frac{2}{3}\right)^8\left(\frac{1}{3}\right)^2$$

For instance, one outcome of the binomial experiment is miss the first free throw, make 8 free throws, and miss the last free throw. The probability of this occurring is

$$\frac{1}{3}\cdot\frac{2}{3}\cdot\frac{2}{3}\cdot\frac{2}{3}\cdot\frac{2}{3}\cdot\frac{2}{3}\cdot\frac{2}{3}\cdot\frac{2}{3}\cdot\frac{2}{3}\cdot\frac{1}{3}=\left(\frac{2}{3}\right)^8\left(\frac{1}{3}\right)^2$$

Another outcome is to miss the second and fourth free throws and make the rest. The probability of this occurring is also $\left(\frac{2}{3}\right)^8\left(\frac{1}{3}\right)^2$. In all there are $C(10,8)$ outcomes of this binomial experiment (why?), and it follows that the probability of making exactly 8 free throws is

$$C(10,8)\left(\frac{2}{3}\right)^8\left(\frac{1}{3}\right)^2$$

Completely analogous reasoning gives the following general result.

Binomial Experiment Probabilities: If a binomial experiment consists of n repetitions of a Bernoulli trial with probability p of success and probability of $q = 1 - p$ of failure, then the probability of exactly r successes is

$$C(n,r)p^r q^{n-r}$$

Example 2: Finding Binomial Experiment Probabilities

A manufacturer has found that 3% of the golf balls it produces are blemished. In a lot of 200 balls

(a) What is the probability that exactly 15 are blemished?

(b) What is the probability that fewer than 5 are blemished?

(c) What is the probability that between 3 and 6 (inclusive) are blemished?

For this problem, $p = .97$ is the probability of success (where success is defined to be getting an unblemished ball) and $q = 1-p = .03$ is the probability of failure. It follows that the answer to (a) is $C(200,185)(.97)^{185}(.03)^{15}$. You can evaluate this expression with CALIPSO's *Calculator* by entering either of the following expressions (or any other acceptable expression)

$$C(200,185)(.97)\hat{}(185)(.03)\hat{}(15)$$
$$C(200,185).97\hat{}185 * .03\hat{}15$$

Observe that for this problem we can also consider a blemished ball as a success and an unblemished one as a failure. In this case the probability of success is $p = .03$ and the probability of failure is $q = 1 - p = .97$. The probability of 15 "successes" is $C(200,15)(.03)^{15}(.97)^{185}$, which, as you can verify with CALIPSO, is the same as the answer obtained before.

In general it does not matter which outcome of a Bernoulli trial you designate to be a success; however, you must be consistent in associating the correct probabilities with success and failure.

For question (b) let success be defined as getting a blemished ball, so that $p = .03$ and $q = 1 - p = .97$. Then the solution of (b) is the sum of the probabilities of 5 outcomes: 0 successes, 1 success, 2 successes, 3 successes, and 4 successes. Since these events are disjoint the solution is

$$\begin{aligned} &C(200,0)(.03)^0(.97)^{200} + C(200,1)(.03)^1(.97)^{199} \\ &+C(200,2)(.03)^2(.97)^{198} + C(200,3)(.03)^3(.97)^{197} \\ &+C(200,4)(.03)^4(.97)^{196} \doteq .28098 \end{aligned} \tag{1}$$

You can evaluate (1) with CALIPSO's *Calculator*; however, it is much easier to use the *Binomial* function that is also found on the *STATISTICS* menu. When you select the *Binomial* function an entry panel asks you to enter the number of trials (n), the probability of success (p), a lower limit (a), and an upper limit (b). The program calculates $P(a \leq X \leq b)$ where X is the event that a success occurs. For $a = 0$ and $b = 4$, $P(0 \leq X \leq 4)$ is the same as expression (1).

Similar reasoning shows that the probability that the answer to (c) is

$$\begin{aligned} &C(200,3)(.03)^3(.97)^{197} + C(200,4)(.03)^4(.97)^{196} \\ &+C(200,5)(.03)^5(.97)^{195} + C(200,6)(.03)^6(.97)^{194} \doteq .54702 \end{aligned} \tag{2}$$

You evaluate (2) with CALIPSO's *Binomial* function as $P(a \leq X \leq b)$ with $n = 200, p = .03, a = 3$, and $b = 6$.

8.4A Exercises: Binomial Experiments

In the following problems express your answer in combination notation or use a hand calculator or CALIPSO's *Calculator* or *Binomial* function to find the appropriate probabilities. In Problems 1 – 4, p is the probability of success and n is the number of trials.

1. § If $p = .3$ and $n = 15$, what is the probability of fewer than 3 successes?

2. If $p = .6$ and $n = 11$, what is the probability of between 4 and 8 (inclusive) successes?

3. § If $p = \frac{1}{3}$ and $n = 17$, what is the probability of fewer than 2 failures or more than 15 failures.

4. If $p = \frac{3}{4}$ and $n = 9$, what is the probability of at least 1 success?

5. § If 60% of a city's population are Democrats and 40% are Republicans, what is the probability that in a sample of 10 people, exactly 4 will be Republicans?

6. Suppose that 2% of the items produced at a factory are defective.

   (a) Find the probability that in a sample of 15 items, at least 1 is defective.

   (b) Find the probability that in a sample of 8 items, either 2 or 3 are defective.

   (c) Find the probability that in a sample of 25 items, none is defective.

7. § You take a multiple choice test consisting of 10 questions. Each question has 4 responses, 1 of which is correct. If you guess the answer to each question, what is the probability that you will get 3 or more correct answers?

8. Suppose 75% of the people who enter a record store will buy at least 1 record. If 30 people come into the store, what is the probability that at least 25 will buy 1 record or more?

9. § Toss a fair coin 10 times. Find the probability that you get exactly 5 heads. Find the probability that you get at least 5 heads.

10. If a traffic light is green 70% of the time, what is the probability of getting a green light at least 8 out of 10 times that you enter the intersection?

11. § Medical records show that 25% of the people afflicted with a certain disease recover and the rest die. A drug company has developed a vaccine and tested it on 20 people suffering from the disease. Sixteen of the people recover and 4 die. If the vaccine were absolutely ineffectual, what is the probability that at least 16 recover? Is the vaccine effective?

12. A traffic control engineer hires people to watch an intersection and count the number of cars that run a stop light. This job is so boring that a watcher only detects 70% of the cars that run the light. If the engineer hires 5 people to independently observe the intersection, what is the probability that a driver running the light is detected by at least 1 observer? What is the probability that at least 3 of the watchers observe that driver running the light?

13. § A company test markets a new recipe for its cola drink. To build interest in its new product, the company buys TV advertising time in which 20 people sample the old and the new drink. Each of them must select one of the drinks as better. Suppose that the old and the new recipes are exactly the same. What is the probability that 12 or more people prefer the new drink?

14. The danger of too much pressure in a certain pipe in a nuclear power generation is so great that 3 gauges are installed to signal an alarm when the pressure exceeds 200 pounds per square inch (psi). Suppose the gauges operate independently and each has a probability of .9 of correctly signaling that the pressure has reached 200 psi. If the pressure reaches 200 psi, what is the probability that at least 1 of the gauges works?

8.5 MARKOV PROCESSES

In common with binomial experiments, *Markov processes* involve a sequence of repetitions of experiments with the same set of outcomes. However, we do not assume that the repeated experiments are independent. We do assume though that the outcomes of the current experiment depend at most on the outcomes of the preceding experiment (that is, the outcomes of the current experiment do not depend on outcomes of experiments prior to the previous experiment).

In the context of Markov processes, the outcomes of the experiments are called *states.* As we repeat the experiment the probabilities of being in any particular state vary and it is the transition of these probabilities that we will study. The original state probabilities form what is often called the *initial distribution.*

Example 1: A Markov Process

Suppose we classify voters into 3 categories (states): Democratic, Republican, Libertarian. Repeated observations have shown:

(a) If a voter votes Democratic in 1 election, then the probability that the voter will vote Democratic in the next election is .75, the probability that the voter will vote Republican in the next election is .15, and the probability the voter will vote Libertarian in the next election is .10.

(b) If a voter votes Republican in 1 election, then the probability that the voter will vote Democratic in the next election is .30, the probability that the voter will vote Republican in the next election is .50, and the probability that the voter will vote Libertarian is .20.

(c) If a voter votes Libertarian in 1 election, then the probability that the voter will vote Democratic in the next election is .25, the probability that the voter will vote Republican in the next election is .15, and the probability that the voter will vote Libertarian in the next election is .60.

The basic "experiment" in this case is to observe how an individual votes in elections year after year (we assume there is an election each year). For any given year we want to determine the probability of a voter being in any of the three "states": Democratic, Republican, or Libertarian. The experiment can be repeated as often as one wants, each year's observations being a repetition of the experiment. Note that the outcome of one year's experiment depends only on the preceding year. Thus, this is an example of a Markov process.

Two fundamental questions arise in connection with Markov processes:

- Given the initial distribution (for example, if in this year's election 40% of the electorate votes Democratic, 50% votes Republican, and

10% votes Libertarian), what is the probability of each state (voting Democratic, voting Republican, or voting Libertarian) in future years? For instance, what percentage of the voters will favor the Democrats, the Republicans, the Libertarians 3 years from now based on how voters voted in this year's election?

- In the "long run" what is the probability of each state? Will, for instance, the percentages of the electorate that vote Democratic, Republican, or Libertarian stabilize over time? We will see that for some Markov processes this is the case, but for others no long-term stabilization of these probabilities occurs.

To get an idea of what lies behind the answer to the first question, we use trees to organize and exhibit the data pertaining to the political parties.

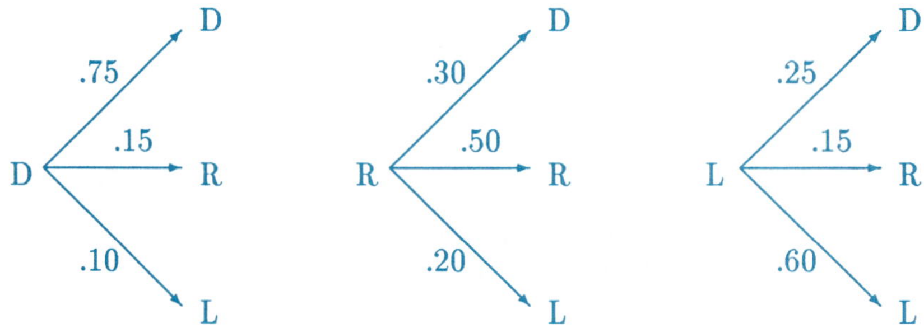

From these trees we form a table whose columns are the probabilities found in each of the trees.

		Current State		
		D	R	L
	D	.75	.30	.25
Next State	R	.15	.50	.15
	L	.10	.20	.60

The *transition matrix* is the matrix T of probabilities:

$$T = \begin{bmatrix} .75 & .30 & .25 \\ .15 & .50 & .15 \\ .10 & .20 & .60 \end{bmatrix} \tag{1}$$

Each of the entries in T represents a conditional probability. For instance the entry .20 is $P(\text{L next} \mid \text{R current})$; that is, .20 is the probability that a voter will vote Libertarian in the next election, given that she voted

Republican in the current election. Similarly, the entry .25 is the probability a voter will vote Democratic in the next election, given that he voted Libertarian in the current election.

Note that the columns of a transition matrix add to 1. Any matrix whose entries are nonnegative and whose columns add to 1 is called a *stochastic matrix.*

A *distribution matrix* (also called a *state matrix*) is a column matrix whose entries represent a probability distribution; that is, each entry is between 0 and 1, inclusive, and the sum of the entries is 1. In our example the distribution matrix

$$M = \begin{bmatrix} .40 \\ .50 \\ .10 \end{bmatrix} \tag{2}$$

indicates that for the current state the probability of voting Democratic is .40, the probability of voting Republican is .50, and the probability of voting Libertarian is .10. Transition matrices allow us to pass from one distribution matrix to another. Suppose, for instance, that the distribution matrix (2) describes the current probabilities of voting Democratic, Republican, or Libertarian. Then, as we will see, multiplying this matrix by the transition matrix T yields a new distribution matrix giving the probabilities of voting Democratic, Republican, or Libertarian in the next election. Since

$$TM = \begin{bmatrix} .75 & .30 & .25 \\ .15 & .50 & .15 \\ .10 & .20 & .60 \end{bmatrix} \begin{bmatrix} .40 \\ .50 \\ .10 \end{bmatrix} = \begin{bmatrix} .475 \\ .325 \\ .200 \end{bmatrix}$$

it follows that the probability a voter will vote Democratic in the next election is .475, the probability a voter will vote Republican is .325, and the probability a voter will vote Libertarian is .20.

To see, for instance, why the matrix multiplication result .475 is the probability a voter will vote Democratic in the next election, consider the tree

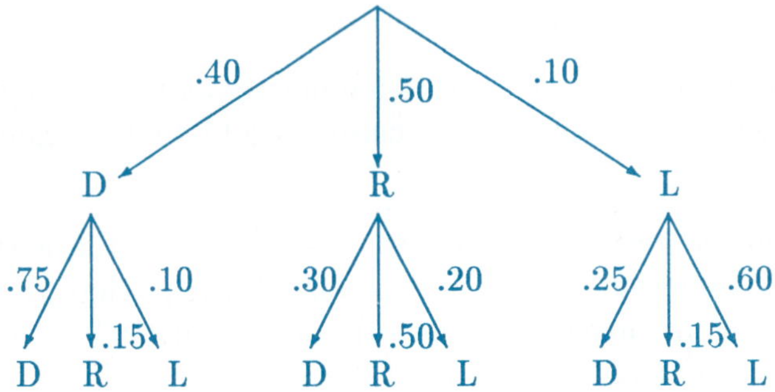

From this tree we see that there are 3 possibilities for voting Democratic in the next election:

(a) Voting Democratic in the current election and voting Democratic in the next election.

(b) Voting Republican in the current election and voting Democratic in the next election.

(c) Voting Libertarian in the current election and voting Democratic in the next election.

The probability that (a) occurs is

$$P(\mathrm{D}_{\mathrm{next}} \cap \mathrm{D}_{\mathrm{current}}) = P(\mathrm{D}_{\mathrm{next}} | \mathrm{D}_{\mathrm{current}}) \cdot P(\mathrm{D}_{\mathrm{current}}) = (.75)(.40)$$

In a like manner you can see that the probability that (b) occurs is (.30)(.50) and the probability that (c) occurs is (.25)(.10). Adding these together (which gives the probability of voting Democratic in the next election) is the same as multiplying the first row of T by the column matrix M. Similarly, you can check using the tree diagram that the probability of voting Republican in the next election is equal to the number obtained by multiplying the second row of T by M. An analogous statement can be made with regard to voting Libertarian in the next election.

The same reasoning shows that multiplying the new distribution matrix

$$\begin{bmatrix} .475 \\ .325 \\ .20 \end{bmatrix}$$

by transition matrix T will give us the probabilities of voting Democratic, Republican, or Libertarian in the election 2 years from now. In other words, $T(TD) = T^2D$ gives us the distribution of the probabilities of the various states after 2 more elections. In general, T^nD gives the voting probabilities in the election held n years from now (where we assume there is an election each year).

Example 2: Finding Future Voting Patterns

Let T be the transition matrix defined in (1). Suppose that in the current election 35% of the electorate voted Democratic, 60% voted Republican, and 5% voted Libertarian. Assume there is an election each year. What is the probability that 6 years from now a person will vote Democratic? From the above discussion we need to find T^6M where $M = \begin{bmatrix} .35 \\ .60 \\ .05 \end{bmatrix}$. You can calculate T^6M by hand or more easily by using the *MATRIX* program of CALIPSO. There are 2 ways to make the calculation with the *Operations* function. One way is to store the answer in matrix X by entering $X = TTTTTTM$. The other way generates successive values of T^kM when you enter the operation $X = T\#M$. After you enter $X = T\#M$ the display of the X matrix shows T^1M. To get the next iteration (T^2M), press [Spacebar]. You can continue pressing [Spacebar] to get T^kM for as many values of k as you want.

Either of these CALIPSO calculations gives $T^6M = \begin{bmatrix} .52207 \\ .23145 \\ .24648 \end{bmatrix}$.

From this matrix we see that about 52% of the people will vote Democratic in an election 6 years from now, about 23% will vote Republican, and about 25% will vote Libertarian.

Example 3: Market Analysis

A soup manufacturer is interested in making a market analysis to see how successful its product, Soup A, will do against competing soups, Soup B, Soup C, and Soup D. Market research has led to the following data whose entries represent the probabilities of the customer switching from one brand of soup to another over a 3-month period.

		From			
		A	B	C	D
	A	.79	.05	.02	.05
To	B	.08	.82	.06	.15
	C	.01	.07	.89	.01
	D	.12	.06	.03	.79

The transition matrix corresponding to these data is

$$T = \begin{bmatrix} .79 & .05 & .02 & .05 \\ .08 & .82 & .06 & .15 \\ .01 & .07 & .89 & .01 \\ .12 & .06 & .03 & .79 \end{bmatrix}$$

In this matrix the entry .08, for instance, is the probability that after a 3-month period, an individual will switch from Soup A to Soup B, that is, $.08 = P(\text{Soup B} \mid \text{Soup A})$. The entry .89 is the probability that after a 3-month period an individual who has bought Soup C will continue to buy Soup C.

Suppose that the distribution matrix $D = \begin{bmatrix} .35 \\ .19 \\ .20 \\ .26 \end{bmatrix}$ gives the current preferences for the various soups. To find the preferences after a year (4 3-month periods) you can use CALIPSO to show that

$$T^4D = \begin{bmatrix} .22352 \\ .31145 \\ .20374 \\ .26129 \end{bmatrix}$$

Thus after a year, about 22% of the customers will prefer soup A, 31% soup B, 20% soup C, and 26% soup D.

In the next section we will treat the "long run" problem. Problem 10 in Exercises 8.5A shows how to use CALIPSO to discover that for the soup problem as n gets larger T^nD approaches the distribution matrix

$$\begin{bmatrix} .16252 \\ .34966 \\ .25816 \\ .22965 \end{bmatrix} \tag{3}$$

This says that in the long run, Soup A will capture about 16% of the market, Soup B's share will be about 35% , and Soup C and Soup D will have about 26% and 23% of the market, respectively. Column matrix (3) is an example of a *stable distribution matrix.* A stable distribution matrix gives us the long-term trend of a Markov process. In the next section we will find ways to tell if a stable distribution exists, and if so how to find it.

The next definition gives a more formal description of the words we introduced in this section to describe Markov processes.

Definitions

- A matrix is *stochastic* if its entries are nonnegative and the sum of the entries in each column is 1.

- A *transition matrix* is a square stochastic matrix that represents a transition from one set of states to the next set of states. Specifically, if the states are $\{s_1, s_2, \ldots, s_n\}$, the matrix

		Current			
		s_1	s_2	$\cdots$	s_n
	s_1	p_{11}	p_{12}	$\cdots$	p_{1n}
Next	s_2	p_{21}	p_{22}	$\cdots$	p_{2n}
	$\vdots$	$\vdots$			
	s_n	p_{n1}	p_{n2}	$\cdots$	p_{nn}

  with entries p_{ij} describes a transition from the probabilities of being in each state of $\{s_1, s_2, \ldots, s_n\}$ to the next set of probabilities of being in each state. For each i, j, $1 \le i, j \le n$,$p_{ij} = P(\text{next state } j|\text{current state } i)$. For each column j, $p_{1j} + p_{2j} + \cdots + p_{nj} = 1$.

- A *distribution matrix* is a stochastic column matrix that gives the probabilities of being in each state.

- For a transition matrix T, the sequence of column matrices TD, T^2D, $T^3D, \ldots$ is called a *Markov chain.*

- If a Markov chain approaches a distribution matrix E, we say E is a *stable distribution matrix* (or a *steady state matrix*) for T. Equivalently (see Problems 9 and 10), E is a stable distribution matrix for T if $TE = E$.

8.5A Exercises: Markov Processes

1. § Which of the following matrices can be transition matrices?

(a) $\begin{bmatrix} .35 & .9 & .2 \\ .5 & .1 & .45 \\ .1 & 0 & .35 \end{bmatrix}$ (b) $\begin{bmatrix} 1.2 & .4 \\ -.2 & .6 \end{bmatrix}$ (c) $\begin{bmatrix} 0 & 1 & 0 \\ 1 & 0 & 1 \end{bmatrix}$

2. Which of the following matrices can be transition matrices?

(a) $\begin{bmatrix} .65 & .80 & .25 \\ .25 & .15 & .50 \\ .10 & .10 & .25 \end{bmatrix}$ (b) $\begin{bmatrix} 2.5 & .7 \\ -1.5 & .3 \end{bmatrix}$ (c) $\begin{bmatrix} .5 & .75 & .25 \\ .5 & .25 & .65 \end{bmatrix}$

3. § Which of the following matrices can be distribution matrices?

(a) $\begin{bmatrix} .4 \\ .3 \\ .1 \end{bmatrix}$ (b) $\begin{bmatrix} 1 \\ 0 \\ 1 \end{bmatrix}$ (c) $\begin{bmatrix} .4 & .1 \\ .2 & 0 \\ .4 & .9 \end{bmatrix}$ (d) $\begin{bmatrix} \frac{1}{3} \\ \frac{1}{3} \\ \frac{1}{3} \end{bmatrix}$

4. Jim and Joy are avid cribbage players and play 1 game against each other every day. They have found that if Jim wins one day, then the probability that he will win the next day is $\frac{2}{3}$.

   (a) Find the transition matrix corresponding to this situation.

   (b) Suppose that Jim won yesterday. What is the probability that he will win tomorrow?

   (c) If Joy won on Sunday, what is the probability she will lose on Wednesday?

5. § United Way estimates that in a certain community 80% of the people who make a donation in one year will make a donation next year. It also estimates that 30% of the people who do not make a donation one year will make a donation next year.

   (a) Find a transition matrix corresponding to this situation.

   (b) Suppose that this year 60% of the people made a donation. What percentage of the people is predicted to donate next year? The year after next?

   (c) If for some reason no one donated last year, what percentage of the people are expected to donate next year?

6. A study of historical weather patterns indicate that if it rains one day, the probability of rain the next day is 0.7. Moreover, if it does not rain one day, the probability of rain the next day is 0.2. Based on these predictions answer the following questions.

   (a) Find a transition matrix corresponding to this situation.

   (b) If it rains today, find the probability that it will not rain 2 days from now.

   (c) If it does not rain on Tuesday, find the probability that it will rain on Saturday.

   (d) If it rains on Wednesday, find the probability that it will not rain during the following weekend.

7. § Pigeons have been trained to peck 3 buttons, A, B, C, to get varying amounts of food. Many observations have shown that if a pigeon pecks button A on one trial, then on the next trial the pigeon will peck button A with probability .5, button B with probability .3, and button C with probability .2. If a pigeon pecks button B on one trial, then on the next trial the pigeon will peck button A with probability .2, button B with probability .6, and button C with probability .2. If a pigeon pecks button C on one trial, then on the next trial the pigeon will peck button A with probability .3, button B with probability .3, and button C with probability .4.

   (a) Find a transition matrix corresponding to this situation.

   (b) Suppose that on an initial trial 25% of the pigeons pecked button A, 40% pecked button B, and 35% pecked button C. Find the percentages for the buttons pecked on the third trial.

   (c) Suppose that a pigeon pecked button B on the first trial. What is the probability that the pigeon will peck button C on the third trial?

8. Suppose that in a certain society it is observed that if the father is left-handed, then the probability that a son will be left-handed is .3, right-handed is .6, and ambidextrous is .1. If the father is right-handed, then the probability that a son will be right-handed is .8, left-handed is .15, and ambidextrous is .05. If the father is ambidextrous, then the probability that a son will be ambidextrous is .2, right-handed is .7, and left-handed is .1.

(a) Find a transition matrix corresponding to this situation.

(b) If a man is right-handed, what is the probability that the son of his son will be right-handed?

9. § Convince yourself that if $T^n D$ approaches E as n gets large, then $TE = E$. (Hint: Note that $T^n D$ and $T^{n+1} D = T(T^n D)$ both approach E.)

Use CALIPSO to solve Problems 10–17.

10. In Problem 9 you saw that if T has a stable distribution matrix E, then both T^n and T^{n+1} will approach E. Therefore, you can use the function $T\#D$ of *MATRIX Operations* to find E. You simply calculate $T^n D$ for higher and higher values of n until the displayed matrix stops changing. The displayed matrix is E (or at least an approximation to E) . Use CALIPSO to check that the distribution matrix for the transition matrix T in Example 3 is the one given in (3).

11. § Given the transition matrix $T = \begin{bmatrix} .2 & .4 & 0 & .1 \\ .3 & .1 & .3 & .1 \\ .2 & .1 & .7 & .4 \\ .3 & .4 & 0 & .4 \end{bmatrix}$ use CALIPSO to show that you get the same stable distribution matrix for the initial distribution matrices

$$\text{(a) } D = \begin{bmatrix} 1 \\ 0 \\ 0 \\ 0 \end{bmatrix} \qquad \text{(b) } D = \begin{bmatrix} .2 \\ .5 \\ .1 \\ .2 \end{bmatrix}.$$

12. It is easy to make a typing error when entering a large matrix of decimals in a computer. As a partial check for typing errors, you can use matrix multiplication and CALIPSO to verify that a transition matrix is stochastic. Think of a way to do this and check that the matrix T in Problem 11 is stochastic.

13. § Suppose that during an epidemic 70% of the people who have the disease one month will continue to have the disease the next month, while 30% of the people with the disease one month will be well the next month. Suppose also that 20% of the people who are well one month will get the disease the next month; 80% of the people who

are well one month remain well the next month. If currently 10% of the population has the disease, what percent of the people will have the disease 3 months from now? In the long run what percent of population will have the disease?

14. Suppose in Problem 13 that 60% of the people who have the disease one month will have the disease the next month; 10% of the people who have the disease one month will die the next month; 30% of the people who have the disease one month will be well the next month. Suppose further that 50% of the people who are well one month will be well the next month; 45% of the people who are well one month will contract the disease the next month; 5% of the people who are well one month will die the next month. Finally suppose that currently 15% of the (living) population has the disease; 85% does not.

    (a) Find the percentage of the current population that will die 4 months from now.

    (b) Find the percentage of the population that will be well 6 months from now.

    (c) Find the percentage of the population that will have the disease a year from now.

15. § Suppose that 15% of the people who own a General Motors car in the current year will buy a Ford the next year; 80% will either keep their present car or buy another General Motors car; 5% will sell their car and buy a bicycle. Suppose that 25% of the people who currently own a Ford will buy a General Motors car next year; 72% will either keep their present car or buy another Ford; 3% will trade in their Ford for a bicycle. Suppose also that 80% of the people with bicycles will either keep their bicycles or buy another bicycle this year; 15% will trade their bicycles in for a Ford; 5% will trade their bicycles in for a General Motors car. If currently, 44% of the people drive a Ford, 48% drive a General Motors car, and 8% ride bicycles, find

    (a) The percentage of people who will drive a Ford 3 years from now.

    (b) The percentage of people who will drive a General Motors car 4 years from now.

    (c) The percentage of people who will ride bicycles 10 years from now.

(d) The long-term driving habits of the population.

16. Long-term weather reports show that if it is sunny at noon today, then the probability it will be sunny tomorrow at noon is .2, the probability it will be rainy is .3, and the probability it will be cloudy tomorrow is .5. If it is raining today at noon, then the probability that it will be sunny tomorrow at noon is .25, the probability it will be cloudy is .4, and the probability it will rain again is .35. Finally if it is cloudy today at noon, then the probability that it will be sunny tomorrow at noon is .3, the probability that it will rain is .3, and the probability that it will be cloudy is .4.

 (a) Suppose it is sunny today at noon. Find the probability that 3 days from today it will be cloudy at noon.

 (b) Suppose it rained yesterday at noon. Find the probability that it will rain a week from today at noon.

 (c) Can you predict a long-term weather trend?

17. § Taxis pick up and deliver passengers in a city that is divided into 6 precincts A, B, C, D, E, and F. The company has determined the probability that a passenger picked up in one precinct will travel to another precinct. These probabilities are given in the table

		From					
		A	B	C	D	E	F
To	A	.32	.12	.12	.05	.30	.15
	B	.12	.15	.06	.15	.15	.31
	C	.09	.22	.19	.45	.22	.03
	D	.24	.43	.21	.02	.11	.15
	E	.18	.03	.32	.13	.06	.12
	F	.05	.05	.10	.20	.16	.24

Suppose that at the start of a day the fraction of the total number of taxis in each precinct is

Precinct	A	B	C	D	E	F
Fraction	.24	.15	.27	.09	.12	.13

 (a) Check that the transition matrix you entered is stochastic by premultiplying it by [1 1 1 1 1 1].

(b) What is the distribution of taxis after each has transported 1 passenger?

(c) What is the distribution of taxis after each has transported 5 passengers?

(d) Is there a stable distribution?

8.6 REGULAR MARKOV CHAINS

At the end of Section 8.5 we introduced the notion of a stable distribution matrix for a transition matrix T. If such a matrix exists, it tells us how a Markov process behaves in the long run. In Example 1 we look at two long-term scenarios.

Example 1: Long-Term Distribution Trends

We consider two transition matrices that lead to quite different long-term results.

(a) Let T and D be the transition and distribution matrices defined by

$$T = \begin{bmatrix} .6 & .2 \\ .4 & .8 \end{bmatrix} \qquad D = \begin{bmatrix} .1 \\ .9 \end{bmatrix}$$

Using CALIPSO you can verify that for large values of n,

$$T^n = \begin{bmatrix} \frac{1}{3} & \frac{1}{3} \\ \frac{2}{3} & \frac{2}{3} \end{bmatrix} \qquad T^n D = \begin{bmatrix} \frac{1}{3} \\ \frac{2}{3} \end{bmatrix} = E$$

In fact, for $n = 15$, CALIPSO gives

$$T^{15} D = \begin{bmatrix} .33333 \\ .66667 \end{bmatrix} = E$$

Moreover, $TE = E$, and thus, E is the stable distribution matrix of T.

Observe also that, for large values of n, the columns of T^n are all equal to E.

(b) Let T and D be the transition and distribution matrices defined by

$$T = \begin{bmatrix} 0 & 1 \\ 1 & 0 \end{bmatrix} \qquad D = \begin{bmatrix} .2 \\ .8 \end{bmatrix}$$

You can verify that

$$TD = \begin{bmatrix} 0 & 1 \\ 1 & 0 \end{bmatrix} \begin{bmatrix} .2 \\ .8 \end{bmatrix} = \begin{bmatrix} .8 \\ .2 \end{bmatrix} \qquad T^2 D = \begin{bmatrix} 1 & 0 \\ 0 & 1 \end{bmatrix} \begin{bmatrix} .2 \\ .8 \end{bmatrix} = \begin{bmatrix} .2 \\ .8 \end{bmatrix}$$

$$T^3 D = \begin{bmatrix} 0 & 1 \\ 1 & 0 \end{bmatrix} \begin{bmatrix} .2 \\ .8 \end{bmatrix} = \begin{bmatrix} .8 \\ .2 \end{bmatrix} \qquad T^4 D = \begin{bmatrix} 1 & 0 \\ 0 & 1 \end{bmatrix} \begin{bmatrix} .2 \\ .8 \end{bmatrix} = \begin{bmatrix} .2 \\ .8 \end{bmatrix}$$

In general,

$$T^n = \begin{cases} \begin{bmatrix} 0 & 1 \\ 1 & 0 \end{bmatrix} & \text{if } n \text{ is even} \\ \\ \begin{bmatrix} 1 & 0 \\ 0 & 1 \end{bmatrix} & \text{if } n \text{ is odd} \end{cases}$$

and

$$T^n D = \begin{cases} \begin{bmatrix} .2 \\ .8 \end{bmatrix} & \text{if } n \text{ is even} \\ \\ \begin{bmatrix} .8 \\ .2 \end{bmatrix} & \text{if } n \text{ is odd} \end{cases}$$

The Markov chain $TD, T^2D, T^3D, T^4D, \ldots$ does not stabilize. The transition matrix T does not have a stable distribution.

The preceding example brings up the question of determining which transition matrices lead to stable (or steady state) distributions. Of equal interest is the question of determining the stable distribution when it exists. A partial answer to the first question is given in terms of regular matrices.

Definition. A transition matrix T is *regular* if T or some power, T^n, of T has no zero entries.

The following result is established in more advanced courses.

Suppose T is a regular transition matrix.

(a) If D is any distribution matrix, then $TD, T^2D, T^3D, \ldots$ will approach a stable distribution matrix E (which does not depend on the distribution matrix D you started with).

(b) The matrices $T, T^2, T^3, \ldots$ approach a matrix F all of whose columns are the same: in fact, each column of F is equal to the stable distribution matrix E.

(c) Algebraically, you can find the stable distribution matrix $E = \begin{bmatrix} e_1 \\ e_2 \\ \vdots \\ e_n \end{bmatrix}$ by solving the system of linear equations

$$\begin{aligned} &e_1 + e_2 + \cdots + e_n = 1 \\ &TE = E \end{aligned} \tag{1}$$

Example 2: A Regular Transition Matrix

The transition matrix

$$T = \begin{bmatrix} \frac{4}{5} & 1 \\ \frac{1}{5} & 0 \end{bmatrix}$$

is regular since the matrix

$$T^2 = \begin{bmatrix} \frac{4}{5} & 1 \\ \frac{1}{5} & 0 \end{bmatrix} \begin{bmatrix} \frac{4}{5} & 1 \\ \frac{1}{5} & 0 \end{bmatrix} = \begin{bmatrix} \frac{21}{25} & \frac{4}{5} \\ \frac{4}{25} & \frac{1}{5} \end{bmatrix}$$

does not contain any 0 entries. We can use (c) above to find the stable distribution matrix corresponding to T by solving the system of equations (1) which for our problem is

$$e_1 + e_2 = 1 \qquad \begin{bmatrix} \frac{4}{5} & 1 \\ \frac{1}{5} & 0 \end{bmatrix} \begin{bmatrix} e_1 \\ e_2 \end{bmatrix} = \begin{bmatrix} e_1 \\ e_2 \end{bmatrix} \tag{2}$$

From (2) we obtain the systems of equations

$$\begin{array}{rcrcl} e_1 & + & e_2 & = & 1 \\ \frac{4}{5}e_1 & + & e_2 & = & e_1 \\ \frac{1}{5}e_1 & & & = & e_2 \end{array} \quad \text{which is equivalent to} \quad \begin{array}{rcrcl} e_1 & + & e_2 & = & 1 \\ -\frac{1}{5}e_1 & + & e_2 & = & 0 \\ \frac{1}{5}e_1 & - & e_2 & = & 0 \end{array} \qquad (3)$$

Solving (3) you will find that $e_1 = \frac{5}{6}, e_2 = \frac{1}{6}$ and so $E = \begin{bmatrix} \frac{5}{6} \\ \frac{1}{6} \end{bmatrix}$ is the stable distribution matrix. From matrix E we see that, in the long term, the probability that you are in state 1 of the Markov process determined by T is $\frac{5}{6}$ and the probability that you are in state 2 is $\frac{1}{6}$. We note once more that these probabilities are the same for *any* initial distribution matrix.

8.6A Hand Exercises: Regular Matrices and Stable Distribution Matrices

In Problems 1 and 2 determine which of the transition matrices are regular. For each regular transition matrix use (c) above to find the corresponding stable distribution matrix.

1. § (a) $\begin{bmatrix} .4 & .2 \\ .6 & .8 \end{bmatrix}$ (b) $\begin{bmatrix} 0 & \frac{1}{4} \\ 1 & \frac{3}{4} \end{bmatrix}$ (c) $\begin{bmatrix} 0 & 0 \\ 1 & 1 \end{bmatrix}$

   (d) $\begin{bmatrix} .5 & 0 \\ .5 & 1 \end{bmatrix}$ (e) $\begin{bmatrix} \frac{1}{5} & \frac{3}{4} \\ \frac{4}{5} & \frac{1}{4} \end{bmatrix}$ (f) $\begin{bmatrix} \frac{1}{3} & 0 \\ \frac{2}{3} & 1 \end{bmatrix}$

2. (a) $\begin{bmatrix} \frac{1}{2} & 0 & \frac{1}{2} \\ 0 & 1 & 0 \\ \frac{1}{2} & 0 & \frac{1}{4} \end{bmatrix}$ (b) $\begin{bmatrix} 0 & 1 & 0 \\ 1 & 0 & 0 \\ 0 & 0 & 1 \end{bmatrix}$ (c) $\begin{bmatrix} .2 & .4 & .1 \\ .5 & .3 & .1 \\ .3 & .3 & .8 \end{bmatrix}$

   (d) $\begin{bmatrix} \frac{1}{3} & 0 & 1 \\ \frac{2}{3} & 1 & 0 \\ 0 & 0 & 1 \end{bmatrix}$ (e) $\begin{bmatrix} \frac{1}{3} & \frac{1}{2} & 1 \\ \frac{1}{3} & \frac{1}{2} & 0 \\ \frac{1}{3} & \frac{1}{2} & 0 \end{bmatrix}$ (f) $\begin{bmatrix} \frac{1}{5} & \frac{1}{3} & 0 \\ \frac{4}{5} & \frac{1}{3} & 0 \\ 0 & \frac{1}{3} & 0 \end{bmatrix}$

Example 3: Checking Regularity with CALIPSO

You can use CALIPSO to verify that the transition matrix

$$T = \begin{bmatrix} 0 & \frac{1}{3} & 0 & \frac{1}{4} \\ \frac{1}{2} & 0 & \frac{1}{4} & \frac{1}{4} \\ 0 & 0 & \frac{1}{4} & \frac{1}{4} \\ \frac{1}{2} & \frac{2}{3} & \frac{1}{2} & \frac{1}{4} \end{bmatrix} \qquad (4)$$

is regular since T^2 has no zero entries.

You can determine if T^n converges to a matrix F as in (b) with CALIPSO. To recall how to raise a matrix to high powers see Example 5 in Section 1.2. Continue to raise T to higher and higher powers until all columns of T^n are equal. (You need to change to decimal mode to do this because fractions usually overflow.) For matrix (4) CALIPSO finds that

$$T^{14} = \begin{bmatrix} .18705 & .18705 & .18705 & .18705 \\ .23741 & .23741 & .23741 & .23741 \\ .14388 & .14388 & .14388 & .14388 \\ .43165 & .43165 & .43165 & .43165 \end{bmatrix}$$

Now let

$$D = \begin{bmatrix} \frac{1}{2} \\ 0 \\ \frac{1}{4} \\ \frac{1}{4} \end{bmatrix}$$

be a distribution matrix. Using CALIPSO to find $T^n D$ for increasing values of n, you will discover that the distribution matrix

$$E = \begin{bmatrix} .18705 \\ .23741 \\ .14388 \\ .43165 \end{bmatrix} \qquad (5)$$

is the stable distribution matrix (steady state matrix) corresponding to T. Note that the columns of T^{14} are all equal to E. You should try another initial distribution matrix D of your own choosing and verify that as n grows larger $T^n D$ also approaches the same stable distribution matrix E. Observe, too, that for large values of n the columns of T^n approach equality to E.

As indicated in (1), we can also find the stable distribution matrix by solving the matrix equation $TE = E$ for E with the additional condition that the sum of the entries in E are 1. From (4) we see that we need to solve the system of equations

$$\begin{array}{rcrcrcrcl} e_1 & + & e_2 & + & e_3 & + & e_4 & = & 1 \\ & & \frac{1}{3}e_2 & & & + & \frac{1}{4}e_4 & = & e_1 \\ \frac{1}{2}e_1 & & & + & \frac{1}{4}e_3 & + & \frac{1}{4}e_4 & = & e_2 \\ & & & & \frac{1}{4}e_3 & + & \frac{1}{4}e_4 & = & e_3 \\ \frac{1}{2}e_1 & + & \frac{2}{3}e_2 & + & \frac{1}{2}e_3 & + & \frac{1}{4}e_4 & = & e_4 \end{array} \tag{6}$$

To solve (6) with CALIPSO we rewrite it as the single matrix equation

$$\begin{bmatrix} 1 & 1 & 1 & 1 \\ -1 & \frac{1}{3} & 0 & \frac{1}{4} \\ \frac{1}{2} & -1 & \frac{1}{4} & \frac{1}{4} \\ 0 & 0 & -\frac{3}{4} & \frac{1}{4} \\ -\frac{1}{2} & \frac{2}{3} & \frac{1}{2} & -\frac{3}{4} \end{bmatrix} \begin{bmatrix} e_1 \\ e_2 \\ e_3 \\ e_4 \end{bmatrix} = \begin{bmatrix} 1 \\ 0 \\ 0 \\ 0 \end{bmatrix} \tag{7}$$

[To obtain the coefficient matrix we subtracted the e_i's from both sides of the last four equations of (6).]

You can easily solve (7) with CALIPSO's *MATRIX Operations*. To do so, start the *MATRIX* program from CALIPSO's main menu. Enter the left matrix in (7) as matrix S and the right-hand side of (7) as matrix D. Then choose *Operations* and enter the formula $E = D/S$. You will find that E is the stable distribution given by (5).

8.6B CALIPSO Exercises: Regular Matrices and Stable Distribution Matrices

1. § Use CALIPSO to determine which of the following transition matrices are regular. For those that are regular find an n such that T^n has all positive entries.

(a) $\begin{bmatrix} .3 & 0 & .2 \\ .7 & 0 & .4 \\ 0 & 1 & .4 \end{bmatrix}$ (b) $\begin{bmatrix} 1 & .5 & 1 \\ 0 & .3 & 0 \\ 0 & .2 & 0 \end{bmatrix}$ (c) $\begin{bmatrix} .1 & .4 & .3 & .2 \\ 0 & .1 & 0 & .1 \\ .6 & .2 & .3 & .1 \\ .3 & .3 & .4 & .6 \end{bmatrix}$

(d) $\begin{bmatrix} .2 & .4 & .4 & .6 \\ 0 & 0 & .2 & 0 \\ .3 & .4 & .2 & 0 \\ .5 & .2 & .2 & .4 \end{bmatrix}$ (e) $\begin{bmatrix} 0 & 1 & .4 \\ 1 & 0 & .2 \\ 0 & 0 & .4 \end{bmatrix}$ (f) $\begin{bmatrix} \frac{1}{4} & 0 & \frac{1}{3} \\ \frac{1}{2} & 1 & \frac{1}{3} \\ \frac{1}{4} & 0 & \frac{1}{3} \end{bmatrix}$

2. Use CALIPSO and the methods of Example 3 to answer the following questions for

$$T = \begin{bmatrix} 0 & .15 & .10 & .75 \\ .25 & 0 & .40 & .10 \\ .35 & .45 & 0 & .15 \\ .40 & .40 & .50 & 0 \end{bmatrix}$$

(a) Show that T is regular.

(b) Find n such that all the columns of T^n are equal.

(c) For the initial distribution matrix $D = \begin{bmatrix} .15 \\ .25 \\ .35 \\ .25 \end{bmatrix}$ find k such that $T^k D$ is a stable distribution matrix. Verify that the columns of T^n that you obtained in part (b) equal $T^k D$.

(d) Find the stable distribution matrix by solving the system of equations (1).

3. Show that even though the transition matrix

$$T = \begin{bmatrix} 0 & 1 \\ 1 & 0 \end{bmatrix}$$

is not regular, the Markov chain $TD, T^2D, T^3D, \ldots$ leads to a stable distribution if

$$D = \begin{bmatrix} \frac{1}{2} \\ \frac{1}{2} \end{bmatrix}$$

Will this be true for any other distribution matrix D?

4. § Liam O'Shaugnessy is the head of a family that has farmed in County Clare for at least 30 generations. He discovers from writings in the family bible that a male child was born to each of these generations, and further notices that an unusual number of first-born children were male. He estimates from the records that if a father was a first-born

son himself, then the probability of his first-born child being male is 0.8. If the father is not himself a first-born son, then the probability that his first child is male is only 0.4. Liam's wife is about to deliver his first child. Without knowing whether Liam is himself a first-born son or not, estimate the probability that the child will be male.

5. A finite mathematics teacher is renowned for his eccentric grading. Unknown to anyone but himself, he uses the following simple scheme in grading the weekly examinations. If a student gets an A one week, then the student gets an A, B, C, D, F next week with respective probabilities $0.5, 0, 0.3, 0.1, 0.1$. A grade of B gets the student A, B, C, D, F with probabilities $0.3, 0.2, 0.5, 0, 0$; a C gets these grades with probabilities $0.2, 0.3, 0.4, 0.1, 0$; a D gets these grades with probabilities $0.3, 0.2, 0.3, 0, 0.2$; and an F gets these grades with probabilities $0, 0.2, 0.5, 0.1, 0.2$. Use CALIPSO to speculate on the grade distribution at the end of a long semester.

6. § Moscow has a small newspaper that cannot afford to subscribe to a national weather forecasting service. Instead the weather editor uses the following secret scheme to predict the weather as clear (C), rainy (R), snowy (S), or foggy (F). If it is clear today, he predicts (C,R,S,F) with probabilities $(0.45, 0.25, 0.20, 0.10)$; if it is rainy today, he predicts with probabilities $(0.25, 0.25, 0.35, 0.15)$; if it is snowy, he predicts with probabilities $(0.50, 0.20, 0.10, 0.20)$; and if it is foggy, he predicts with probabilities $(0.15, 0.15, 0.60, 0.10)$. If he has used this scheme all year, use CALIPSO to speculate on his weather predictions at the end of the year.

7. In order to stimulate interest in her meals, Josephine, the dorm cook, announces that she will vary the carbohydrates she serves in the following interesting way. She serves either pasta, potatoes, bread, or hominy grits according to the following plan. If she serves pasta one day, then she serves (pasta, potatoes, bread, hominy) the next day with probability $(0.30, 0.20, 0.40, 0.10)$; if she serves potatoes one day, she determines the next days' servings by using the probabilities $(0.15, 0.10, 0.25, 0.50)$; the day after bread, she uses the probabilities $(0.40, 0.50, 0.05, 0.05)$; and the day after hominy grits, she uses the probabilities $(0.45, 0.25, 0.20, 0.10)$. Use CALIPSO to find the probability of having hominy grits on the last day of the semester?

8.7 ABSORBING MARKOV CHAINS

In applications it may happen that once a particular state is reached, it becomes impossible to leave that state. In other words, the probability of going from that state to the same state is 1. Such a state is called an *absorbing state*. Transition matrices that lead to absorbing states are called *absorbing matrices*, and the corresponding Markov chain is an *absorbing Markov chain*. Somewhat more formally we have the following definition.

Definition. A transition matrix T for a Markov process is an *absorbing* matrix if

(a) There is at least one absorbing state.

(b) It is possible to move from any state to an absorbing state either directly or by passing through intermediate states.

In Problem 5 of Exercises 8.7A you are asked to show that if a transition matrix T has a column with a 1 on the *main diagonal* (the diagonal running from the upper left to the lower right) and 0's elsewhere in the column, then this column will remain unchanged in the matrices $T^2, T^3, \ldots$. For instance, if $T = \begin{bmatrix} * & 0 & * \\ * & 1 & * \\ * & 0 & * \end{bmatrix}$, then, for each n, T^n has the same form. It follows that, in this case, once you enter state 2, you will remain in that state.

The next example illustrates these ideas.

Example 1: An Absorbing Matrix and Markov Chain

Suppose you are tempted to play the following game. You start with 3 dollars. You roll a die and and each time 1 or 2 comes up you win a dollar; if 3, 4, 5, or 6 appears you lose a dollar. The game ends when you either have no money or you have 4 dollars. We will use a Markov process to determine various probabilities for this game. First note that the states you can be in are 0, 1, 2, 3, 4 (dollars). The probabilities are given by the transition matrix

		Current				
		0	1	2	3	4
	0	1	$\frac{2}{3}$	0	0	0
	1	0	0	$\frac{2}{3}$	0	0
Next	2	0	$\frac{1}{3}$	0	$\frac{2}{3}$	0
	3	0	0	$\frac{1}{3}$	0	0
	4	0	0	0	$\frac{1}{3}$	1

(1)

(To see, for instance, that the probability of going from state 2 to state 3 is $\frac{1}{3}$, note that currently you have \$2; there are 2 possibilities out of 6 of winning a dollar to enter state 3.)

To verify that (1) is an absorbing matrix, we check the two conditions (a) and (b) of the definition.

(a) States 0 and 4 are absorbing states, since if you are in state 0 (that is, you have 0 dollars) the game ends and so the probability that you remain in this state is 1; similarly, if you are in state 4 (you have won 4 dollars), the game also ends, and so the probability that you remain in this state is also 1. As noted above an easy way to find absorbing states from the transition matrix is to look along the main diagonal. A column corresponds to an absorbing state if it has a 1 for its diagonal entry and 0's elsewhere.

(b) We must check each of the nonabsorbing states (1, 2, and 3) to see that some sequence of play can put us in an absorbing state (0 or 4). From state 1 you can move immediately to state 0 with probability $\frac{2}{3}$. From state 2 you can move to state 1 with probability $\frac{2}{3}$, and from state 1 you can move to state 0 with probability $\frac{2}{3}$. From state 3 you can move directly to state 4 with probability $\frac{1}{3}$.

It is convenient to rearrange a transition matrix with absorbing states so that all of the absorbing states occupy the left-most columns of the matrix and the 1's remain on the diagonal. You can always do this by interchanging rows and columns that correspond to absorbing states. For example, matrix (2) below is obtained from (1) by interchanging the columns for states 1 and 4 and the rows for states 1 and 4. The rearranged absorbing transition matrix is said to be in *standard form.*

$$T = \quad \text{Next} \begin{array}{c|ccccc} & \multicolumn{5}{c}{\text{Current}} \\ & 0 & 4 & 2 & 3 & 1 \\ \hline 0 & 1 & 0 & 0 & 0 & \frac{2}{3} \\ 4 & 0 & 1 & 0 & \frac{1}{3} & 0 \\ 2 & 0 & 0 & 0 & \frac{2}{3} & \frac{1}{3} \\ 3 & 0 & 0 & \frac{1}{3} & 0 & 0 \\ 1 & 0 & 0 & \frac{2}{3} & 0 & 0 \end{array} \tag{2}$$

Observe that (2) provides exactly the same information as did the (1). For instance, in both matrices we have that the probability of going from state 3 to state 2 to is $\frac{2}{3}$.

Next we use CALIPSO to obtain T^2 and T^3. (Note that you enter only the matrix in (2) in CALIPSO, not the row and column labels.)

$$T^2 = \quad \text{Next} \begin{array}{c|ccccc} & \multicolumn{5}{c}{\text{Current}} \\ & 0 & 4 & 2 & 3 & 1 \\ \hline 0 & 1 & 0 & \frac{4}{9} & 0 & \frac{2}{3} \\ 4 & 0 & 1 & \frac{1}{9} & \frac{1}{3} & 0 \\ 2 & 0 & 0 & \frac{4}{9} & 0 & 0 \\ 3 & 0 & 0 & 0 & \frac{2}{9} & \frac{1}{9} \\ 1 & 0 & 0 & 0 & \frac{4}{9} & \frac{2}{9} \end{array}$$

$$T^3 = \quad \text{Next} \begin{array}{c|ccccc} & \multicolumn{5}{c}{\text{Current}} \\ & 0 & 4 & 2 & 3 & 1 \\ \hline 0 & 1 & 0 & \frac{4}{9} & \frac{8}{27} & \frac{22}{27} \\ 4 & 0 & 1 & \frac{1}{9} & \frac{11}{27} & \frac{1}{27} \\ 2 & 0 & 0 & 0 & \frac{8}{27} & \frac{4}{27} \\ 3 & 0 & 0 & \frac{4}{27} & 0 & 0 \\ 1 & 0 & 0 & \frac{8}{27} & 0 & 0 \end{array}$$

We can interpret these results as follows. From T^2 we see, for instance, that the if you start with \$2, then the probability that you will lose all of your money after playing the game twice is $\frac{4}{9}$. From T^3 we see that if you begin with \$1, then the probability that you will win (obtain \$4) after playing the game 3 times is $\frac{1}{27}$.

Using CALIPSO to find larger and larger powers of T (switch to decimal mode to do this), you will see that eventually the entries tend to stabilize.

In fact, the matrix

		Current				
		0	4	2	3	1
	0	1	0	.8	.53333	.93333
	4	0	1	.2	.46667	.066667
Next	2	0	0	0	0	0
	3	0	0	0	0	0
	1	0	0	0	0	0

(3)

is the stable matrix associated with T, since raising T to higher powers does not lead to any changes in these entries.

Properties of Absorbing Matrices

(a) Taking higher powers of any absorbing matrix will eventually lead to a stable matrix.

(b) Whatever your initial state you will always end in an absorbing state. For instance, in the above example, if you started in state 2 (began the game with \$2), then the probability that you will lose (end up in state 0) is .8 and the probability that you will win (end up with \$4) is .2. The probability of ending up in any nonabsorbing state is 0.

(c) Contrary to the case of regular matrices, the final state matrix depends on the initial state. For example, the probability of winning if you start with \$3 is different than if you started with \$1. Recall that for regular matrices, all initial states eventually lead to the same long-term distribution.

(d) If the original matrix has been put in the standard form

$$T = \begin{bmatrix} I & A \\ \mathbf{0} & B \end{bmatrix} \tag{4}$$

where I is an identity matrix, $\mathbf{0}$ denotes a submatrix consisting of all 0's, and A and B are submatrices corresponding to the nonabsorbing states, then it can be shown that the stable matrix corresponding to T is

$$\begin{bmatrix} I & A(I-B)^{-1} \\ \mathbf{0} & \mathbf{0} \end{bmatrix}$$

Observe that the entries of $A(I-B)^{-1}$ are the probabilities of ending up in an absorbing state given that you started in a nonabsorbing state.

(e) Suppose the original matrix has been put in standard form (4). Form the matrix $(I-B)^{-1}$, called the *fundamental* matrix, and label the columns of $(I-B)^{-1}$ in the same way the columns of B are labelled. The sum of the entries in column j of $(I-B)^{-1}$ give the average number of transitions it will take a process beginning in the jth state to enter an absorbing state. Thus, for instance, in Example 1 you can find how many moves on the average it will take to enter an absorbing state when you begin with \$2.

Example 2: Absorbing Matrices

For the absorbing transition matrix T given in (2), the submatrices A and B in (4) are

$$A = \begin{array}{c|ccc} & 2 & 3 & 1 \\ \hline 0 & 0 & 0 & \frac{2}{3} \\ 4 & 0 & \frac{1}{3} & 0 \end{array} \qquad B = \begin{array}{c|ccc} & 2 & 3 & 1 \\ \hline 2 & 0 & \frac{2}{3} & \frac{1}{3} \\ 3 & \frac{1}{3} & 0 & 0 \\ 1 & \frac{2}{3} & 0 & 0 \end{array}$$

and so

$$I - B = \begin{bmatrix} 1 & 0 & 0 \\ 0 & 1 & 0 \\ 0 & 0 & 1 \end{bmatrix} - \begin{bmatrix} 0 & \frac{2}{3} & \frac{1}{3} \\ \frac{1}{3} & 0 & 0 \\ \frac{2}{3} & 0 & 0 \end{bmatrix} = \begin{bmatrix} 1 & -\frac{2}{3} & -\frac{1}{3} \\ -\frac{1}{3} & 1 & 0 \\ -\frac{2}{3} & 0 & 1 \end{bmatrix}$$

You can use CALIPSO to find that

$$(I-B)^{-1} = \begin{bmatrix} \frac{9}{5} & \frac{6}{5} & \frac{3}{5} \\ \frac{3}{5} & \frac{7}{5} & \frac{1}{5} \\ \frac{6}{5} & \frac{4}{5} & \frac{7}{5} \end{bmatrix} \qquad A(I-B)^{-1} = \begin{bmatrix} \frac{4}{5} & \frac{8}{15} & \frac{14}{15} \\ \frac{1}{5} & \frac{7}{15} & \frac{1}{15} \end{bmatrix}$$

Thus from (d) above the stable matrix associated with T is

		Current				
		0	4	2	3	1
	0	1	0	$\frac{4}{5}$	$\frac{8}{15}$	$\frac{14}{15}$
	4	0	1	$\frac{1}{5}$	$\frac{7}{15}$	$\frac{1}{15}$
Next	2	0	0	0	0	0
	3	0	0	0	0	0
	1	0	0	0	0	0

(5)

Note that the matrix (3) that we obtained with CALIPSO by taking powers of T is the decimal approximation to (5).

From (5) we see, for instance, that the probability of ending up in absorbing state 4 having started in state 3 is $\frac{7}{15}$. Similarly, the probability of ending up in absorbing state 0 if you started in state 1 is $\frac{14}{15}$.

Using the last column of $(I-B)^{-1}$ we see by property (e) that the average number of transitions needed to get to an absorbing state from state 1 is $\frac{3}{5}+\frac{1}{5}+\frac{7}{5}=\frac{11}{5}$. Similarly if you begin in state 2, then on the average it will take $\frac{9}{5}+\frac{3}{5}+\frac{6}{5}=\frac{18}{5}$ transitions to get to an absorbing state.

8.7A Hand Exercises: Absorbing Matrices and Markov Processes

In Problems 1 and 2

(a) Determine if the transition matrix is absorbing.

(b) If the matrix is not absorbing, explain why not; if the matrix is absorbing, rewrite the matrix in standard form.

1.§ $\begin{bmatrix} 0 & \frac{3}{4} \\ 1 & \frac{1}{4} \end{bmatrix}$ 2. $\begin{bmatrix} \frac{1}{2} & 0 & 0 \\ \frac{1}{2} & 1 & \frac{1}{2} \\ 0 & 0 & \frac{1}{2} \end{bmatrix}$

The transition matrices found in Problems 3 and 4 are absorbing. Write each of these matrices in standard form (4) and

(a) Find the corresponding stable matrix.

(b) Find the probability of each nonabsorbing state being absorbed by each absorbing state.

(c) Find the probability of each nonabsorbing state being absorbed after three transitions.

(d) Find the average number of transitions it will take for each nonabsorbing state to reach an absorbing state.

3.§ $\begin{bmatrix} \frac{1}{2} & 0 & \frac{1}{2} \\ \frac{1}{2} & 1 & 0 \\ 0 & 0 & \frac{1}{2} \end{bmatrix}$

4. $\begin{bmatrix} \frac{1}{3} & 0 & 0 & 0 \\ \frac{1}{3} & 1 & 0 & \frac{1}{4} \\ \frac{1}{3} & 0 & 1 & \frac{1}{4} \\ 0 & 0 & 0 & \frac{1}{2} \end{bmatrix}$

5. Explain why if a transition matrix T has a column with 1 on the main diagonal and 0's elsewhere in that column, then this column remains unchanged in each of the powers $T^2, T^3, \ldots$.

The next example introduces a common application of absorbing matrices and offers some suggestions on how to use CALIPSO to calculate the quantities we have introduced.

Example 3: Analyzing a Maze with CALIPSO

On February 29 of each leap year in the village of Dogpatch marriageable bachelors participate in a competition to wed available spinsters. The village builds a maze of 12-ft high fences with strategically placed ladders. The spinster sits in a room inside the maze and the bachelor is placed in another room. At the sound of the cannon the bachelor begins climbing ladders in a random manner in the hope of either finding the spinster or the single exit to escape from the maze. The maze below is a typical arrangement in which the spinster is situated in Room 4, the bachelor starts in Room 1, and the outside world is designated Room 8. The diamonds show the location of the ladders, and it is assumed that the bachelor is equally likely to try any of the ladders leading out of a room.

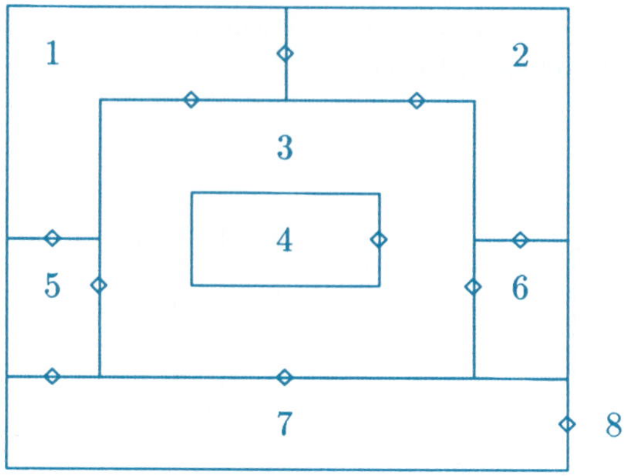

Our task is to analyze the probable result of this competition when many bachelors and spinsters try out the maze. To this end we regard each room as a state and form the absorbing transition matrix

$$
T = \quad \text{Next} \quad
\begin{array}{c|cccccccc}
 & \multicolumn{8}{c}{\text{Current}} \\
 & 1 & 2 & 3 & 4 & 5 & 6 & 7 & 8 \\
\hline
1 & 0 & \frac{1}{3} & \frac{1}{6} & 0 & \frac{1}{3} & 0 & 0 & 0 \\
2 & \frac{1}{3} & 0 & \frac{1}{6} & 0 & 0 & \frac{1}{2} & 0 & 0 \\
3 & \frac{1}{3} & \frac{1}{3} & 0 & 0 & \frac{1}{3} & \frac{1}{2} & \frac{1}{3} & 0 \\
4 & 0 & 0 & \frac{1}{6} & 1 & 0 & 0 & 0 & 0 \\
5 & \frac{1}{3} & 0 & \frac{1}{6} & 0 & 0 & 0 & \frac{1}{3} & 0 \\
6 & 0 & \frac{1}{3} & \frac{1}{6} & 0 & 0 & 0 & 0 & 0 \\
7 & 0 & 0 & \frac{1}{6} & 0 & \frac{1}{3} & 0 & 0 & 0 \\
8 & 0 & 0 & 0 & 0 & 0 & 0 & \frac{1}{3} & 1
\end{array}
\qquad (6)
$$

When you enter a matrix like (6) in CALIPSO you should use decimal mode because fraction mode is likely to overflow when you make many calculations; however, you should enter the data in the original matrix as fractions and let CALIPSO convert the fractions to decimals. In this way CALIPSO will carry many more digits of accuracy than it displays on the screen.

Next we need to convert (6) to standard form. This is not such an easy thing to do by hand because we must make two row interchanges and two column interchanges to accomplish it. Here is a trick from linear algebra that makes it easy to do it with CALIPSO. Enter an 8×8 identity matrix in CALIPSO. Choose *By Command* from the *MATRIX* menu and interchange row 1 and row 4. Then interchange row 2 and row 8. Suppose the result is in matrix E. (Note that we performed the same row operations on the identity matrix that we would have performed on T in the process of putting it in standard form.) Now use *Operations* to compute $S = ETE$. You will see that S is the desired standard form matrix. What happened is that multiplying T on the left by E interchanged rows 1 and 4 and rows 2 and 8. Multiplying on the right by E interchanged columns 1 and 4 and columns 2 and 8. The standard form matrix S is

$S =$ (7)

		Current							
		4	8	3	1	5	6	7	2
	4	1	0	$\frac{1}{6}$	0	0	0	0	0
	8	0	1	0	0	0	0	$\frac{1}{3}$	0
	3	0	0	0	$\frac{1}{3}$	$\frac{1}{3}$	$\frac{1}{2}$	$\frac{1}{3}$	$\frac{1}{3}$
Next	1	0	0	$\frac{1}{6}$	0	$\frac{1}{3}$	0	0	$\frac{1}{3}$
	5	0	0	$\frac{1}{6}$	$\frac{1}{3}$	0	0	$\frac{1}{3}$	0
	6	0	0	$\frac{1}{6}$	0	0	0	0	$\frac{1}{3}$
	7	0	0	$\frac{1}{6}$	0	$\frac{1}{3}$	0	0	0
	2	0	0	$\frac{1}{6}$	$\frac{1}{3}$	0	$\frac{1}{2}$	0	0

Now we compute the stable matrix approached by $S, S^2, S^3, \ldots$. Choose *Operations* and enter $R = S\hat{}$. (Of course, you must use different letters in these commands if you have named your matrices differently.) S^2 appears on the screen; press (or hold down) *Spacebar* until all of the entries below the first two rows are displayed as 0.

Note: If you print a copy of a matrix such as R that is created by using operations of CALIPSO, you may be surprised to see that the zeros are printed as small numbers like $1.234E-09$. Numbers sometimes are displayed as 0 on the screen and as a nonzero, but very small number, in print. This is done to give better accuracy in the printed output than can be easily read on the screen. In fact, a high degree of accuracy of numbers reported by a computer after many calculations is always doubtful. For most calculations, CALIPSO is usually accurate to only about 8 decimal places. Also, it is worth noting that the rows in the chain $S, S^2, S^3, \ldots$ *approach* 0, though they probably never actually get there.)

It takes 201 iterations to get the stable matrix

		Current							
		4	8	3	1	5	6	7	2
	4	1	0	.61806	.58333	.52778	.61111	.38194	.60417
	8	0	1	.38194	.41667	.47222	.38889	.61806	.39583
	3	0	0	0	0	0	0	0	0
Next	1	0	0	0	0	0	0	0	0
	5	0	0	0	0	0	0	0	0
	6	0	0	0	0	0	0	0	0
	7	0	0	0	0	0	0	0	0
	2	0	0	0	0	0	0	0	0

(8)

Since our bachelor started in room 1, the probability of his finding the spinster is .58333 and the probability of his escaping is .41667. Other columns give the probabilities for bachelors who start in other rooms.

Finally, we calculate the fundamental matrix. We need to extract the matrix called B in (4) from the standard matrix (7). We do this by locating the standard matrix on the screen and selecting *Extract Block* from the menu. When you do this, CALIPSO asks for the row and column numbers of the upper left corner and of the lower right corner of the block to extract. We respond with (3,3) and (8,8). Suppose the extracted block is in matrix B. To calculate $(I - B)^{-1}$ we first form a 6×6 identity matrix I. To get the fundamental matrix we use *Operations* and enter $F = (I - B)\hat{}$. We get

		3	1	5	6	7	2
	3	3.7083	3.5	3.1667	3.6667	2.2917	3.625
	1	1.75	3	2	2	1.25	2.25
$F =$	5	1.5833	2	2.6667	1.6667	1.4167	1.75
	6	1.2222	1.3333	1.1111	2.4444	.77778	1.6667
	7	1.1458	1.25	1.4167	1.1667	1.8542	1.1875
	2	1.8125	2.25	1.75	2.5	1.1875	3.1875

(9)

We can find the average time it takes to get from Room j to an absorbing state by adding the entries in the columns of (9). To do this with CALIPSO we enter a 1×6 row matrix containing all 1's. Call this matrix U. Then we use *Operations* to calculate UF and get

	3	1	5	6	7	2
$UF =$	11.22	13.333	12.111	13.444	8.7778	13.667

(10)

From (10) we see, for example, that on the average it will take 8.7778 moves to get from Room 7 to an absorbing state (either Room 4 or 8).

8.7B CALIPSO Exercises: Absorbing Matrices and Markov Processes

In Problems 1 - 6

(a) Determine if the transition matrix is absorbing.

(b) If the matrix is not absorbing, explain why not; if the matrix is absorbing, rewrite the matrix in standard form.

1.§ $\begin{bmatrix} 1 & 0 & 0 & 0 & 0 \\ 0 & \frac{1}{3} & \frac{1}{2} & \frac{1}{2} & \frac{1}{2} \\ 0 & 0 & 0 & \frac{1}{4} & \frac{1}{2} \\ 0 & \frac{1}{3} & \frac{1}{2} & 0 & 0 \\ 0 & \frac{1}{3} & 0 & \frac{1}{4} & 0 \end{bmatrix}$

2. $\begin{bmatrix} .3 & 1 & .4 & 0 \\ .4 & 0 & .2 & 0 \\ .3 & 0 & .4 & 0 \\ 0 & 0 & 0 & 1 \end{bmatrix}$

3.§ $\begin{bmatrix} 0 & \frac{1}{10} & 0 & 0 \\ \frac{1}{10} & \frac{1}{10} & 0 & \frac{1}{2} \\ \frac{4}{5} & \frac{2}{5} & 1 & \frac{1}{5} \\ \frac{1}{10} & \frac{2}{5} & 0 & \frac{3}{10} \end{bmatrix}$

4. $\begin{bmatrix} 1 & 0 & 0 \\ 0 & .3 & .8 \\ 0 & .7 & .2 \end{bmatrix}$

5.§ $\begin{bmatrix} .2 & 0 & .3 & 0 & 0 \\ 0 & 1 & 0 & 0 & 0 \\ .3 & 0 & .1 & .6 & 0 \\ .5 & 0 & .6 & 0 & .6 \\ 0 & 0 & 0 & .4 & .4 \end{bmatrix}$

6. $\begin{bmatrix} .25 & .3 & 0 & 0 \\ .25 & .3 & 0 & 0 \\ .25 & .2 & 1 & 0 \\ .25 & .2 & 0 & 1 \end{bmatrix}$

The transition matrices found in Problems 7–9 are absorbing. Write each of these matrices in standard form (4) and

(a) Find the corresponding stable matrix.

(b) Find the probability of each nonabsorbing state being absorbed by each absorbing state.

(c) Find the probability of each nonabsorbing state being absorbed after 3 transitions.

(d) Find the average number of transitions it will take for each nonabsorbing state to reach an absorbing state.

7.§ $\begin{bmatrix} \frac{1}{4} & 0 & 0 & \frac{1}{3} & 0 \\ \frac{1}{4} & 1 & 0 & 0 & 0 \\ 0 & 0 & 1 & \frac{1}{6} & \frac{1}{3} \\ \frac{1}{2} & 0 & 0 & 0 & \frac{1}{3} \\ 0 & 0 & 0 & \frac{1}{2} & \frac{1}{3} \end{bmatrix}$

8. $\begin{bmatrix} \frac{1}{4} & 0 & 0 & 0 \\ 0 & 1 & \frac{1}{2} & \frac{1}{3} \\ \frac{3}{4} & 0 & 0 & 0 \\ 0 & 0 & \frac{1}{2} & \frac{2}{3} \end{bmatrix}$

9. $\begin{bmatrix} 1 & 0 & .5 & 0 & 0 & 0 & 0 \\ 0 & 0 & 0 & .5 & 0 & 0 & 0 \\ 0 & .5 & .5 & 0 & 0 & 0 & 0 \\ 0 & 0 & 0 & .5 & 0 & .5 & 0 \\ 0 & .5 & 0 & 0 & 1 & 0 & .5 \\ 0 & 0 & 0 & 0 & 0 & .5 & 0 \\ 0 & 0 & 0 & 0 & 0 & 0 & .5 \end{bmatrix}$

10. Change the rules of the game played in Example 1 as follows. Each play consists of selecting a number between 1 and 10 (inclusive) from a hat. You win \$1 if the number is a prime; otherwise you lose \$1. You begin playing with \$3 and will quit if you have either \$0 or \$5.

   (a) Construct the appropriate absorbing transition matrix.

   (b) Rewrite in standard form the matrix obtained in (a).

   (c) Find the stable (steady state) matrix associated with the matrix obtained in (b).

   (d) What is the probability that you will lose all of your money?

   (e) What is the average number of times you can play (the average number of transitions) before you lose your money or win \$5?

   (f) What is the probability of winning \$5 after you have played the game twice?

11. § A certain four-year college wants to predict attrition and progression rates. They have data for the past decade of the percentages of students from each class who have dropped out, progressed to the next class, or graduated. To analyze the data, they assign the following

states: State 1 is the set of drop-outs; State 2 is the set of graduates; and States 3, 4, 5, and 6 are the freshman, sophomore, junior, and senior classes, respectively. Their records show that 22% of the freshman class drops out and the rest become sophomores; 13% of the sophomore class drops out and the rest become juniors; 9% of the junior class drops out and the rest become seniors; 6% of the senior class drops out and the rest graduate.

(a) Describe this process with an absorbing transition matrix in standard form.

(b) Find the stable matrix.

(c) Determine the long-term probability that an entering freshman will graduate.

12. Verify that transition matrix T in (2) satisfies properties (a) – (e) of the Properties of Absorbing Matrices. Also obtain the stable matrix by calculating $A(I - B)^{-1}$.

13. § A rat placed in the following maze wanders from room to room passing through doors (indicated by gaps) with equal likelihood. However, once the rat enters chamber 6 or 8 it is caught in a trap and wanders no more.

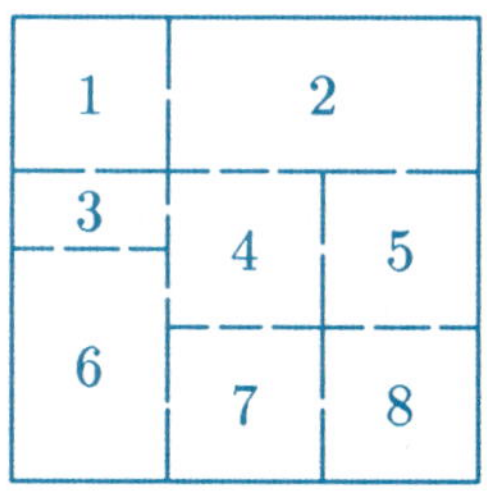

(a) Find an appropriate absorbing transition matrix for this situation and then write the matrix in standard form.

(b) If the rat begins in chamber 1, what is the probability that it will eventually end up in chamber 6? chamber 8?

(c) If the rat begins in chamber 5, what is the probability that it will eventually end up in chamber 6? chamber 8?

(d) If the rat begins in chamber 4, what is the average number of doors it will pass through before being trapped?

(e) If the rat begins in chamber 3, what is the probability that it is in chamber 5 after passing through 5 doors?

14. A rat placed in the following maze wanders from room to room passing through doors (indicated by gaps) with equal likelihood. If the rat enters chamber 8 it is caught in a trap. If the rat passes through the door near numeral 10, it escapes the maze and runs away.

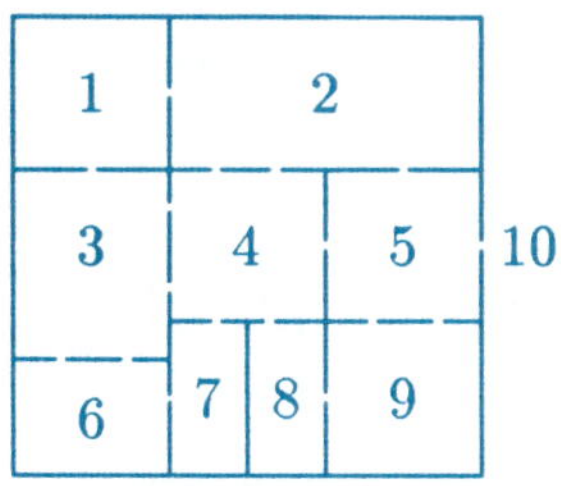

(a) Find an appropriate absorbing transition matrix for this situation and then write the matrix in standard form.

(b) If the rat begins in chamber 1, what is the probability that it will be trapped? What is the probability it will get free?

(c) If the rat begins in chamber 7, what is the probability that it will be trapped? What is the probability it will get free?

(d) If the rat begins in chamber 1, what is the average number of doors it will pass through before it escapes?

(e) If the rat begins in chamber 1, what is the probability that it is in chamber 6 after passing through 6 doors?

9 Statistics

In this chapter we will see how to organize and describe data in ways that are easy to understand and interpret.

9.1 FREQUENCY DISTRIBUTIONS, PROBABILITY DISTRIBUTIONS, AND HISTOGRAMS

The following situation illustrates some of the basic ideas of this section. Sixty people are surveyed to determine the average number of hours they spend watching television each day. Table 1 gives the results of the survey.

Table 1

12	7	3	4	4	6	0	3	2	8
8	2	7	4	0	2	1	5	9	0
8	9	3	5	0	0	5	1	1	6
8	4	2	5	3	1	7	4	6	4
9	12	0	6	3	4	5	9	10	7
6	8	4	4	2	3	6	3	7	6

It is not particularly easy to assimilate the data in Table 1. In Table 2 we retabulate the data as a *frequency distribution* (the second column) and a *relative frequency distribution* (the third column). This table gives a much better idea of the results of the survey. Note that the relative frequency of a particular occurrence is the ratio of the number of times the occurrence happened to the total number of occurrences. Also observe that if you add all of the entries in the frequency distribution column of Table 2 you will get 60 (the number of people surveyed); if you add the entries of the relative frequency distribution column you get 1.

Table 2: Relative Frequency of Watching TV

Hours Spent	Number of Occurrences (Frequency)	Proportion of Occurrences (Relative Frequency)
0	6	$\frac{6}{60}$
1	4	$\frac{4}{60}$
2	5	$\frac{5}{60}$
3	7	$\frac{7}{60}$
4	9	$\frac{9}{60}$
5	5	$\frac{5}{60}$
6	7	$\frac{7}{60}$
7	5	$\frac{5}{60}$
8	5	$\frac{5}{60}$
9	4	$\frac{4}{60}$
10	1	$\frac{1}{60}$
12	2	$\frac{2}{60}$

More generally, suppose an experiment has outcomes $x_1, x_2, \ldots, x_k$ and that for each $i, 1 \leq i \leq k$, the outcome x_i occurs f_i times. Let $n = f_1 + f_2 + \cdots + f_k$ be the total number of occurrences. Table 3 gives the frequency distribution and the relative frequency distribution for this data.

Table 3

Outcome	Frequency	Relative Frequency
x_1	f_1	$\frac{f_1}{n}$
x_2	f_2	$\frac{f_2}{n}$
$\vdots$		
x_k	f_k	$\frac{f_k}{n}$

It is often helpful to use visual means to summarize data. In particular, *histograms* (*bar graphs*) provide an easy way for describing visually a frequency distribution. To construct a histogram corresponding to the data found in Table 2, simply draw the "bars" indicated in Figure 1; the height of each bar corresponds to the frequency of occurrences of the given outcome. Note that each bar corresponds to the left-hand value of the unit interval forming its base.

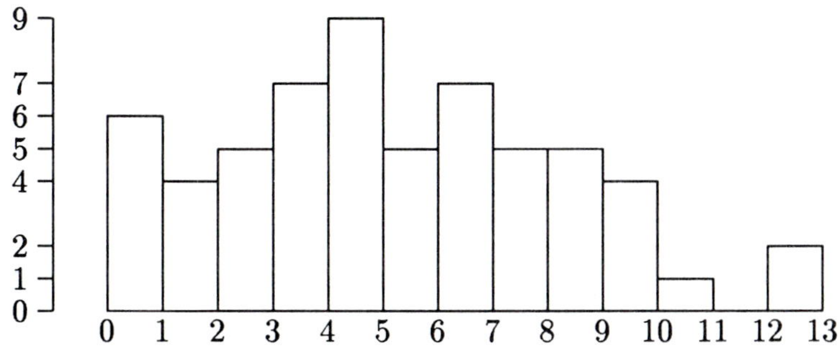

Figure 1

It is often convenient to group the data; this is especially true if there are many outcomes. For instance, for the data in Table 2, we might consider groups or classes having bar width 3, where the interval corresponding to the first bar is $0 \leq x < 3$. That is, we lump all of the occurrences for the outcomes $0, 1, 2$ together; we do the same for outcomes $3, 4, 5$; $6, 7, 8$; $9, 10, 11$; $12, 13, 14$. Table 4 gives the corresponding frequency and relative frequency information.

Table 4: Distribution of Data in Table 2

Interval	Frequency	Relative Frequency
$0 \leq x < 3$	15	$\frac{15}{60}$
$3 \leq x < 6$	21	$\frac{21}{60}$
$6 \leq x < 9$	17	$\frac{17}{60}$
$9 \leq x < 12$	5	$\frac{5}{60}$
$12 \leq x < 15$	2	$\frac{2}{60}$

Figure 2 gives the resulting histogram.

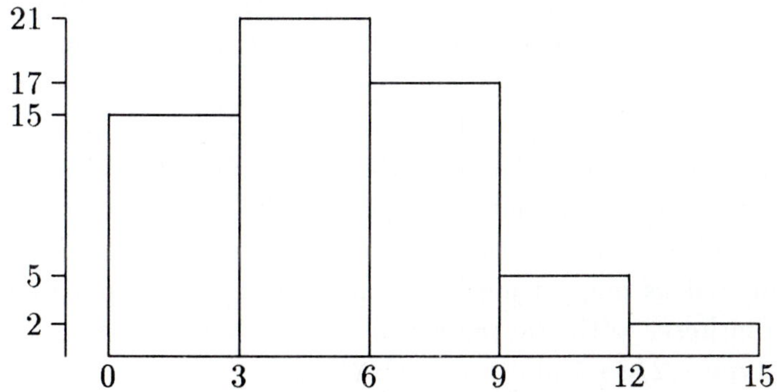

Figure 2

For the example just discussed we can make other choices as well. We can, for instance, use bar width 4, beginning with the interval $-1 \leq x < 3$. Figure 3 gives the resulting histogram.

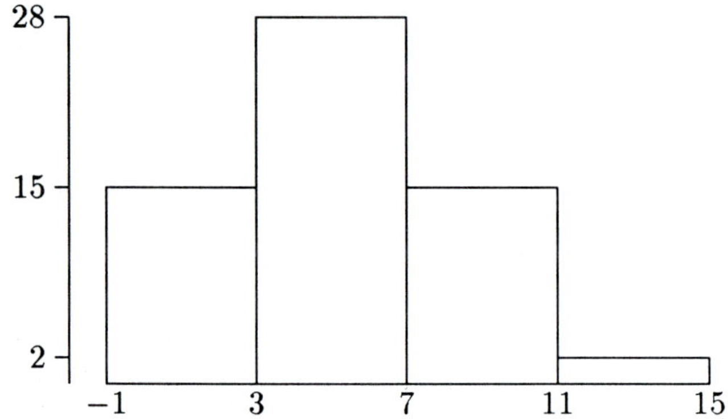

Figure 3

CALIPSO

You draw histograms with CALIPSO by choosing the *STATISTICS* application from the main CALIPSO menu. You then choose *Descriptive* from the *STATISTICS* menu. The *Descriptive* option also calculates mean, standard deviation, and median. We will discuss these concepts in Sections 9.2 and 9.3. You can ignore them for now.

When you press *Enter* with the cursor on *Descriptive* a panel appears that allows you to enter data. You can enter the 60 numbers from Table 1. To do so, you type a number in one field, move to the next field with ↓, and type the next number. If you make a mistake, you can correct it with the *Backspace* key or by returning to any field with an arrow key and overtyping your previous entry. Be sure to type your entries in sequence. If you leave any entry field blank, CALIPSO will assume that is the end of your data set.

Alternatively, you can save typing effort by using the repetition feature. Note for instance that there are six 0's in the data set of Table 2. You can enter six 0's at once by typing $0*6$. Similarly, you can enter five 7's by typing $7 * 5$. The number following the $*$ is a repetition factor; the number before the $*$ is repeated as many times as is specified by the repetition factor. Do not use parentheses with the repetition feature; for instance, to enter $-2/3$ five times, type $-2/3 * 5$ and not $(-2/3) * 5$.

Some of the exercises ask you to process the same data set several times in order to draw several histograms or answer several questions. You can recall the last data set to the data entry panel by pressing F2 when the data entry panel first appears. This saves you the trouble of typing it again.

When you have typed all your data, press *Enter*. The instructions at the top of the screen will change. You now enter the number of classes (intervals) in the histogram. Note that you cannot choose more than 20 classes; this limit is imposed because more intervals than that cannot be drawn legibly on the screen. Next you enter the width of each class interval and the starting value of the first interval. Alternatively, you can let CALIPSO pick the width of the class intervals and the starting value by simply leaving these entries blank. The histogram in Figure 2 was drawn by CALIPSO. To reproduce it, specify 5 classes, a class width of 3, and a starting value of 0. To obtain the histogram in Figure 3 use 4 intervals, a width of 4, and a starting value of -1.

The histogram appears on the screen when you press *Enter*. As advertised on the screen, you can print the histogram by typing a Y. The histogram will not be printed immediately. Rather an entry beginning "Graph Data ..." is made in the worksheet. If you print the worksheet (either immediately or after making more calculations), the histogram will be printed in the proper place in the worksheet. Note that you should be selective about printing histograms since it takes about 3 minutes to print one on a dot matrix printer. Also, recall that you must run the DOS program GRAPHICS before printing any graphics. Usually you run this program every time you start your computer by placing an appropriate entry in the AUTOEXEC.BAT file. If you choose not to do that, you must remember to run it *before* you start CALIPSO. See Appendix A for a complete discussion of printing graphics.

9.1A Exercises: Frequency, Relative Frequency, and Histograms

Each exercise asks you to process the same data set several times. You can do this most easily by using the F2 key to recall the last data set to the data entry panel as we described in the instructions above for using CALIPSO.

For Problems 1 - 3

(a) Set up a frequency distribution table similar to Table 2.

(b) For the given numbers of classes, class widths, and left endpoint, set up frequency distribution tables (similar to Table 4) and draw the corresponding histograms by hand.

(c) Use CALIPSO to check the histograms you drew in (b).

1. 7, 10, 15, 15, 7, 7, 11, 4, 10, 10, 26, 12, 12, 14, 32, 31, 11, 13, 13, 20.

   (a) Use 10 classes of width 3 and left endpoint 3.

   (b) Use 6 classes of width 5 and left endpoint 4.

2. −4, −4, −4, −2, −2, 0, 0, 0, 5, 5, 5, 5, 6, 6, 8, 8, 13, 13, 14, 15, 15, 17, 17, 17, 20, 20, 20.

   (a) Use 7 classes of width 4 and left endpoint −6.

   (b) Use 5 classes of width 5 and left endpoint −5.

3. 12, 14, 17, 28, 28, 36, 36, 36, 42, 51, 52, 57, 57, 79, 83, 83, 112, 120, 120.

   (a) Use 19 classes of width 6 and left endpoint 10.

   (b) Use 11 classes of width 10 and left endpoint 12.

4. For each minute of a one-hour period, a surveyor counted the number of customers waiting at a supermarket checkout counter. Table 5 gives the survey results.

Table 5: Survey of Customers

8	6	9	2	2	6	8	3	8	5
6	3	6	3	3	5	9	5	7	4
2	9	7	5	2	2	7	2	5	2
1	4	5	1	5	9	2	6	4	6
0	7	3	0	7	5	5	4	4	9
5	7	1	6	9	6	3	1	6	1

   (a) Use CALIPSO to draw a histogram of these data that has 5 classes. Choose an appropriate starting interval.

   (b) Draw a histogram with 3 classes. Let CALIPSO choose the bar width and starting interval.

5. A traffic engineer obtains a count of the number of cars that pass through an intersection for 100 one-hour periods. She arranges the data in Table 6. The first row of Table 6 gives the various counts that occurred during the one-hour period. The second row gives the number

of one-hour periods that this same count occurred. For instance, the first column of Table 6 means that 35 cars were counted during 8 one-hour periods.

Table 6

Count	35	36	40	50	55	60	70	75	80	86	90	93	95
Frequency	8	9	9	8	6	4	8	8	6	8	3	9	4

Use CALIPSO to draw a histogram of these data with 7 classes.

Probability Distributions and Random Variables

Random variables play a basic role in probability and statistics.

Definition. A *random variable* is a function (a rule) that assigns a number to each outcome in a sample space S. We often let X denote this function.

Example 1: A Random Variable

Let E be the experiment of tossing a coin 6 times (typical outcomes are HHTHHT, THTHTH, etc.). Let X be the random variable that assigns to each outcome of E the number of heads that occurs. Thus, for instance, if $s =$ HTHHHT, then $X(s) = 4$; and if $s =$ TTTTTT, then $X(s) = 0$.

In what follows we will be interested in the number of times (the frequency) that a random variable X can take on a given value x. For instance, in Example 1, if $x = 4$, then it follows from the counting methods of Chapter 7 that the number of outcomes s with the property that $X(s) = 4$ is $C(6,4) = 15$. You can verify that Table 7 is the frequency distribution table for Example 1 that gives the number of times the random variable X assumes the given value x.

Table 7

x	0	1	2	3	4	5	6
Frequency	1	6	15	20	15	6	1

Of equal interest is the *probability distribution table* associated with a random variable. For a random variable X we let $P(X = x)$ denote the probability that the value of the random variable X is x. For instance, if X is the random variable given in Example 1, then $P(X = 2)$ is the probability, $\frac{C(6,2)}{64} = \frac{15}{64}$, that exactly 2 heads will appear. (Note that $2^6 = 64$ is the number of possible outcomes of the experiment E.) Table 8 gives the frequency and probability distributions for this experiment.

Table 8

x	0	1	2	3	4	5	6
Frequency	1	6	15	20	15	6	1
$P(X = x)$	$\frac{1}{64}$	$\frac{6}{64}$	$\frac{15}{64}$	$\frac{20}{64}$	$\frac{15}{64}$	$\frac{6}{64}$	$\frac{1}{64}$

Definition. A *probability histogram* is a graph of the probabilities that the values x are assumed by a random variable X. For each possible value x you draw a bar whose height is proportional to the probability that x will occur.

We first describe how to draw a probability histogram with CALIPSO and then give such a histogram in Figure 4 for the data in Table 8.

CALIPSO

To use CALIPSO to draw a probability histogram, choose *Probability Histogram* from the *STATISTICS* menu. The data entry screen allows you to enter a value for x and a corresponding probability $P(X = x)$. When you have made all your entries from Table 8 press Enter to see the probability histogram.

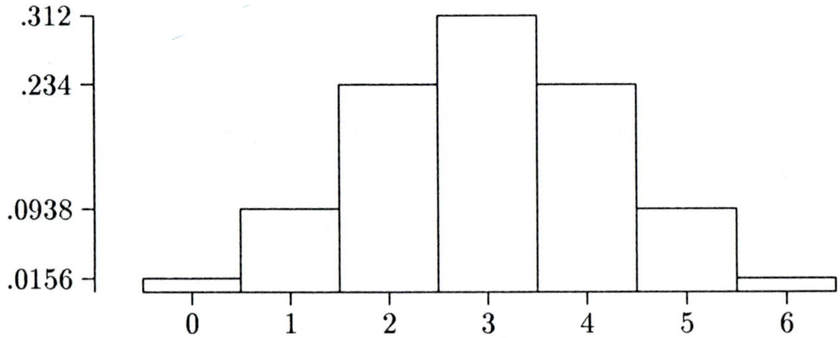

Figure 4

Observe that you can add up appropriate entries in Table 8 (or in the histogram in Figure 4) to calculate probabilities such as $P(3 \leq X \leq 5)$, which is the probability that the random variable X is equal to 3, 4, or 5. This probability is the sum

$$\begin{aligned} P(X = 3) + P(X = 4) + P(X = 5) &= \tfrac{20}{64} + \tfrac{15}{64} + \tfrac{6}{64} = \tfrac{41}{64} \\ &\doteq .312 + .234 + .0938 = .6398 \end{aligned}$$

From Table 8 we see also that $P(X \geq 4) = \frac{22}{64}$ since

$$P(X \geq 4) = P(X = 4) + P(X = 5) + P(X = 6) = \frac{15}{64} + \frac{6}{64} + \frac{1}{64}$$

Example 2: Using CALIPSO to Construct a Probability Distribution Table and Histogram for a Binomial Distribution

Suppose that the probability of making a sales (a success) to a prospective client is $p = 0.3$ (and, consequently, the probability of failing to make a sale is $q = 0.7$). Let $n = 7$ be the number of clients you contact and let X be the random variable that gives the number of successes. To construct the probability distribution table for X you need to find

$$P(X = 0), P(X = 1), \ldots, P(X = 7).$$

CALIPSO will construct Table 9 and the corresponding histogram for you if you choose *Binomial Distribution* from the *STATISTICS* menu. Specify $n = 7, p = 0.3, a = 0$, and $b = 7$. In general, you use any integers a and b satisfying $0 \leq a \leq b \leq n$ to calculate $P(X = a), \ldots, P(X = b)$.

Table 9: Probability Distribution Table

Number of Successes (r)	$P(X = r) = C(7, r)(0.3)^r(0.7)^{7-r}$
0	0.082354
1	0.247060
2	0.317650
3	0.226890
4	0.097240
5	0.025005
6	0.003572
7	0.000219

Note: You may sometimes encounter the terms *probability density function* (or *density function*) associated with a random variable. The probability density function for a random variable X is the function that assigns a probability p_i to each value x_i of X. The probabilities p_i are those found in a probability distribution table.

9.1B Exercises: Probability Distributions; Histograms

You can draw the histograms required in these problems by hand or with CALIPSO. Histograms are not given with the answers because you can easily obtain them with CALIPSO.

1. Draw a probability histogram of the relative frequency distribution in Table 2.

2. Draw a probability histogram of the relative frequency distribution in Table 4.

3. Table 10 gives income levels of the families of children attending an elementary school. Each range of incomes includes the lower limit and excludes the upper limit. Use CALIPSO to present the data in a probability histogram.

Table 10: Income Levels

Group	Income Level	Percent
1	0 – 5,000	1
2	5,000 – 10,000	3
3	10,000 – 15,000	2
4	15,000 – 20,000	8
5	20,000 – 25,000	18
6	25,000 – 30,000	15
7	30,000 – 35,000	21
8	35,000 – 40,000	15
9	40,000 – 50,000	10
10	50,000 – 55,000	6

4. A coin is tossed 5 times. X is the random variable that assigns the number of tails occurring for each outcome.

   (a) Formulate a probability distribution table for X and draw the corresponding histogram.

   (b) Use your table to calculate $P(X < 3)$ and $P(1 \leq X \leq 4)$.

5. § A die is rolled. If the number appearing is an even number, you win \$2. If the number is odd, you win twice the number appearing.

(a) Express this situation in terms of a random variable X. Set up a probability distribution table for X and draw the corresponding histogram.

(b) Use your table to calculate $P(X = 2), P(X \geq 3)$, and $P(2 < X \leq 10)$.

6. A manufacturer knows that 2% of the bicycles it produces are defective. Let X be the random variable that gives the number of nondefective bicycles in a shipment of 10 bicycles.

(a) Set up a probability distribution table for X and draw the corresponding histogram.

(b) Use your table to calculate $P(X = 10), P(2 \leq X \leq 10)$, and $P(X \geq 9)$.

7. § An unfair die is such that

$$P(1) = .1, P(2) = .2, P(3) = .3, P(4) = .2, P(5) = .1, \text{ and } P(6) = .1.$$

Roll the die twice and let X be the random variable that assigns the sum of the numbers appearing.

(a) Set up the probability distribution table for X and draw the corresponding histogram.

(b) Use your table to calculate $P(X = 7 \text{ or } 11), P(X \geq 10)$, $P(4 < X < 9), P(X \leq 5)$.

8. Suppose you take a multiple choice test consisting of 20 questions. Each question has 5 possible responses, exactly 1 of which is correct. Assume that you failed to study for the test and must guess the answer to each question. Let X be the random variable that assigns the number of answers you get right (out of 20 questions).

(a) Set up a probability distribution table and draw the corresponding histogram.

(b) Use your table to calculate $P(X = 4), P(X \leq 2), P(X \geq 7)$, and $P(3 < X < 8)$.

9. § An urn contains 3 red, 2 blue, and 5 orange balls. An experiment consists of drawing 3 balls at random, observing the number of balls of each color, and returning the balls to the urn.

(a) Determine the probability distribution of this experiment.

(b) Draw the probability histogram.

10. A marksman can hit a beer can with probability $\frac{1}{4}$. He shoots until he hits the can or takes 4 shots.

    (a) Determine the probability distribution of the number of shots taken.

    (b) What is the probability that the marksman fails to hit the can in 4 attempts?

11. § An experiment consists of repeated trials in which 7 cards are drawn from a 52-card deck and the number of hearts is recorded. After each trial the cards are returned to the deck and the deck is shuffled. Draw a frequency histogram for the experiment.

9.2 FINDING THE CENTER: MEDIANS, MEANS, AND EXPECTED VALUES

To a large extent statistics involves the study of populations. A *population* is a set of objects. Examples of populations include the set of all Americans, the set of white mice used in an experiment, the set of new car buyers in New York, the set of chairs produced by a manufacturer, etc.

If a population is large, it may be impossible to gather data about each member of the population. Instead, use is made of samples of the population; a *sample* is a subset of a population. A *statistic* is a numerical measurement of a sample. A *parameter* is a numerical measurement of a population. In this and subsequent sections we will examine a variety of statistics and parameters. We begin with the mean of a set of numbers.

Definition. The *mean* (or *average*) $\bar{x}$ of a set of n numbers $x_1, x_2, \ldots, x_n$ is defined by

$$\bar{x} = \frac{x_1 + x_2 + \cdots + x_n}{n} \tag{1}$$

Example 1: Mean of a Set of Numbers

The mean of the numbers $3, 10, 5, 3, 7, 7, 7, 16, 19$ is

$$\bar{x} = \frac{3 + 10 + 5 + 3 + 7 + 7 + 7 + 16 + 19}{9} \doteq 8.5556$$

In Example 1 observe that we added three 7's; obviously the same result is attained if we multiply 7 by 3. Thus, to compute the mean of a set of numbers we can first form the frequency distribution of the numbers and then use the formula

$$\bar{x} = \frac{f_1x_1 + f_2x_2 + \cdots + f_kx_k}{n} \tag{2}$$

where $x_1, x_2, \ldots, x_k$ is the complete list of the different numbers in the set, f_1 is the frequency that the number x_1 occurs, f_2 is the frequency that the number x_2 occurs, etc; n is the total number of entries in the set.

Example 2: Calculating a Mean from a Frequency Distribution

The frequency distribution of the set of numbers $4, 7, 7, 10, 3, 3, 3, 8, 10, 7, 2, 2, 3, 3, 2$ is

Number	2	3	4	7	8	10
Frequency	3	5	1	3	1	2

Using (2) we find that the mean of this set of numbers is

$$\bar{x} = \frac{3 \cdot 2 + 5 \cdot 3 + 1 \cdot 4 + 3 \cdot 7 + 1 \cdot 8 + 2 \cdot 10}{15} \doteq 4.9333$$

You should distinguish between the mean of a sample and the mean of an entire population. The *mean of a sample* (sometimes referred to as the *sample mean*) is given by (1) or (2) where the values x_i have been assigned to members of some sample of a population. It is often impossible to find the mean of a population because of the size of the population. The formula for the population mean is the same, though it is usually denoted by the Greek letter μ instead of $\bar{x}$. In many instances a sample mean $\bar{x}$ is used to estimate the population mean μ, where, of course, the sample is drawn from the population whose mean you want to find.

The median provides another way of describing the "center" of a set of numbers.

Definition. The *median* of a set of numbers that are listed in increasing (or decreasing) order is the middle entry if the number of entries is odd, and is the mean of the two middle entries if the number of entries is even.

Example 3: The Median of a Set of Numbers

The median of the set of numbers $3, 5, 5, 8, 11, 13, 13$ is the middle number 8. The median of the set of numbers $6, 6, 8, 13, 14, 17, 20, 22, 22, 22$ is the mean, 15.5, of the two middle numbers 14 and 17.

Related to the idea of a mean or an average is the notion of *expected value.* To motivate this concept suppose you are considering playing the following game. A hat contains 24 cards. Six of the cards are imprinted with \$20, 4 with \$−24, 8 with \$15, and 6 with \$−12. You pay \$5 to draw a card from the hat and either win the positive amount shown or lose the negative amount shown on the card you draw. Is it advantageous to play this game? To answer this question we calculate the expected value of the game and compare it to what it costs to play. The expected value is essentially what on average you expect to win (or lose) each time you play. We assume the game is "fair"; that is, the cards are well shuffled and the player cannot see a card before it is drawn.

Since there are 24 cards, 6 of which are \$20 cards, you expect to draw a \$20 card 1 out of every 4 times. Therefore, it is reasonable to expect a gain of $20(\frac{1}{4})$ each time you play as the \$20 card's "contribution" to your winnings. Similarly, since you expect to draw a \$−24 card one out of every 6 times, it is reasonable to expect a loss of $24(\frac{1}{6})$ each time you play as the "contribution" of \$−24 cards to the game. Applying similar reasoning to the other 2 kinds of cards, you expect that "on average" the payoff for each play is

$$20 \cdot \frac{1}{4} - 24 \cdot \frac{1}{6} + 15 \cdot \frac{1}{3} - 12 \cdot \frac{1}{4} = 3 \tag{3}$$

Since you must pay \$5 to enter the game it is not worth your while to play because on the average you will lose \$2 each time you play.

In a slightly more formal way we can view this game in terms of a random variable X that takes on values x equal to $20, -12, 15, -24$ and having probablity distribution

x	20	-12	15	-24
$P(X=x)$	$\frac{1}{4}$	$\frac{1}{4}$	$\frac{1}{3}$	$\frac{1}{6}$

Note that we can write the expected value of the game [given by (3)] as

$$x_1 \cdot P(X=x_1) + x_2 \cdot P(X=x_2) + x_3 \cdot P(X=x_3) + x_4 \cdot P(X=x_4)$$

where the values x_i and $P(X=x_i)$ come from the probability distribution table. This idea leads to the following important definition.

Definition. If X is a random variable with possible values $x_1, x_2, \ldots, x_k$ and if $P(X=x_1)=p_1, P(X=x_2)=p_2, \ldots, P(X=x_k)=p_k$, then the *expected value* of X is

$$E(X) = x_1p_1 + x_2p_2 + \cdots + x_kp_k$$

Example 4: The Expected Value of a Random Variable

The expected value of the random variable X having probability distribution

x	7	-3	10	2	5
$P(X=x)$	.2	.1	.2	.4	.1

is $E(X) = 7(.2) + (-3)(.1) + 10(.2) + 2(.4) + 5(.1) = 4.4$.

To find the expected value of a random variable when you are given frequencies instead of probabilities you first convert the frequencies to probabilities.

Example 5: Finding the Expected Value from Frequencies

Table 1 gives the frequency that scores between 1 and 10 were attained on a nationwide test given to 35,000 students.

Table 1: Test Scores for 35,000 Students

Scores	1	2	3	4	5	6	7	8	9	10
Frequency	956	1845	2671	4321	5769	6003	5812	4211	2250	1162

Let X be the random variable that assigns the scores obtained on the test. From Table 1 we see that the expected value $E(X)$ (the average test score) is

$$\begin{aligned} E(X) = \ & 1 \cdot \tfrac{956}{35{,}000} + 2 \cdot \tfrac{1845}{35{,}000} + 3 \cdot \tfrac{2671}{35{,}000} + 4 \cdot \tfrac{4321}{35{,}000} \\ & + 5 \cdot \tfrac{5769}{35{,}000} + 6 \cdot \tfrac{6003}{35{,}000} + 7 \cdot \tfrac{5812}{35{,}000} + 8 \cdot \tfrac{4211}{35{,}000} \\ & + 9 \cdot \tfrac{2250}{35{,}000} + 10 \cdot \tfrac{1162}{35{,}000} \doteq 5.74 \end{aligned}$$

In standard statistics texts it is shown that if X is random variable for a binomial distribution, where X assigns the number of successes in a binomial experiment consisting of n trials and probability of success p (see Section 8.4), then

$$E(X) = np \tag{4}$$

Example 6: Expected Value of a Binomial Distribution

Suppose you perform a binomial experiment with $n = 30$ trials and probability $p = 0.4$ of success on each trial. Let X be the random variable that gives the number of successes each time you perform the experiment. Then $E(X) = np = 0.4 \cdot 30 = 12$ is the expected number of successes when you perform the experiment.

CALIPSO

You can use *Calculator* from the *STATISTICS* menu to calculate means and expected values. For instance, to calculate the mean of the data in Example 2, you enter

$$(3 * 2 + 5 * 3 + 1 * 4 + 3 * 7 + 1 * 8 + 2 * 10)/15$$

You can also use the *Descriptive* option from the *STATISTICS* menu to calculate the mean and the median automatically. You enter data as you did in Section 9.1 to draw histograms. The mean, median, and standard deviation appear in the worksheet. (We discuss the standard deviation in the next section.)

9.2A Exercises: Calculating Means, Medians, and Expected Values

In Problems 1 – 3 find the mean and median of the given sets of numbers.

1. § 4, 6, 19, 23, 23, 32, 35, 35, 35, 49.

2. −20, −10, −10, 0, 0, 4, 4, 4, 6, 6, 6, 6, 10, 12, 16, 16.

3. § 71, 123, 24, 16, 71, 71, −80, 24, 24, −14, 16, 16, 24, 16.

In Problems 4 – 6 find the mean and median of the numbers found in the distribution tables. If you use CALIPSO's *Descriptive* option, recall that you use the repetition factor to denote the frequency of a number; for instance, to enter −2/3 with frequency 15, type −2/3 ∗ 15. [Do not enter (−2/3) ∗ 15 as you would do to multiply −2/3 by 15.]

4.

Number	4	8	10	22	25	40
Frequency	3	7	1	5	10	4

5. §

Number	$-\frac{2}{3}$	2	$\frac{6}{5}$	4	5
Frequency	15	29	50	3	16

6.

Number	22	30	76	120	152
Frequency	245	123	82	17	21

In Problems 7 – 10 find $E(X)$ corresponding to the given probability distribution tables. You can use CALIPSO's *Calculator* or a hand calculator.

7. §

x	12	−12	4	6	10
$P(X = x)$	.2	.05	.3	.15	.3

8.

x	−10	−6	0	3	4	20
$P(X = x)$	.4	.05	.1	.25	.05	.15

9. §

x	5	8	15	20	25
Frequency	17	5	10	6	12

10.

x	−12	−8	−4	0	6	10
Frequency	15	5	15	20	10	10

11. § You play the following simple game. You roll a die. If an even number appears you win 3 times (in dollars) the amount shown; if an odd number appears, you lose twice the amount shown. What is the expected value of this game? If it costs $7 to play, would you play?

12. You roll 2 dice. If the sum of the dice is odd, you win 5 times the amount of the sum (in dollars); if the sum is even you lose 4 times the sum. Find the expected value of this game. If it costs $5 to play, would you play?

13. § If your probability of making a free throw in basketball is .8, what is the expected number of made free throws if you shoot 80 times?

14. You take a true-false test consisting of 20 questions. If you make the correct answer you get 2 points; if you make the wrong answer you lose a point. Assuming you try to answer each question but are always guessing the answer, what is your expected score?

9.3 MEASURING THE SPREAD: VARIANCE AND STANDARD DEVIATION

Though the mean and (and median) can provide useful information they often do not give a very complete picture of the given data. For instance, the mean of the numbers $8, 8, 8, 9, 10, 10, 10$ is the same as the mean of the numbers $-154, -30, -30, 13, 20, 36, 115$, yet clearly the range of these sets of numbers is quite different. To help identify differences between such sets of numbers we can use other statistics such as the variance and the standard deviation. Both of these statistics essentially indicate the amount of spread of the numbers from the mean.

Definition. Let $x_1, x_2, \ldots, x_n$ be a set of n values with mean μ. Then the *variance* σ^2 of these values is defined by

$$\sigma^2 = \frac{(x_1 - \mu)^2 + (x_2 - \mu)^2 + \cdots + (x_n - \mu)^2}{n} \tag{1}$$

and the *standard deviation* of these values is $\sigma = \sqrt{\sigma^2}$.

The standard deviation of a set of values is simply the square root of the variance. Note that both of these parameters measure in some way to what extent the given values vary from the mean. A small variance (or standard deviation) indicates that the values tend to be close to the mean; the larger the variance (or standard deviation) the more some or all of the values tend to be far from the mean.

Example 1: Calculating a Variance and Standard Deviation

The mean of the values $-6, 3, 4, 4, 9, 10$ is $\mu = 4$. The variance is

$$\sigma^2 = \frac{(-6-4)^2 + (3-4)^2 + (4-4)^2 + (4-4)^2 + (9-4)^2 + (10-4)^2}{6} = 27 \tag{2}$$

and the standard deviation is $\sigma = \sqrt{27} \doteq 5.196$.

If the data are given in a frequency format (that is, value x_1 occurs with frequency f_1, x_2 occurs with frequency $f_2, \ldots, x_k$ occurs with frequency f_k), then it is easy to verify that the following formula for the variance σ^2 is valid

$$\sigma^2 = \frac{f_1(x_1-\mu)^2 + f_2(x_2-\mu)^2 + \cdots + f_k(x_k-\mu)^2}{n}$$

where μ is the mean of the data and $n = f_1 + f_2 + \cdots + f_k$. As before, the standard deviation is the square root of the variance.

Example 2: Calculating the Variance and Standard Deviation of Data

The mean of the data

Value	-3	-1	15	7	10
Frequency	7	2	1	4	6

(3)

is $\mu = 4$. The variance is

$$\sigma^2 = \frac{7(-3-4)^2 + 2(-1-4)^2 + 1(15-4)^2 + 4(7-4)^2 + 6(10-4)^2}{20} \doteq 39.95$$

and the standard deviation is $\sigma = \sqrt{39.95} \doteq 6.1887$.

For reasons made clear in more advanced statistics courses, the variance of a sample is defined in a slightly different way than the variance of data arising from an entire population. The variance (and standard deviation) of data derived from a population is that given by (1) and (2). However, for a sample we have the following definition.

Definition. Let $x_1, x_2, \ldots, x_n$ be a sample of n values and let

$$\bar{x} = \frac{x_1 + x_2 + \cdots + x_n}{n}$$

be the sample mean of these values. The *sample variance* s^2 of this sample is defined by

$$s^2 = \frac{(x_1 - \bar{x})^2 + (x_2 - \bar{x})^2 + \cdots + (x_n - \bar{x})^2}{n - 1}$$

and the *sample deviation* is $s = \sqrt{s^2}$.

Next we define the variance and standard deviation of a probability distribution.

Definition. Let X be a random variable that assumes values $x_1, x_2, \ldots, x_n$ with corresponding probabilities $p_1, p_2, \ldots, p_n$. (That is, $P(X = x_i) = p_i$, for $i = 1, 2, ..., n$.) Let μ be the mean (expected value) of this probability distribution. Then the *variance* σ^2 is defined by

$$\sigma^2 = (x_1 - \mu)^2 p_1 + (x_2 - \mu)^2 p_2 + \cdots + (x_n - \mu)^2 p_n$$

and the *standard deviation* is $\sigma = \sqrt{\sigma^2}$.

Example 3: Calculating the Variance and Standard Deviation of a Probability Distribution

The mean of the probability distribution for a random variable X

x	-5	4	10	7
$P(X = x)$	.40	.25	.15	.20

is $\mu = -5(.40) + 4(.25) + 10(.15) + 7(.20) = 1.9$. The variance is $\sigma^2 = (-5 - 1.9)^2(.40) + (4 - 1.9)^2(.25) + (10 - 1.9)^2(.15) + (7 - 1.9)^2(.20) = 35.19$ and the standard deviation is $\sigma = \sqrt{35.19} \doteq 5.9321$.

CALIPSO

You can use CALIPSO's *Calculator* to make the calculations in the examples above. You enter expressions in the same way as we did in those examples except for the operation of squaring. To square a number you use the exponent operator ˆ. For instance, to calculate 0.5^2 you enter 0.5ˆ2.

It is much easier to use the *Descriptive* option to make these calculations, though you will not get practice using the formulas if you do so. To use this feature, select *Descriptive* from the *STATISTICS* menu. Enter your data in the table. If you are using a frequency distribution, recall that you enter multiple copies of a datum with the $*$ operation. For instance, to enter 9 copies of 6, type $6 * 9$.

Example 4: Using CALIPSO to Calculate Variations and Standard Deviations

We illustrate a CALIPSO calculation with the data (3) from Example 2. After choosing *Descriptive*, enter the data in the table as

$$\begin{array}{l} -3 * 7 \\ -1 * 2 \\ 15 \\ 7 * 4 \\ 10 * 6 \end{array}$$

Press [Enter]. Since *Descriptive* always draws a histogram, you are asked for specifications for it. The population mean and standard deviation and the sample mean and standard deviation appear in the worksheet.

To use *Calculator* to compute the population mean you type an expression like

$$(7(-3) + 2(-1) + 15 + 4 * 7 + 6 * 10)/20$$

The answer 4 appears in the worksheet. To compute the population standard deviation, type

$$((7(-3-4)\hat{}2+2(-1-4)\hat{}2+1(15-4)\hat{}2+4(7-4)\hat{}2+6(10-4)\hat{}2)/20)\hat{}(1/2)$$

Example 5: The Mean and Standard Deviation of a Binomial Distribution

In statistics texts it is shown that the mean and standard deviation of a binomial distribution (n trials and probability p of success) are given by $\mu = np$ and $\sigma = \sqrt{np(1-p)}$. For instance, if $n = 40$ and $p = .2$, then $\mu = 40(.2) = 8$ and $\sigma = \sqrt{40(.2)(.8)} = 2.530$.

The Chebychev inequality gives an estimate of the probability that an outcome lies within a given distance from the mean. You can find a proof of the validity of this inequality in most statistics texts.

Chebychev Inequality. Suppose that μ is the mean and σ is the standard deviation of a probability distribution. Let c be a positive number. Then the probability that an outcome x lies between $\mu - c$ and $\mu + c$ is at least $1 - \frac{\sigma^2}{c^2}$. That is, the Chebychev inequality

$$P(\mu - c \leq x \leq \mu + c) \geq 1 - \frac{\sigma^2}{c^2}$$

is valid.

Example 6: Applying the Chebychev Inequality

(a) Suppose a probability distribution has mean $\mu = 4$ and standard deviation $\sigma = 2$. Then the probability that an outcome x lies between $4 - 3 = 1$ and $4 + 3 = 7$ is at least $1 - \frac{2^2}{3^2}$. That is,

$$P(1 \leq x \leq 7) \geq 1 - \frac{2^2}{3^2} = \frac{5}{9}$$

(b) Suppose that a certain brand of tires has a tread-life expectancy of 40,000 miles with standard deviation 2000. Using the Chebychev inequality we can estimate the probability that the tread life of a tire is between 37,000 and 43,000 miles to be $1 - \frac{2000^2}{3000^2} \doteq .555556$.

9.3A Exercises: Calculating Variances and Standard Deviations

For Problems 1 – 4, calculate:

(a) The sample mean ($\bar{x}$), sample variance (s^2), and sample standard deviation (s) of the given data.

(b) The population mean (μ), population variance (σ^2), and population standard deviation (σ) of the given data.

1. § $-2, 3, 3, 5, 6$.

2. $-6, -10, 5, 2, 22, 5$.

3. $-5, 5.2, -3.4, 6.1, 7.3$.

4. § $12, 12, 15, 12, 11, 9, 8, 7, 5, 1$.

5. $9, 10, 10, 10, 11, 11, 11, 11, 13, 14$.

6. $.7, 1.2, 3.6, 4.4, 5.1$.

7. § Given a probability distribution with mean $\mu = 5$ and standard deviation $\sigma = 1.5$ use the Chebychev inequality to estimate the probability that an outcome lies between 3 and 7.

8. Suppose a probability distribution has mean $\mu = -3$ and standard deviation $\sigma = 2$. Use the Chebychev inequality to estimate the probability that an outcome is less than -6 or greater than 0.

9. § Suppose a probability distribution has mean $\mu = 10$ and standard deviation $\sigma = 3$. Find the number c such that the probability that an outcome lies between $10 - c$ and $10 + c$ is at least $\frac{2}{3}$.

10. Suppose that a certain toy will on the average endure 200 hours of use with standard deviation 8. Use the Chebychev inequality to estimate the probability that a toy holds up between 175 and 225 hours.

11. § Find the mean and standard deviation of a binomial distribution with $n = 100$, $p = .35$.

12. Find the mean and standard deviation of a binomial distribution with $n = 120$, $p = .46$.

13. § Find the mean and standard deviation of a binomial distribution with $n = 2000$, $p = .25$.

14. § Often an alternate formula, $\sigma^2 = E(X^2) - \mu^2$, is also given to simplify calculations of the variance σ^2 of a random variable X. Apply this formula to Problems 1 – 4 to calculate σ^2. For instance, the random variable in Problem 1 takes on the values $-2, 3, 3, 5, 6$ and $E(X^2) = \frac{(-2)^2+3^2+3^2+5^2+6^2}{5}$. Answers to Problems 1 and 3 are given.

9.4 NORMAL DISTRIBUTIONS

Thus far we have dealt with what are often referred to as *discrete random variables.* These are random variables that take on only a finite number of values. In practice, however, it is frequently necessary to work with random variables that can assume an infinite number of values. For instance, if X is a random variable that gives the average number of bushels of wheat per acre on an Iowa farm, then X can take on an infinite number of values such as 360.15, 324.238, etc. Such a random variable is called a *continuous random variable.*

A moment's thought should indicate to you why we cannot use probability histograms to graph the probabilities of continuous random variables. However, we are able to associate curves with these variables and use areas under the curves to calculate probabilities.

We will leave it to you to pursue in more advanced courses the general theory underlying continuous random variables. In this section we will focus our attention on attributes of an important class of these variables.

Data obtained from studying IQ's, product lifetimes, weights of humans, etc., often lead to what are called *normal distributions* or *normal random variables.* Typically associated with such a distribution is a bell-shaped curve, called a *normal curve.* Figure 1 illustrates one such curve.

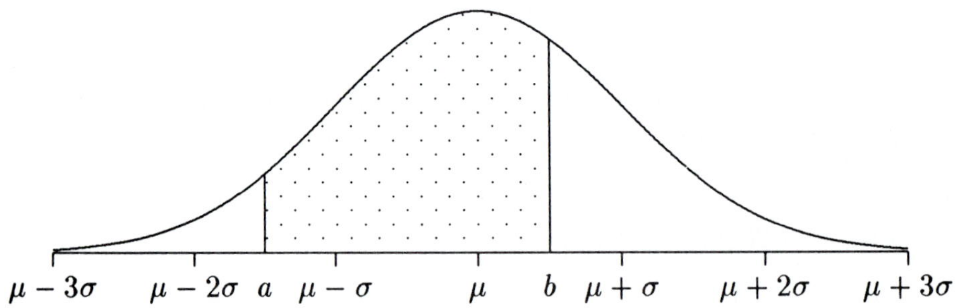

Figure 1

Though we will not make use of this fact, a normal curve is the graph of the function $y = \frac{1}{\sigma\sqrt{2\pi}}e^{-\frac{(x-\mu)^2}{2\sigma^2}}$ where μ is the mean of the normal distribution and σ is the standard deviation.

The usefulness of this curve is derived from the following properties. If the curve corresponds to a normal distribution X with mean μ and standard deviation σ, then it can be shown that

1. The area under the curve is 1.
2. The curve is symmetric about the vertical line $x = \mu$.
3. The probability that X is between a and b, $P(a \leq X \leq b)$, is the area under the curve between the vertical lines $x = a$ and $x = b$.
4. The probability that X is greater than b, $P(X \geq b)$, is the area under the curve to the right of the vertical line $x = b$. The probability that X is less than a, $P(X \leq a)$, is the area under the curve to the left of the vertical line $x = a$.

Note that it follows from 3 that the probability that X is equal to any particular number is 0, since the area of a vertical line segment is 0.

Example 1: Areas Under Normal Curves

Let X be a normal random variable with mean $\mu = 50$ and standard deviation $\sigma = 10$. The areas of the shaded regions in Figure 2 are the probabilities that X is in the indicated ranges.

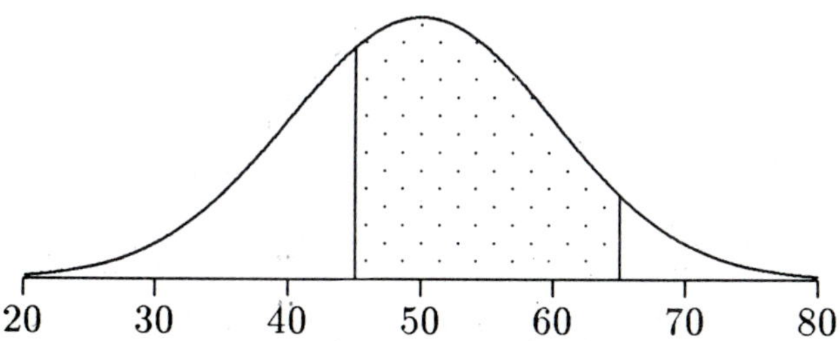

Figure 2a: $P(45 \leq X \leq 65)$

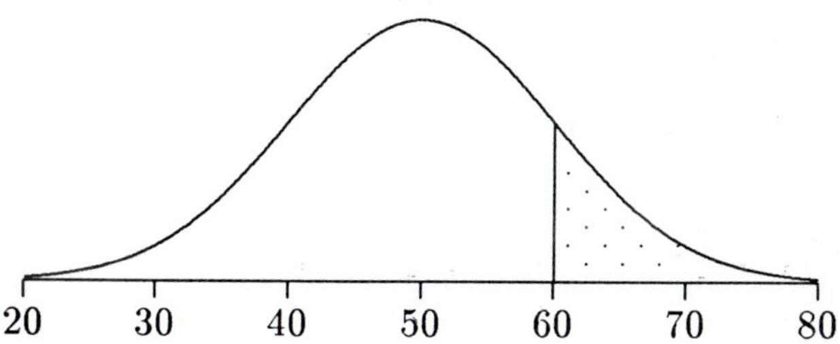

Figure 2b: $P(60 \leq X)$

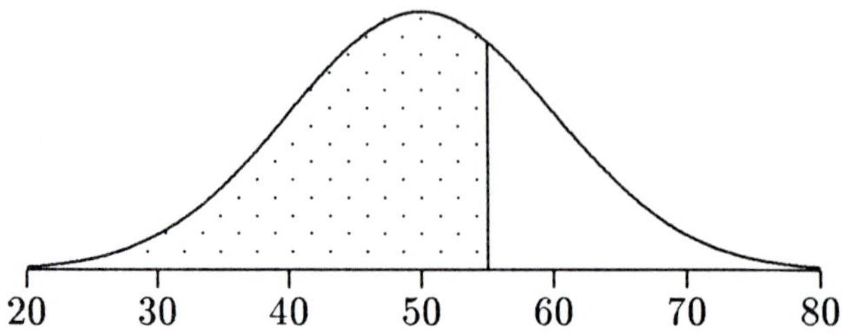

Figure 2c: $P(X \leq 55)$

The obvious question now is how to calculate the areas of particular regions under a normal curve. CALIPSO provides an easy means for making these calculations.

CALIPSO

To calculate a probability such as $P(a \leq X \leq b)$ for a random variable X with mean μ and standard deviation σ go to the *STATISTICS* menu and select *Normal Distribution*. You are asked to enter the mean (μ), the standard deviation (σ), the value of a, and the value of b. CALIPSO draws the normal distribution curve and shades the area between a and b. The probability $P(a \leq X \leq b)$ is displayed under the curve and is also given in the worksheet for later reference.

Example 2: Using CALIPSO

Suppose the average speed of cars traveling along a portion of Highway 101 is 62 mph and the standard deviation is 3.6. Find

(a) The probability that a car is traveling between 63 and 67 mph.

(b) The probability that a car is traveling faster than 60 mph.

(c) The probability that a car is traveling slower than 70 mph.

Solution. Execute *Normal Distribution* with $\mu = 62, \sigma = 3.6$ and with

(a) $a = 63, b = 67$ to find that $P(63 \leq X \leq 67) \doteq .3081$;

(b) $a = 60$ and b blank (which effectively sets it to ∞) to find $P(60 \leq X) \doteq .7107$;

(c) a is blank (which effectively sets it to $-\infty$) and $b = 70$ to get $P(X \leq 70) \doteq .9869$.

Using Tables to Find Areas Under a Normal Curve

Although CALIPSO gives a particularly easy means of finding probabilities associated with normal distributions, from time to time you may need to use tables for these calculations. Tables generally are given for a *standard normal distribution*, which is a normal distribution with mean $\mu = 0$ and standard deviation $\sigma = 1$. Table 1, located at the end of this section, is a small version of such a table. As indicated in the accompanying figure, Table 1 gives the values $A(z) = P(Z \leq z)$ where Z is a standard normal distribution and $A(z)$ is the area under the curve to the left of the vertical line $x = z$.

Example 3: Using Table 1 for Standard Normal Distributions

We use Table 1 to calculate the following probabilities for a standard normal distribution Z.

- To find $P(Z \leq 1.15)$ you simply locate 1.15 in the z column and see that $P(Z \leq 1.15) = A(1.15) \doteq .8749$.
- To find $P(-.35 \leq Z \leq 1.80)$ observe that the corresponding area under the standard normal curve is equal to $A(1.80) - A(-.35)$, which from Table 1 is $.9641 - .3632 = .6009$.
- To find $P(Z \geq -.65)$, note that the corresponding area under the curve is equal to $1 - A(-.65) \doteq 1 - .2578 = .7428$.

You can also use Table 1 to find probabilities associated with nonstandard normal distributions. Suppose that X is a normal random variable with mean μ and standard deviation σ. To find $P(a \leq X \leq b)$ observe that the following inequalities are equivalent:

$$\begin{array}{ccccc} a & \leq & X & \leq & b \\ a-\mu & \leq & X-\mu & \leq & b-\mu \\ \dfrac{a-\mu}{\sigma} & \leq & \dfrac{X-\mu}{\sigma} & \leq & \dfrac{b-\mu}{\sigma} \end{array}$$

It can be shown that the continuous random variable $Z = \frac{X-\mu}{\sigma}$ is a standard normal variable and so you can use Table 1 to find

$$P\left(\frac{a-\mu}{\sigma} \leq Z \leq \frac{b-\mu}{\sigma}\right)$$

which is equal to $P(a \le X \le b)$.

Moreover, using the same arguments you can show that if X is distributed normally with mean μ and standard deviation σ, then

$$\begin{aligned} P(X \le b) &= P\left(\tfrac{X-\mu}{\sigma} \le \tfrac{b-\mu}{\sigma}\right), \text{and} \\ P(a \le X) &= 1 - P(X \le a) = 1 - P\left(\tfrac{X-\mu}{\sigma} \le \tfrac{a-\mu}{\sigma}\right) \end{aligned}$$

As before, $Z = \frac{X-\mu}{\sigma}$ is a standard normal distribution.

Example 4: Using Table 1 to Find Probabilities of Nonstandard Normal Distributions

Let X be a normal random variable with mean $\mu = 3$ and standard deviation $\sigma = 2$.

- Find $P(1.5 \le X \le 4.2)$.

  $$P(1.5 \le X \le 4.2) = P\left(\frac{1.5-3}{2} \le Z < \frac{4.2-3}{2}\right) = P(-.75 \le Z \le .6)$$

  where $Z = \frac{X-3}{2}$ is a standard normal random variable. From Table 1 we find

  $$\begin{aligned} P(1.5 \le X \le 4.2) = P(-.75 \le Z \le .6) &= A(.6) - A(-.75) \\ &\doteq .7257 - .2266 = .4991. \end{aligned}$$

- Find $P(X \le -.65)$.

  $$P(X \le -.65) = P\left(Z \le \frac{-.65-3}{2}\right) = P(Z \le -1.825)$$

  where Z is a standard normal random variable. From Table 1 we find (with interpolation)

  $$P(X \le -.65) = P(Z \le -1.825) = A(-1.825) \doteq .0340$$

- Find $P(2.3 \le X)$.

  $$P(2.3 \le X) = P\left(\frac{2.3-3}{2} \le \frac{X-3}{2}\right) = P(-.35 \le Z) = 1 - P(Z \le -.35)$$

  It follows from Table 1 that

  $$P(2.3 \le X) = 1 - P(Z \le -.35) = 1 - A(-.35) \doteq 1 - .3632 = .6368$$

Normal distributions are often used to approximate discrete distributions. We study one kind of such approximations in the next section. Example 5 examines another kind of approximation of a discrete distribution by a continuous distribution.

Example 5: Analyzing the Distribution of Test Scores

It is common to assume that scores on a test such as the ACT are normally distributed by a continuous random variable. Actually, the test scores assume only integer values and are not continuously distributed; however, little error is introduced by assuming a continuous distribution when the number of scores being analyzed is very large. The assumption does lead to some curious conclusions. If X gives the test scores, then $P(X \geq 90) = P(X > 90)$ because the area below the normal curve and to the right of $X = 90$ is the same whether the boundary line $X = 90$ is included in the area or not. Of course, the only way you can get a score X greater than 90 is to get at least a 91, and the actual probability $P(X > 90)$ is strictly less than the probability $P(X \geq 90)$.

Suppose the ACT scores for a given year are normally distributed with mean 72 and standard deviation 12. What is the probability that a score lies between 76 and 85? This question is vaguely stated since it is not clear if 76 and 85 are to be included in the range or not. However, that does not matter if we are assuming a continuous distribution of scores since the area under the normal curve between 76 and 85 is the same whether we include the endpoints or not. By using Table 1 or *Normal Distribution* from the *STATISTICS* menu and entering $\mu = 72, \sigma = 12, a = 76$, and $b = 85$ you will find that $P(76 \leq X \leq 85) \doteq .22997$.

9.4A Exercises: Normal Distributions

In Problems 1–4 suppose that Z is a standard normal distribution. Use Table 1 to find the given probabilities.

1. § $P(Z \leq -.80)$, $P(Z \geq -.80)$, $P(-.80 \leq Z \leq 1.20)$

2. $P(1.10 \leq Z)$, $P(Z \leq 1.65)$, $P(-.45 \leq Z \leq 3.30)$

3. § $P(Z \leq -2.55)$, $P(Z \geq -2.55)$, $P(-2.55 \leq Z \leq 1.20)$

4. $P(1.60 \leq Z)$, $P(Z \geq 1.65)$, $P(1.45 \leq Z \leq 3.50)$

In Problems 5–9 you are given the mean μ and standard deviation σ of a normal distribution X. Use Table 1 and the method of Example 2 to calculate the given probabilities. Then check your answers with CALIPSO's *Normal Distribution* option.

5. § $\mu = 6, \sigma = 2.1$. Find $P(5 \leq X \leq 8), P(X \geq 4), P(X \leq 5)$.

6. $\mu = -5, \sigma = 1$. Find $P(-3 \leq X \leq 9), P(X \geq -1), P(X \leq -7)$.

7. § $\mu = 120, \sigma = 12$. Find $P(110 \leq X \leq 115), P(X \geq 105), P(X \leq 120)$.

8. $\mu = 50, \sigma = 15$. Find $P(12 \leq X \leq 18), P(X \geq 10), P(X \leq 80)$.

9. § $\mu = 20, \sigma = 6$. Find $P(10 \leq X \leq 15), P(X \geq 25), P(X \leq 12)$.

10. If X is a normal distribution with $\mu = 7$ and $\sigma = 2$, find the probability that X is greater than 9 or less than 6; that is, find

$$P(X \geq 9 \text{ or } X \leq 6).$$

Solve the problem first with Table 1 and then with CALIPSO.

11. § Suppose that SAT scores are normally distributed with mean 480 and standard deviation 92.

    (a) What is the probability that a score is less than or equal to 440?

    (b) What is the probability a score lies between 400 and 500?

    (c) What percentage of the students taking the test will score better than 650?

12. Suppose that the number of raisins in a box of cereal is normally distributed with $\mu = 25$ and $\sigma = 2.3$. [Note that this is an unrealistic assumption because it assumes that the number of raisins can be any real number. In the next section we discuss approximating a random variable that assumes discrete values (in the case of numbers of raisins, integer values) with a continuous distribution.]

    (a) What is the probability that a box will contain 27 or more raisins? More than 27 raisins?

    (b) What is the probability that a box will contain fewer than 21 raisins?

(c) What is the probability that a box will contain exactly 25 raisins? How do you explain your answer?

13. § The quantity of oil that Quik-Lube uses each month is normally distributed with $\mu = 3500$ and $\sigma = 125$. If Quik-Lube has 3800 quarts on hand at the beginning of the month, find the probability that it will not need to order more oil before the end of the month.

14. A grazing field contains about 1000 grass plants per square meter. A scientist samples the field and estimates that the grass plants have a mean height of 4 cm and a standard deviation of 0.2 cm. Assuming the heights are normally distributed, estimate how many plants per square meter have height between 4.8 and 5.4 cm.

15. § The Potlatch stud mill manufactures studs that are nominally 8 feet long. Because of inaccuracies in cutting and variations due to shrinkage while drying, the finished product has a mean length of 8 feet and a standard deviation of 1/2 inch. Assuming the lengths are normally distributed, estimate the proportion of studs with lengths between 95 in and 97 in. What proportion of the studs are shorter than 94 in?

16. A timber cruiser estimates that each acre of the Kelly Creek drainage contains 42 cedar trees. By measuring the diameter of a random sample of trees at one foot above the ground she estimates that the mean diameter of the trees in the drainage is 220 inches with a standard deviation of 50 inches. Assuming the diameters are uniformly distributed, estimate the proportion of trees in each acre with diameter between 200 and 250 inches. What proportion of the trees on an acre are expected to be larger than 275 inches?

Table 1: $A(z) = P(Z \le z)$ is the area under the normal curve from $-\infty$ to z.

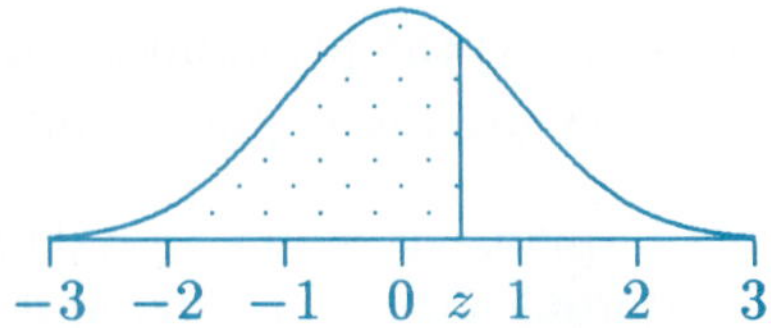

z	$A(z)$	z	$A(z)$	z	$A(z)$	z	$A(z)$
−3.50	.0002	−1.70	.0446	0.10	.5398	1.90	.9713
−3.45	.0003	−1.65	.0495	0.15	.5596	1.95	.9744
−3.40	.0003	−1.60	.0548	0.20	.5793	2.00	.9772
−3.35	.0004	−1.55	.0606	0.25	.5987	2.05	.9798
−3.30	.0005	−1.50	.0668	0.30	.6179	2.10	.9821
−3.25	.0006	−1.45	.0735	0.35	.6368	2.15	.9842
−3.20	.0007	−1.40	.0808	0.40	.6554	2.20	.9861
−3.15	.0008	−1.35	.0885	0.45	.6736	2.25	.9878
−3.10	.0010	−1.30	.0968	0.50	.6915	2.30	.9893
−3.05	.0011	−1.25	.1056	0.55	.7088	2.35	.9906
−3.00	.0013	−1.20	.1151	0.60	.7257	2.40	.9918
−2.95	.0016	−1.15	.1251	0.65	.7422	2.45	.9929
−2.90	.0019	−1.10	.1357	0.70	.7580	2.50	.9938
−2.85	.0022	−1.05	.1469	0.75	.7734	2.55	.9946
−2.80	.0026	−1.00	.1587	0.80	.7881	2.60	.9953
−2.75	.0030	−0.95	.1711	0.85	.8023	2.65	.9960
−2.70	.0035	−0.90	.1841	0.90	.8159	2.70	.9965
−2.65	.0040	−0.85	.1977	0.95	.8289	2.75	.9970
−2.60	.0047	−0.80	.2119	1.00	.8413	2.80	.9974
−2.55	.0054	−0.75	.2266	1.05	.8531	2.85	.9978
−2.50	.0062	−0.70	.2420	1.10	.8643	2.90	.9981
−2.45	.0071	−0.65	.2578	1.15	.8749	2.95	.9984
−2.40	.0082	−0.60	.2743	1.20	.8849	3.00	.9987
−2.35	.0094	−0.55	.2912	1.25	.8944	3.05	.9989
−2.30	.0107	−0.50	.3085	1.30	.9032	3.10	.9990
−2.25	.0122	−0.45	.3264	1.35	.9115	3.15	.9992
−2.20	.0139	−0.40	.3446	1.40	.9192	3.20	.9993
−2.15	.0158	−0.35	.3632	1.45	.9265	3.25	.9994
−2.10	.0179	−0.30	.3821	1.50	.9332	3.30	.9995
−2.05	.0202	−0.25	.4013	1.55	.9394	3.35	.9996
−2.00	.0228	−0.20	.4207	1.60	.9452	3.40	.9997
−1.95	.0256	−0.15	.4404	1.65	.9505	3.45	.9997
−1.90	.0287	−0.10	.4602	1.70	.9554	3.50	.9998
−1.85	.0322	−0.05	.4801	1.75	.9599	3.55	.9998
−1.80	.0359	−0.00	.5000	1.80	.9641	3.60	.9998
−1.75	.0401	0.05	.5199	1.85	.9678	3.65	.9999

9.5 APPROXIMATING BINOMIAL RANDOM VARIABLES WITH NORMAL RANDOM VARIABLES

In Section 9.1, we discussed binomial random variables and showed how to use CALIPSO to calculate probabilities involving these variables. However, for large values of n (the number of trials) even CALIPSO may not be inclined to perform the necessary calculations. For instance, C(1000,500) is beyond the capacity of CALIPSO. Pocket calculators are even more limited in finding such values. We can, nevertheless, use normal random variables to approximate binomial distributions, and then take advantage of CALIPSO's normal curves (or tables) to calculate the desired probabilities to a good degree of accuracy.

To understand why this possible, consider the binomial distribution with $n = 12, p = .5$. The mean of this distribution is $\mu = np = 6$ and the standard deviation is $\sigma = \sqrt{np(1-p)} = \sqrt{3} = 1.732$. Figure 1 gives a histogram describing this distribution.

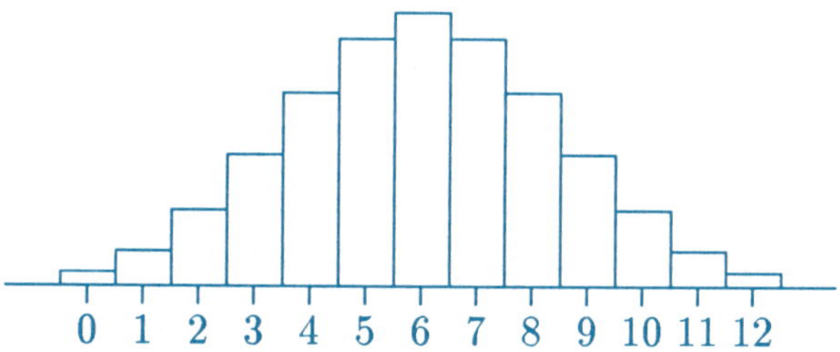

Figure 1

In Figure 2 we have superimposed a normal curve associated with the normal distribution with mean $\mu = 6$ and standard deviation $\sigma = 1.732$.

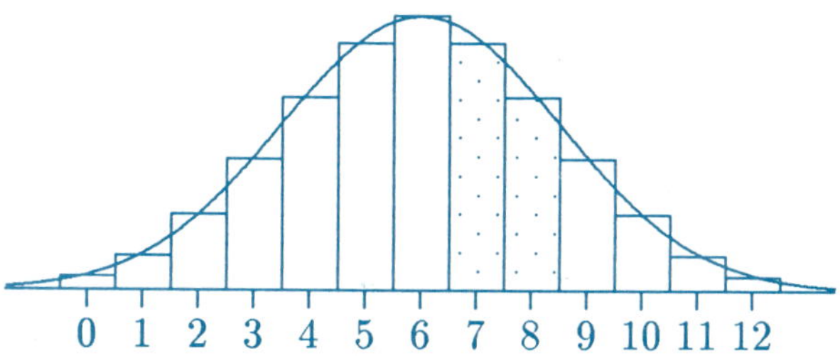

Figure 2

Note that this normal distribution gives a reasonably good approximation of the binomial distribution. We take advantage of this observation as follows. Suppose for the binomial distribution with $n = 12$ and $p = .5$, you want to find $P(8 \leq X \leq 9) = P(X = 8) + P(X = 9)$ To approximate this value use the normal random variable X with $\mu = 6$ and $\sigma = 1.732$. The area below the normal curve in Figure 2 between 7.5 and 9.5 is approximately equal to $P(8) + P(9)$ because the portions of the rectangles that lie above the normal curve approximately equal the portions cut off by the curve. Using CALIPSO's *Normal Distribution* function you will find that $P(7.5 \leq X \leq 9.5) \doteq 0.1716$. Using CALIPSO's *Binomial Distribution* function you will find that that the exact value is $P(8) + P(9) \doteq .1746$. The normal distribution gives a fair estimate even for this small value of n. The estimate improves as n becomes large, and in practice you would only approximate a binomial distribution with the area under the normal curve when n is large. The next example illustrates this.

Example 1: Approximating a Binomial Distribution

We use a normal distribution to approximate the solution of the following problem. Suppose it is known that 3% of a manufacturer's light bulbs are defective. In a shipment of 5000 bulbs, what is the probability that 125 or more bulbs are defective? You could obtain the exact answer by calculating $P(125) + P(126) + P(127) + \cdots + P(5000)$; however, this is an immense calculation, well beyond the capability of CALIPSO or almost any statistical program. To use a normal distribution to approximate this value, note that $\mu = 5000(.03) = 150$ and $\sigma = \sqrt{5000(.03)(.97)} \doteq 12.06$. For a normal distribution X with $\mu = 150$ and $\sigma = 12.06$, you can use CALIPSO to find that $P(X \geq 124.5) \doteq .9828$. To find the probability that fewer than 140 bulbs are defective you can use CALIPSO to find $P(X \leq 139.5) \doteq .192$.

9.5A Exercises: Approximating a Binomial Distribution

In the following problems use an appropriate normal distribution to approximate the true probability. In Problems 1 and 2 calculate the true probability by hand and compare this answer to your approximation. Then check your answer with CALIPSO's *Binomial Distribution* function. Problems 3 and 4 are too large to calculate the true probabilities.

1. § $P(2 \leq X \leq 5)$ where X is binomial distribution with $n = 250$ and $p = .01$.

2. $P(42 \le X \le 45)$ where X is a binomial distribution with $n = 100$ and $p = .4$.

3. § $P(200 \le X \le 240)$ where X is a binomial distribution with $n = 10,000$ and $p = .023$.

4. $P(160 \le X \le 240)$ where $n = 15,000$ and $p = .01$.

5. § A university has noted that .3% of the transcripts it issues contain errors. If 6000 transcripts are prepared, what is the probability that fewer than 4 contain errors? What is the probability that at least 2 contain errors?

6. Suppose that in Idaho 60% of the drivers use seat belts on the highway. In a statewide check of 1500 drivers what is the probability that more than 950 of these drivers do not use seat belts? What is the probability that between 800 and 900 (inclusive) of these drivers do use seat belts?

7. § A man, who has little better to do, tosses a coin 2000 times. What is the probability he records no more than than 990 heads? What is the probability he records at least 850 and no more than 950 heads?

8. A new drug promises a success rate of 45% in growing hair on bald men. A group of 300 bald men are invited to test the drug. What is the probability that more than half of the men will grow hair? What is the probability that fewer than 125 of the men will grow hair?

9.6 LINEAR REGRESSION *

Many experiments take the form of gathering data as a set of pairs. For instance, at each hour of an 8-hour day we can record the number of transactions that have occurred up to that time on the New York Stock Exchange. Or the experiment might consist of measuring the humidity each time the temperature changes by 1 degree at a given site. Since such experiments result in a set of pairs of numbers, we can plot these pairs on a graph to get a visual representation of the results. The problem we want to study arises in the attempt to fit such experimental data to a smooth curve that in some sense represents a "best fit." In general, one can attempt to fit a given data set by curves from classes of functions such as quadratics or exponentials. In this introduction we consider only data sets that can reasonably be approximated by lines. With that restriction the curve fitting problem is usually called *linear regression.*

To illustrate the basic idea underlying linear regression we consider the following situation. Table 1 gives the rainfall (in inches) observed in a tropical forest on 4 days in the first week of January. Based on this data we want to predict the rainfall for other days during the first week of January.

Table 1

Day	1	2	3	5
Rainfall	0	2	1	4

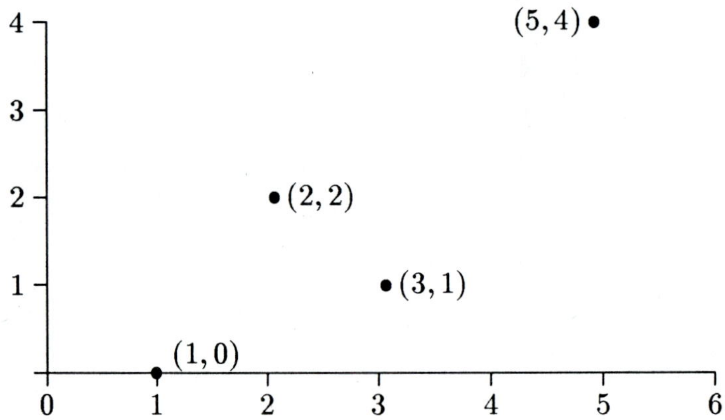

Figure 1

First we plot the data by letting the x coordinate correspond to the days, and the y coordinate to the rainfall amounts. Figure 1 indicates the points $(1,0),(2,2),(3,1),(5,4)$.

As you can see, the points in Figure 1 do not lie on a line. Suppose, however, that for purposes of estimation we believe it reasonable to approximate these points with a line. Thus we try to find a line that "fits" the data set in a way that approximates the rainfall that actually occurred. The value of finding such an approximating line is that we can use it to predict the rainfall that occurred on days in which we made no measurements. Figure

2 shows a good possibility.

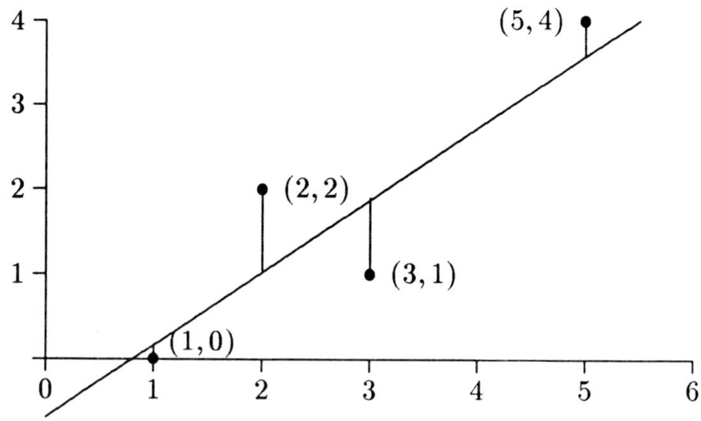

Figure 2

If you study more statistics you will learn that the line that fits the given data most usefully is the one with the property that the sum of the squares of the vertical distances from the data points to the line is the least. This line is called the *regression line* or the *least squares regression line.* For the data set in Table 1, Figure 2 shows these vertical distances.

We will show you how to calculate the equation of the line that minimizes the sum of the squares of these distances. We will not explain why the calculations work because either calculus or linear algebra techniques are required to do that. If you take a statistics or linear algebra course, you will learn how to justify these calculations.

Any line that is not vertical has an equation of the form $y = mx + b$ where m and b are constants. A point (x, y) lies on the line if and only if x and y satisfy the equation $y = mx + b$. We first show you how to plot regression lines with CALIPSO and then consider how to use matrix algebra to calculate m and b.

CALIPSO

Choose the *STATISTICS* application from the CALIPSO menu and then choose *Regression* from the *STATISTICS* menu. You will see a data entry panel that accepts pairs of values for x and y. You can enter the data in Table 1 as four pairs (x, y) where x corresponds to a day and y corresponds to the rainfall on that day. Table 2 gives the pairs to enter in CALIPSO.

Table 2

x	y
1	0
2	2
3	1
5	4

After entering the data you press *Enter* to see a plot of the data points and the regression line. If you want to print a copy of the plot, press Y; if not, press *Spacebar*. The (x, y) pairs that you entered are printed in the worksheet and the equation of the regression line is given. For our example, the equation is $y = 0.886 * x + -0.686$. This line is usually written without $*$ and $+$, that is, as $y = 0.886x - 0.686$. CALIPSO has substituted the value 0.886 that it calculated for m and the value -0.686 that it calculated for b in the expression $y = m * x + b$. Note that we can use the equation $y = .886x - .686$ to approximate the rainfall on a day in which we did not measure it, say day 4. The approximation is $y = .886(4) + .686 = 4.23$.

If you typed a Y earlier to obtain a printed copy of the plot, another entry appears in the worksheet with some numbers that CALIPSO will use to print. To actually obtain the printed copy, you must print the worksheet.

Example 2: Using CALIPSO to Plot a Regression Line

Table 3 gives the average time (in minutes) it takes to catch a fish on a river according to the temperature (in degrees Celsius) of the water.

Table 3: Time to Catch a Fish

Temperature	2	4	6	8	10	12	14
Time	10	13	12	15	20	23	27

Using CALIPSO you can verify that the least squares regression line is $y = 1.41x + 5.86$. From this line, we can estimate the time required to catch a fish if the water temperature is 16 degrees to be $y = 1.41(16) + 5.86 = 28.42$. The estimated time for catching a fish if the water temperature is 9 degrees is $y = 1.41(9) + 5.86 = 18.55$.

9.6A Exercises: Least Squares Regression

In Problems 1 – 4 use CALIPSO to find the least squares regression line corresponding to the given data. Use your line to estimate the indicated values.

1. §

x	1	4	6	8	12
y	3	0	3	5	7

(a) $x = 14, y = ?$ (b) $x = ?, y = 6$ (c) $x = ?, y = 3$

2.

x	1	2	3	4	5	6	7	8	9	10
y	3	6	6	12	11	12	16	21	19	24

(a) $x = 7, y = ?$ (b) $x = 8, y = ?$ (c) $x = 2.5, y = ?$

3. §

x	3	5	2	4	10	12	8	7	6
y	13	19	8	15	43	45	30	30	25

(a) $x = 7, y = ?$ (b) $x = 8.5, y = ?$ (c) $x = ?, y = 5$

4.

x	2	4	8	16	32	64	128
y	124	110	92	69	54	45	12

(a) $x = 50, y = ?$ (b) $x = 0, y = ?$ (c) $x = 14, y = ?$

5. § Table 4 gives the air particulate levels in Moscow corresponding to outside temperatures in December. Estimate the air particulate level if the temperature is (a) 32, (b) −10, (c) 60.

Table 4: Particulate Levels

Temperature	0	10	20	30	40	50
Level	182	150	135	110	85	48

6. Table 5 gives median values of houses for the years 1985 – 1990. Based on this information estimate the median value of a house in the year 1993.

Table 5: Median Value of Houses

Year	1985	1986	1987	1988	1989	1990
Value	62,000	61,500	63,500	68,000	73,500	79,000

Note: Table 5 will produce an unsatisfactory graph if you use the years 1985–1990 as the x variable because the x axis will extend from 0 to 1990. All of the data points will be crowded in a small interval at the extreme right. Such situations often arise in graphing a regression line. You can obtain a useful graph by counting years beyond 1985 rather than beyond the year 0. To do this, substitute 0 for 1985, 1 for 1986, ..., 5 for 1990.

7. § Amounts of raisins in cereal boxes (even of the same size) tend to vary. Three 5-, 15-, and 20-ounce boxes are checked for their raisin content. Based on the information in Table 6, estimate the number of raisins in a 25-ounce box.

Table 6: Number of Raisins in a Box

Size	5 ounce	10 ounce	15 ounce	20 ounce
Raisins	26	58	72	90

8. A common technique used by an auditor in examining the inventory of a company is to compare the value of items listed in the inventory to current prices for these items. The auditor does not expect the cost of an item in inventory to be the same as its current cost since the item in inventory was purchased at an earlier time and the price is likely to have changed; however, it is reasonable to expect that there is a linear relationship between the current and inventory values. Plot the regression line for the data in Table 7. Do you think the inventory values bear a linear relationship with current prices? Which items do you think are incorrectly valued in the inventory list?

Table 7: Inventory Values Versus Current Prices

Item Number	1	2	3	4	5	6	7	8	9	10
Current Price (x_i)	12	14	9	25	50	72	36	90	60	54
Inventory Value (y_i)	24	15	8	36	49	69	45	89	85	56

We now discuss how to use matrix algebra to find the regression line. We will not be able to prove that our method works because the linear algebra needed to do that is beyond the scope of this book.

Consider the data in Table 2. If we substitute the four values for x and the corresponding values for y in the equation $mx + b = y$ we get the system of linear equations

$$\begin{array}{rcccl} m & + & b & = & 0 \\ 2m & + & b & = & 2 \\ 3m & + & b & = & 1 \\ 5m & + & b & = & 4 \end{array} \tag{1}$$

We already observed that geometrically these data points do not lie on a line. We conclude the same thing algebraically by observing that the system (1) does not have a solution for m and b.

We can write (1) in matrix notation as

$$\begin{bmatrix} 1 & 1 \\ 2 & 1 \\ 3 & 1 \\ 5 & 1 \end{bmatrix} \begin{bmatrix} m \\ b \end{bmatrix} = \begin{bmatrix} 0 \\ 2 \\ 1 \\ 4 \end{bmatrix} \tag{2}$$

If we let A and X denote the two matrices on the left side of equation (2) and B denote the matrix on the right side, we can write (2) in the shorter form

$$AX = B \tag{3}$$

It is quite surprising that although (3) has no solutions, the equation

$$A^T AX = A^T B \tag{4}$$

obtained by multiplying both sides of (3) by A^T does have a solution. A^T, the *transpose* of A, is the matrix obtained from A by interchanging rows and columns; that is, the first row of A becomes the first column of A^T, the second row of A becomes the second column of A^T, etc. For our A matrix

$$A^T = \begin{bmatrix} 1 & 2 & 3 & 5 \\ 1 & 1 & 1 & 1 \end{bmatrix}$$

You can easily check that

$$A^T A = \begin{bmatrix} 1 & 2 & 3 & 5 \\ 1 & 1 & 1 & 1 \end{bmatrix} \begin{bmatrix} 1 & 1 \\ 2 & 1 \\ 3 & 1 \\ 5 & 1 \end{bmatrix} = \begin{bmatrix} 39 & 11 \\ 11 & 4 \end{bmatrix}$$

and

$$A^T B = \begin{bmatrix} 1 & 2 & 3 & 5 \\ 1 & 1 & 1 & 1 \end{bmatrix} \begin{bmatrix} 0 \\ 2 \\ 1 \\ 4 \end{bmatrix} = \begin{bmatrix} 27 \\ 7 \end{bmatrix}$$

Thus, we can write (4) as

$$\begin{bmatrix} 39 & 11 \\ 11 & 4 \end{bmatrix} \begin{bmatrix} m \\ b \end{bmatrix} = \begin{bmatrix} 27 \\ 7 \end{bmatrix} \tag{5}$$

As you can check, (5) has a unique solution because its leftmost matrix is invertible. Another surprising thing is that this solution gives the values of m and b for the regression line.

In general, you can find the coefficients of the regression line as follows. Suppose the data are given by the pairs

$$(x_1, y_1), (x_2, y_2), \ldots, (x_n, y_n)$$

Form the matrices

$$A = \begin{bmatrix} x_1 & 1 \\ x_2 & 1 \\ \vdots & \\ x_n & 1 \end{bmatrix} \quad X = \begin{bmatrix} m \\ b \end{bmatrix} \quad B = \begin{bmatrix} y_1 \\ y_2 \\ \vdots \\ y_n \end{bmatrix}$$

The system of equations

$$A^T AX = A^T B \tag{6}$$

is called the *normal equation* for the data set. It can be shown $A^T A$ is an invertible matrix except in the trivial case where all of the (x_i, y_i) are equal. The solution to (6) gives the coefficients of the regression line.

CALIPSO

You can easily solve the normal equation (6) with CALIPSO's *MATRIX* application. First enter the A and B matrices. Next choose *Operations* from the *MATRIX* menu and enter either of the expressions

$$C = A'B/(A'A) \quad \text{or} \quad C = (A'A)\hat{}A'B \tag{7}$$

The C matrix will always be a 2×1 column matrix whose first entry is the value of m and whose second entry is the value of b in the equation $y = mx + b$ for the regression line.

The A' in (7) is CALIPSO's notation for A^T and ˆ is its notation for the inverse. You do not need to memorize these notations since the *Operations* screen in which you enter expressions describes the notation for you. If you are solving several problems on one worksheet, you may need to use other letters than C, A, and B to enter the normal equation.

Example 3: Finding a Regression Line with Matrices

To find the regression line for the problem given in Example 2, enter the following matrices in CALIPSO's *MATRIX* application.

$$A = \begin{bmatrix} 2 & 1 \\ 4 & 1 \\ 6 & 1 \\ 8 & 1 \\ 10 & 1 \\ 12 & 1 \\ 14 & 1 \end{bmatrix} \qquad B = \begin{bmatrix} 10 \\ 13 \\ 12 \\ 15 \\ 20 \\ 23 \\ 27 \end{bmatrix}$$

Next choose *Operations* and enter the expression $C = A'B/(A'A)$. The answer is

$$C = \begin{bmatrix} \frac{79}{56} \\ \frac{41}{7} \end{bmatrix} \doteq \begin{bmatrix} 1.41 \\ 5.86 \end{bmatrix}$$

The equation of the regression line is $y = \frac{79}{56}x + \frac{41}{7} \doteq 1.41x + 5.86$.

9.6B Exercises: Using Matrices to Find the Regression Line

1–8. § Use matrices and CALIPSO to solve Problems 1–8 of Exercises 9.6A.

9. § In the introductory example we found the regression line for the data in Table 2 to be $y = 0.886x - 0.686$. Use the *MATRIX* application to find the exact values of m and b as fractions.

10. A physics class investigates whether the pressure exerted by oxygen in an enclosed tank varies linearly with the temperature. To accomplish this task the class makes the 7 measurements given in Table 8. Due to inaccuracies in measurement, the pairs (temperature, pressure) do not fall exactly on a line. The students decide to calculate the regression line for this data as the best fit. Calculate the regression line for them and use it to interpolate the pressure when the temperature is 125 degrees. Also extrapolate the temperature if the pressure is 3 atm.

Table 8: Temperature (^{o}C) versus Pressure (atm)

Temperature	0	50	100	150	200	250	300
Pressure	1.25	1.95	1.74	1.96	2.25	2.46	2.70

11. § To test a new drug designed to lower cholesterol levels a research team injects the drug into a volunteer and measures his cholesterol level the next day. This procedure is performed on the same person with varying dosages in each of 6 successive weeks. Table 9 summarizes the experimental results. Find the regression line for the data.

Table 9: Dosages and Cholesterol Levels

Dosage (mg)	0	2	4	6	8	10
Cholesterol Level	292	270	248	232	212	190

12. The average number of years of life left to a 65-year-old woman has increased from 12.2 years in 1900 to 19.2 years in 1990. Table 10 gives data on this survival time for each decade with decade 0 representing 1900-1909 and decade 9 representing 1980-1989. Find the regression line and predict the life expectancy of a 65-year-old woman for decade 10 representing 1990-1999.

Table 10: Survival Time of 65-Year-Old Women after Age 65

Decade	0	1	2	3	4	5	6	7	8	9
Years	12.2	12.6	12.8	13.0	13.2	15.5	16.6	17.5	18.1	19.2

13. § A chain store wants to determine the demand for plastic lawn chairs as a function of price. It offers the chairs for sale in each of its 8 stores at various prices. Table 11 gives the prices and demands (numbers sold). The marketing manager decides to compute the regression line to use as a graphic demonstration of the experiment. What is the regression line?

Table 11: Demand Versus Price

Price ($)	11	13	15	17	19	21	23
Demand	32	45	40	35	24	15	10

14. A food scientist studies the growth rate of yeast in bread dough as a function of temperature. His technique is to add 25 gm of yeast to 1 kg of dough and measure the amount of yeast present after one hour. Table 12 summarizes his measurements. Find the regression line.

Table 12: Yeast Growth versus Temperature

Temperature	16	18	20	22	24	26	28	30
Yeast (gm)	30	34	38	42	54	60	62	65

10 Mathematics of Finance

10.1 SIMPLE AND COMPOUND INTEREST

In this section we develop formulas for computing earnings that arise from depositing money in accounts offering simple or compound interest.

The initial amount you deposit in a bank is called the *principal.* Suppose you deposit P dollars in a bank that offers an annual *simple interest rate* r. Then after t years you will have earned interest amounting to

$$I = Prt$$

dollars and your final balance B will be

$$B = P + Prt = P(1 + rt)$$

Example 1: Simple Interest

If you buy a money-market bond for \$25,000 (the principal) from a bank offering an annual simple interest rate of 7%, then after 4 years you will have earned interest

$$I = 25,000(.07)(4) = 7,000$$

and your ending balance will be

$$B = 25,000 + 7,000 = 32,000$$

After $6\frac{1}{2}$ years your balance will be

$$B = 25,000(1 + .07 \times 6.5) = 36,375$$

More commonly, banks offer *compound interest.* Compound interest allows you to earn interest on interest already received. For instance, suppose a bank offers an (annual) interest rate r compounded annually and your principal is P. After 1 year you will have a balance of $P + Pr$ dollars. During the second year you earn interest on this new amount (rather than just on the principal P). Table 1 shows how your money grows when the interest is compounded annually.

Table 1: Compound Interest

Original Deposit	P
Balance after 1 yr	$P + Pr = P(1+r)$
Balance after 2 yr	$P(1+r) + P(1+r)r = P(1+r)(1+r) = P(1+r)^2$
Balance after 3 yr	$P(1+r)^2 + P(1+r)^2 r = P(1+r)^2(1+r) = P(1+r)^3$
⋮	
Balance after n yr	$B = P(1+r)^n$

Example 2: Calculating Compound Interest

If you deposit \$10,000 in a bank offering 6% interest compounded annually, then after 10 years your balance will be

$$B = 10,000(1 + .06)^{10} = 17,908.48. \tag{1}$$

To use CALIPSO's *Calculator* to calculate B, enter (1) as

10000(1.06)^10

If you have a hand calculator that computes exponents, you can easily evaluate (1). While calculator formats vary, the most common way to exponentiate uses a key labelled x^y. To calculate 1.06^{10} with such a key, you enter 1.06 in the register (1.06 is x), type 10, and press x^y. If you have a financial calculator, it may have a special key to calculate $(1 + r)^n$.

If your calculator only adds, subtracts, multiplies, and divides, it will not be of much use for financial calculations. In the sequel, when we say you can make a calculation on a hand calculator, we assume you will use one that at least has an exponential function.

Many banks compound interest quarterly (every three months), monthly, or even daily. If r is the annual interest rate and the interest is compounded q times a year, then you are actually earning interest at the rate $\frac{r}{q}$ for each of the q periods of time; we call $i = \frac{r}{q}$ the *periodic interest rate*. Also, leaving the money in the bank t years is equivalent to leaving the money for $n = tq$ periods. Thus, for calculating the balance B after t years formula (1) above becomes

$$B = P\left(1 + \frac{r}{q}\right)^{tq} = P(1+i)^n \tag{2}$$

where $i = \frac{r}{q}$ is the periodic interest rate and $n = tq$ is the total number of periods.

Example 3: Calculating Interest Compounded Periodically

Suppose you deposit \$800 in a bank offering a 9% annual interest rate compounded monthly. Then the periodic interest rate is $i = \frac{.09}{12} = .0075$. You can you use (2) and CALIPSO's *Calculator* (or a hand calculator) to show that after 5 years ($n = 60$ periods) your balance will be

$$B = 800(1 + .0075)^{60} = 800(1.0075)^{60} = 1,252.54$$

Bank regulatory laws often require that a customer be told the simple annual interest rate that yields the same return as the compound interest rate. In this context, the annual compound interest rate is called the *nominal rate* and the annual simple interest rate is called the *effective rate.* If the nominal rate is r and interest is compounded q times a year, then the value of an investment of P dollars after 1 year is $P(1+\frac{r}{q})^q$. The effective interest rate s must satisfy

$$P\left(1 + \frac{r}{q}\right)^{q} = P(1+s) \tag{3}$$

To find s we cancel P from both sides of (3) and solve for s to get

$$s = \left(1 + \frac{r}{q}\right)^{q} - 1$$

Example 4: Calculating an Effective Interest Rate

If a bank pays 5% compounded quarterly on its passbook savings accounts, the effective annual rate is

$$s = \left(1 + \frac{.05}{4}\right)^{4} - 1 = .050945 \doteq 5.09\%$$

The next example shows how you can determine the principal you need to deposit to achieve a desired amount of money after a given number of years. In this context the principal is sometimes called the *present value* of the future balance.

Example 5: Calculating the Present Value of a Future Balance

A bank offers a 7% annual interest rate compounded quarterly. You want to determine the principal P you should deposit so that after 10 years you will have a balance of \$5000. The principal P is the present value of the future balance, \$5000.

To make this calculation simply solve for P in formula (2) with $i = \frac{.07}{4} = .0175, n = 10 \cdot 4 = 40$, and $B = 5000$ to obtain

$$P = \frac{5000}{(1 + .0175)^{40}} = 2498.00$$

On occasion a bank will advertise that interest is *compounded continuously*. Though this notion is a bit disturbing, it can be understood if you think of continually decreasing the length of the time periods over which interest is compounded. For instance, you can compound interest (at least in theory) daily, hourly, each second, each microsecond, etc. The "limit" of this idea is interest compounded continuously or compounded somehow each "instant" of time. To find a formula for the balance resulting from continuously compounding interest first observe that the compound interest formula

$$B = P\left(1 + \frac{r}{q}\right)^{tq}$$

gives the balance after t years no matter how small the time periods over which the interest is compounded. Note, further, that as the length of the time periods decreases, the number of periods q per year increases. In a calculus course it is shown that as q grow larger, the values $(1 + \frac{r}{q})^{q/r}$ approach the number e where e is an irrational number approximately equal to 2.7182818. Therefore if we write $B = P(1 + \frac{r}{q})^{tq}$ as

$$B = P\left(1 + \frac{r}{q}\right)^{\frac{q}{r}rt} = P\left(\left(1 + \frac{r}{q}\right)^{\frac{q}{r}}\right)^{rt}$$

you can see that as q gets larger and larger, B approaches the value

$$B = Pe^{rt} \tag{4}$$

This is the formula for calculating the balance after t years if your account earns interest compounded continuously.

Example 6: Continuously Compounded Interest

If you deposit \$20,000 in an account that pays an annual interest rate of 8% compounded continuously, then after 15 years your balance will be

$$B = 20,000e^{(.08)15} = 66,402.34$$

In the exercises you are invited to compare the virtues of the various methods for earning interest.

10.1A Exercises: Simple and Compound Interest

In Problems 1 – 3 use CALIPSO or a hand calculator to calculate your balance after (a) 5 years, (b) 36 months, and (c) 40 months given the principal P and the simple interest rate r.

1. § $P = 5000; r = .08$.

2. $P = 10,000; r = .06$.

3. § $P = 2000; r = .15$.

In Problems 4 – 9 use CALIPSO or a hand calculator to calculate the balance after (a) 5 years, (b) 36 months, and (c) 50 months given the principal P and the annual interest r compounded as indicated. Assume that interest is only paid when the amount has been on deposit for the entire compounding period; for instance, if you withdraw your money after 50 months and interest is compounded annually, you receive no interest for the 49th and 50th month.

4. $P = 5000; r = .08$, compounded quarterly.

5. § $P = 5000; r = .08$, compounded monthly.

6. $P = 10,000; r = .06$, compounded annually.

7. § $P = 10,000; r = .06$, compounded daily.

8. $P = 2000; r = .15$, compounded semiannually (twice a year).

9. § $P = 2000; r = .15$, compounded monthly.

10. If $P = 5000$ and $r = .08$, find the balance after 30 years if the interest is compounded continuously.

11. § If $P = 10,000$ and $r = .06$, find the balance after 30 years if the interest is compounded continuously.

12. If $P = 5000$ and $r = .08$, compare the balance after one year if the interest is compounded annually, monthly, weekly, daily, and continuously.

13. § If $P = 10,000$ and $r = .06$, compare the balance after 1 year if the interest is compounded annually, monthly, weekly, daily, and continuously.

14. Calculate the effective rate corresponding to the nominal rate $r = .12$ if the interest is compounded (a) quarterly and (b) monthly.

15. § Calculate the effective rate corresponding to the nominal rate $r = .15$ if the interest is compounded (a) monthly and (b) weekly.

16. What should your initial deposit (principal) be in an account paying 10% interest compounded monthly if you want to have a balance of $15,000 after 8 years?

17. § What should your initial deposit be in an account paying 8% interest compounded quarterly if you want to have a balance of $100,000 after 20 years?

10.2 ANNUITIES

Annuities involve making equal deposits (or withdrawals) at regular intervals over some period of time. For instance, you may choose to deposit $60 each month for a period of 8 months at an annual interest rate of 10%. We find a formula that allows you to compute your ending balance.

For simplicity, assume that you deposit your money at the end of each month. It follows from (2) in Section 10.1 that at the end of the 8 months your first month's deposit of \$60 will be worth

$$B_1 = 60\left(1 + \frac{0.1}{12}\right)^7$$

dollars since during the 8 months the interest wiil have been compounded 7 times (no interest was earned the first month because the money was deposited at the end of the month). Interest on your second month's deposit of \$60 will be compounded 6 times. Therefore

$$B_2 = 60\left(1 + \frac{0.1}{12}\right)^6$$

is the amount the second deposit contributes after 8 months. Similarly, your remaining \$60 deposits will contribute after 8 months

$$B_3 = 60\left(1 + \tfrac{0.1}{12}\right)^5 \qquad B_6 = 60\left(1 + \tfrac{0.1}{12}\right)^2$$
$$B_4 = 60\left(1 + \tfrac{0.1}{12}\right)^4 \qquad B_7 = 60\left(1 + \tfrac{0.1}{12}\right)$$
$$B_5 = 60\left(1 + \tfrac{0.1}{12}\right)^3 \qquad B_8 = 60$$

(Observe that your last deposit will not generate any interest.)

Adding these amounts gives the total ending balance. Note that we can write the sum

$$B_1 + B_2 + \cdots + B_8 = 60\left(1 + \frac{0.1}{12}\right)^7 + 60\left(1 + \frac{0.1}{12}\right)^6 + \cdots + 60\left(1 + \frac{0.1}{12}\right) + 60$$

in the form

$$a + as + as^2 + as^3 + as^4 + as^5 + as^6 + as^7$$

where $a = 60$ and $s = 1 + \frac{0.1}{12}$. In Problem 13 of Exercises 10.2A you are asked to show that for any positive integer k

$$a + as + as^2 + \cdots + as^{k-1} = a\frac{s^k - 1}{s - 1} \qquad (1)$$

So, if $a = 60$ and $s = 1 + \frac{0.1}{12} = 1.0083333$, you can use CALIPSO or a hand calculator to show that after 8 months your annuity is worth

$$60\frac{s^8 - 1}{s - 1} = 60\frac{1.0083333^8 - 1}{.0083333} = 494.24$$

You can derive the following general formula for calculating the balance of an annuity by using the arguments given for the specific case just discussed.

Annuity Balance. If periodic deposits of A dollars are made q times per year, and if the annual interest r is compounded q times per year, then after t years the balance B will be

$$B = A\frac{(1+i)^n - 1}{i} \tag{2}$$

where $i = \frac{r}{q}$ is the periodic interest and $n = tq$ is the number of periods.

Example 1: Calculating an Annuity

Suppose you decide to deposit \$500 quarterly in an account that pays 9% interest compounded quarterly. What will your balance be after 15 years? The periodic interest is $i = \frac{.09}{4} = .0225$ and the number of periods is $n = 15 \cdot 4 = 60$. Using (2) and CALIPSO or a hand calculator you can verify that the answer is

$$B = 500\frac{(1.0225)^{60} - 1}{.0225} = 62,225.22$$

You can also solve formula (2) for A to determine the amount you should deposit monthly to achieve a desired balance after a given period of time. The result is

$$A = \frac{B}{\frac{(1+i)^n - 1}{i}} \tag{3}$$

CALIPSO

The denominator in (3) occurs so often in financial calculations that it is given the special notation

$$s_{\overline{n}|i} = \frac{(1+i)^n - 1}{i} \quad \text{read "}s\text{ sub }n\text{ angle }i\text{"}$$

Values of $s_{\overline{n}|i}$ are tabulated in many handbooks of mathematical tables. Some financial calculators have a special key to calculate it. The next example shows how to use the expression $S(n, i)$ to calculate $s_{\overline{n}|i}$ with CALIPSO's *Calculator*.

Example 2: Determining Payments

You plan to make monthly deposits in an account that pays an annual interest of 7% compounded monthly. How much should you deposit each month so that after 6 years you will have a $10,000 balance? To determine this amount solve (2) for A to get

$$A = \frac{B}{\frac{(1+i)^n - 1}{i}} = \frac{B}{s_{\overline{n}|i}}$$

where $B = 10,000, i = \frac{.07}{12}$ is the periodic interest, and $n = 6 \cdot 12 = 72$ is the number of periods. CALIPSO uses the expression $S(n,i)$ for $s_{\overline{n}|i}$. To see that $A = 112.1567$ with CALIPSO, enter either of the expressions

$$10000/S(72, .07/12) \quad \text{or} \quad 10000/S(72, 7/1200)$$

in CALIPSO's *Calculator.*

Next we consider the notion of the present value of an annuity. The *present value* of an annuity is the amount of a lump sum payment that will yield the same amount of money as would periodic payments of an annuity over the same period of time at the same interest. (As usual, the interest is compounded over each payment period.) To calculate the present value P of an annuity over n periods at compounded periodic interest i we set the amount

$$B = A\frac{(1+i)^n - 1)}{i}$$

generated by the annuity equal to the amount

$$B = P(1+i)^n$$

generated by the lump sum payment P over the same period of time. Thus, we have

$$A\frac{(1+i)^n - 1}{i} = P(1+i)^n$$

from which we see that

$$P = A\frac{(1+i)^n - 1}{i(1+i)^n} \tag{4}$$

CALIPSO

The factor of A in (4) occurs so often in financial calculations that it is given the special notation

$$a_{\overline{n}|i} = \frac{(1+i)^n - 1}{i(i+1)^n} \quad \text{read "}a\text{ sub }n\text{ angle }i\text{"}$$

As is the case for $s_{\overline{n}|i}$, values of $a_{\overline{n}|i}$ are tabulated in mathematical tables to high accuracy. Some financial calculators have a special key to calculate it. The next example shows how to use the expression $A(n, i)$ to calculate $a_{\overline{n}|i}$ with CALIPSO's *Calculator*.

Example 3: Calculating a Present Value

Suppose your annuity involves quarterly payments of \$400 for 10 years. The annual interest r is 13%, and so the periodic interest i is $\frac{13}{4} = 3.25\% = .0325$. From (4) we find that the present value of this annuity is

$$P = 400\frac{(1.0325)^{60} - 1}{.0325(1.0325)^{60}} = 400a_{\overline{60}|.0325} = 10,501.46$$

To calculate P with CALIPSO's *Calculator*, enter

$$400A(60, 13/400),\ 400A(60, 0.13/4), \text{or } 400A(60, 0.0325)$$

Frequently, you may find yourself in a situation where you have a given amount of money, which you place in an account, and then make periodic withdrawals from (rather than deposits to) this account. Such a situation is referred to as a *decreasing annuity*. The following formula allows you to make the appropriate calculations.

Decreasing Annuity Balance. Suppose for t years, q equal withdrawals of W dollars are made each year from an account that pays an annual interest rate r compounded q times per year. Let $i = \frac{r}{q}$ be the periodic interest and $n = qt$ the total number of periods. Then after t years the ending balance will be 0 if the initial deposit D is

$$D = W\frac{(1+i)^n - 1}{i(1+i)^n} = Wa_{\overline{n}|i} \qquad (5)$$

Notice that (5) is the same as expression (4) for the present value of an annuity with D replacing P and W replacing A. This should come as no surprise since the two concepts are the same except that for a decreasing annuity the bank pays you, and for an increasing annuity you pay the bank.

Example 4: Calculating a Decreasing Annuity

Suppose you want to withdraw \$1500 every 6 months from an account that pays 14% interest compounded twice a year. You want to make these withdrawals over a 5-year period. Then by (5) you should make a one-time deposit

$$D = 1500\frac{1.07^{10} - 1}{.07 \cdot 1.07^{10}} = 1500a_{\overline{10}|.07}$$

You can calculate D with a financial calculator or with CALIPSO by entering $1500A(10, 0.07)$ to find that $D = \$10,535.37$.

You can use (5) to solve a variety of other kinds of problems. The following example illustrates this.

Example 5: Applying the Decreasing Annuity Formula

You have deposited your \$20,000 lottery winnings in an account that pays 8% interest compounded quarterly. To help pay your college expenses you want to make equal withdrawals from this account every 3 months for a period of 5 years (and end with a balance of 0). To find how much you can withdraw each quarter solve (5) for W to obtain

$$W = \frac{D}{a_{\overline{n}|i}} = \frac{20,000}{a_{\overline{20}|.02}}$$

You can calculate W with a financial calculator or with CALIPSO by entering $20000/A(20, 0.08/4)$ to find that $W = \$1,223.13$.

10.2A Exercises: Annuities

1. § If you deposit \$100 monthly in an account paying 7% interest compounded monthly, what will your balance be after 5 years?

2. If you deposit \$1000 quarterly in an account paying 12% interest compounded quarterly, what will your balance be after 20 years?

3. § If you deposit \$10 monthly in an account paying 10% interest compounded monthly, what will your balance be after 15 years?

4. Your account pays 8% interest compounded monthly. What should your monthly payments be if you want to have a balance of \$20,000 after 10 years?

5. § Your account pays 15% interest interest compounded annually. What should your yearly payments be if you want to attain a balance of \$50,000 after 20 years?

In Problems 6 – 8 calculate the present value of the given annuity.

6. Payment: \$300, monthly
   Interest: 8%, compounded monthly
   Time period: 6 years

7. § Payment: \$150, quarterly
   Interest: 12%, compounded quarterly
   Time period: 15 years

8. Payment: \$25, monthly
   Interest: 18%, compounded monthly
   Time period: 10 years

9. § Find the amount of the deposit you should make for a decreasing annuity that will pay \$200 monthly for 8 years, if the interest rate is 10% compounded monthly.

10. Find the amount of the deposit you should make for a decreasing annuity that will pay \$3000 annually for 25 years, if the interest rate is 15% interest compounded annually.

11. § You deposit \$25,000 in an account paying 12% interest compounded quarterly. You want to make equal withdrawals from this account each quarter for 10 years, after which time the account is completely depleted. How much can you withdraw each time?

12. You deposit \$15,000 in an account paying 8% interest compounded monthly. You want to equal withdrawals from this account each month for 12 years, at which time the account is completely depeleted. How much can you withdraw each month?

13. § Prove the identity in (1) by carrying out the following steps.

    (a) Explain why

    $$(1 + s + s^2 + \cdots + s^{k-1}) - s(1 + s + s^2 + \cdots + s^{k-1}) = 1 - s^k$$

    (b) Write the left-hand side of the expression in part (a) as

    $$(1 - s)(1 + s + s^2 + \cdots + s^{k-1})$$

    (c) Explain why

    $$1 + s + s^2 + \cdots + s^{k-1} = \frac{1 - s^k}{1 - s}$$

    (d) Explain why (1) follows from part (c).

14. Congress estimates that the deficit in five years will be \$425,000,000,000. Rather than incur that deficit, it decides to deposit money monthly in a sinking fund (a *sinking fund* is another name for an annuity). If the annual interest on the fund is 14%, how much should congress deposit each month to raise the required amount after 5 years of payments?

10.3 AMORTIZATION

Amortization is closely related to the notion of decreasing annuities. Paying off a loan is a prime example of amortization. Suppose, for instance, that you have taken out a \$12,000 loan to pay for an automobile. You are to make a sequence of equal payments at regular time intervals to pay off the loan. The interest on the loan is compounded periodically, where the length of each period is equal to the payment intervals. Various questions arise. How much is each payment? What is the original loan amount? How much do you owe after making some of the payments? How much interest have you paid in retiring the loan?

Observe that an amortization can be considered as a decreasing annuity problem where equal payments (withdrawals) are made periodically until the loan (the final balance) is paid off in full (is 0). Thus, the decreasing annuity formula

$$D = W\frac{(1+i)^n - 1}{i(1+i)^n} = Wa_{\overline{n}|i} \tag{1}$$

applies to amortizations as well as to decreasing annuities. Moreover, solving (1) for W

$$W = \frac{D}{a_{\overline{n}|i}} \tag{2}$$

allows you to calculate the periodic payments (which are equal to W).

Example 1: Calculating Monthly Payments

To purchase a boat you take out a 2-year loan of $30,000 at 16% interest, compounded monthly. You make your payments at the end of each month. Then

$$W = \frac{30,000}{a_{\overline{24}|.16/12}} = \$1,468.89$$

is your monthly payment over the 2-year (24 payment) period.

Example 2: Calculating a Loan Amount

You want to take out a 36-month loan with monthly payments of $300. The annual interest rate is 15% compounded monthly. What is the amount of the loan? This amount is the value D in (1), where $i = \frac{.15}{12} = .0125, W = 300$, and $n = 36$. You can verify that

$$D = 300a_{\overline{36}|.0125} = \$8,654.18$$

It is often of interest to know the balance you owe on a loan after having made a certain number of payments. In order to develop a general formula to calculate the unpaid balance we consider a short-term loan for which we can easily examine the sequence of payments.

Example 3: A Short-Term Loan

A teenager buys a bicycle for $350 and agrees to pay for it in 6 monthly installments at 12% annual interest compounded monthly. By (2) his monthly payments are

$$W = \frac{350}{a_{\overline{6}|.01}} = \$60.39$$

Table 1 shows a complete analysis of the amortization of the loan.

Table 1: Analysis of an Amortization

Payment number	Amount	Interest	Applied to principal	Unpaid balance
1	60.39	3.50	56.89	293.11
2	60.39	2.9311	57.4589	235.6511
3	60.39	2.3565	58.0335	177.6176
4	60.39	1.7762	58.6138	119.0038
5	60.39	1.1900	59.2000	59.8038
6	60.39	0.5980	59.7920	−.0119

Note that it is common for an amortization analysis to end with a final balance that differs from zero by a few pennies. Part of this error comes from rounding in the interest calculations, but most of it derives from the fact that the monthly payment must be rounded to the nearest penny, whereas the actual payment to retire the loan exactly is unlikely to be accurate to the penny.

For the short-term loan in Example 3 it was easy to find the unpaid balance at the end of each payment period by creating Table 1. However, if the loan had been for a 30-year mortgage on a house, it would be onerous to determine the unpaid balance at the end of 24 years of monthly payments in this way.

Let us try to use an annuity formula to calculate the unpaid balance on the bicycle loan after the fourth payment (it is 119.0038 in Table 1). This unpaid balance must be the same as the amount we need to place in a decreasing annuity that runs for 2 months and yields \$60.39 at the end of each month. We use formula (5) from Section 10.2 to calculate the unpaid balance D for the withdrawal (payment) amount $W = 60.39$.

$$D = Wa_{\overline{n}|.01} = 60.39a_{\overline{2}|.01} = 118.99$$

The small discrepancy between 118.99 and 119.0038 is again due to the slight error in Table 1 caused by the need to make payments to the nearest penny. You can easily see how to extend this analysis to obtain the following generalization.

Balance of an Amortization. Suppose a loan runs for n periods at periodic interest rate i with periodic payments W. The unpaid balance after m payments is the same as the amount D that must be placed in an annuity that yields a payment of W for $n - m$ periods. That is, the unpaid balance is

$$D = Wa_{\overline{n-m}|i} \tag{3}$$

Example 4: Calculating an Unpaid Balance

Suppose you buy a house for $80,000 on a 30-year mortgage at 12%. What is your unpaid balance after 10 years?

We calculate the monthly payment by the same method used in Example 1 to be $W = \frac{80{,}000}{a_{\overline{360}|.01}}$. We use (3) to calculate the unpaid balance. In this case, $n = 30 \cdot 12 = 360$ and $m = 10 \cdot 12 = 120$, and so $n - m = 240$. The unpaid balance is

$$D = Wa_{\overline{240}|.01} = 80{,}000\frac{a_{\overline{240}|.01}}{a_{\overline{360}|.01}} \doteq \$74{,}734.40$$

This is pretty discouraging progress for 10 years' work; however, the rate of progress does improve. You should use CALIPSO or a financial calculator to check the entries in Table 2.

Table 2: Some Unpaid Balances on a House Loan

Payment Number	Number of Years	Unpaid Balance
120	10	74,734.40
240	20	57,355.87
300	25	36,993.06
336	28	17,480.97
348	29	9,261.69

Example 5: Calculating the Interest Paid on a Loan

Suppose you take out a 10-year, $15,000 loan at 15% interest, compounded monthly. How much interest have you paid by the end of the 10 years? If you use (2) and CALIPSO to calculate the payments W you get

$$W = \frac{15{,}000}{a_{\overline{120}|.15/12}} = 242.00$$

Since you must make 120 payments of $242.00, the total amount you have paid is $29,040. Subtracting $15,000 from this amount you find that the interest paid is $14,040.

10.3A Exercises: Amortization

1. § You take out a 5-year loan of \$10,000 at 15% interest, compounded quarterly. What are your quarterly payments?

2. You take out a 3-year loan of \$5000 at 18% interest, compounded monthly. What are your monthly payments?

3. § You take out a 25-year loan of \$75,000 at 9% interest compounded annually. What are your annual payments?

4. You want to take out a loan at 12% interest compounded monthly. You are willing to make monthly payments of \$200 for 3 years. What is the amount of the loan?

5. § You want to take out a loan at 15% interest compounded quarterly. You can afford quarterly payments of \$500 for 4 years. What is the amount of the loan?

6. You want to take out a loan at 16.5% interest compounded monthly. You can make monthly payments of \$150 for 3 years. What is the amount of the loan?

7. § You take out a 4-year loan of \$8000 at 12% interest compounded monthly. After 1 year, how much do you still owe on the loan?

8. You take out a 15-year loan of \$30,000 at 10% interest compounded semiannually. After 8 years how much do you still owe on the loan?

9. § You take out a 5-year loan of \$20,000 at 15% interest rate compounded quarterly. After $3\frac{1}{2}$ years how much do you still owe on the loan?

10. You take out a 8-year loan of \$15,000 at 12% interest compounded quarterly. If you make quarterly payments, how much interest will you have paid by the end of the 8 years?

11. § You take out a 30-year loan of \$60,000 at 10% interest compounded monthly. If you make monthly payments, how much interest will you have paid by the end of the 30 years?

12. Suppose you take out a 6-year, \$12,000 loan at 14% compounded monthly. If you make monthly payments, what is the total amount you have paid by the end of the 6 years?

13. § Suppose you take out a 25-year, \$80,000 loan at 10% interest compounded monthly. If you make monthly payments, what is the total amount you have paid by the end of the 25 years?

14. Suppose you take out a 15-year, \$10,000 loan at 12% interest compounded quarterly. How much do you pay in interest?

15. § You buy a house for \$90,000 and make a 10% down payment. You finance the balance with a 25-year mortgage at 12.5% compounded monthly.

    (a) What are your monthly payments?

    (b) What is the total amount you pay for the house over 25 years?

    (c) If you sell the house at the end of 10 years of payments, what is the balance to be paid off?

    (d) How much of your 121st payment goes toward the principal and how much is for interest?

16. You buy a house for \$120,000 and make a 15% down payment. You finance the balance with a 30-year mortgage at 14.2% compounded monthly.

    (a) What are your monthly payments?

    (b) What is the total amount you pay for the house over 30 years?

    (c) If you sell the house at the end of 6 years of payments, what is the balance to be paid off?

    (d) How much of your 100th payment goes toward the principal and how much is for interest?

17. § You buy a car for \$15,000 and make a 10% down payment. You finance the balance with a 42-month loan at 14.5% compounded monthly.

    (a) What are your monthly payments?

    (b) What is the total interest you pay over the 42-month period?

    (c) If you trade in the car after 2 years, what is the balance to be paid off?

(d) How much would your payments be if you made a 20% down payment?

18. On December 31, 1991, the national debt was 3,801,698 million dollars ($3.801698 \cdot 10^{12}$). If congress decides in January 1992 to pay off this debt over the next 20 years, what will the monthly payments be at an annual interest rate of 8.5%? What will the monthly payments be to pay it off in 50 years? Will they ever pay it off?

10.4 MODELING FINANCIAL PROBLEMS WITH LINEAR PROGRAMMING

Financial problems often involve linking a series of investments into a plan that maximizes yield over a period of time. Example 1 illustrates this kind of problem.

Example 1: Saving for College

A woman wants to set up a fund to pay for her granddaughter's college education. The girl is now 8 years old and is expected to start college 10 years from now. The grandmother wants to calculate the amount of money she must invest now so that her granddaughter will have $25,000 available to her at the start of each of her 4 college years. She has identified three potential investments.

- She can purchase a municipal bond at 45% of its face value. The bond will be redeemable at its face value in 12 years.
- She can purchase a security that pays 7.5% effective annual interest and is redeemable at its face value in 10 years.
- She can deposit money at 4.7% effective annual interest in a savings account.

The grandmother lets b denote the face value of the bond, s denote the face value of the security, and $d_0, d_1, \ldots, d_{12}$ denote the amounts deposited in savings at the start of year 0 through year 12. She plans to deposit d_0 dollars in savings initially, and to deposit any extra cash realized from interest or redemptions in subsequent years.

Except for years in which the bond or the security is redeemed, the cash income consists of a 7.5% return from the security and a 4.7% return on the previous year's savings. For the years prior to the start of college, this cash will be redeposited in savings according to the formula

$$d_i = 0.075s + 1.047d_{i-1}$$

She enters the following LP model in CALIPSO's *LP SOLVER*. (Since *LP SOLVER* does not take subscripts, she enters d_i as *di*.)

```
MIN .45b + s + d0;
d1 = 0.075s + 1.047d0;                {Year 1}
d2 = 0.075s + 1.047d1;                {Year 2}
d3 = 0.075s + 1.047d2;                {Year 3}
d4 = 0.075s + 1.047d3;                {Year 4}
d5 = 0.075s + 1.047d4;                {Year 5}
d6 = 0.075s + 1.047d5;                {Year 6}
d7 = 0.075s + 1.047d6;                {Year 7}
d8 = 0.075s + 1.047d7;                {Year 8}
d9 = 0.075s + 1.047d8;                {Year 9}
d10 + 25000 = 1.075s+ 1.047d9;        {Start college. Security matures}
d11 + 25000 = 1.047d10;               {Year 11}
d12 + 25000 = b + 1.047d11;           {Bond matures}
25000 = 1.047d12;                     {Graduation year}
```

CALIPSO finds the solution to be $b = 48,877.75, s = 25,322.15, d0 = 0$ and $z = 47,317.13$.

It is possible to solve the problem of Example 1 by using compound interest and annuity formulas. However, you can see that such a solution will be complicated and that it would be easy to make a mistake. The LP solution is clear and simple. It does involve a good deal of tedious typing, especially if you wanted to solve a problem over many periods. Here are two editing features of *LP SOLVER* that can help reduce typing:

- The first 9 equations in the LP model for Example 1 are the same except for the changes to the subscript i in *di*. You can type the first of these equations $d1 = 0.075s + 1.047d0$; and use the *F*6 key to make 8 copies. You can then work through the copies changing the subscripts. (See the help screens of *LP SOLVER* for additional useful features.)

- If you are interested in modeling large problems with CALIPSO, you should examine CALIPSO's inequality generation features. You do this by starting *LP SOLVER* and reading the help messages that advertise inequality generation. To help you get started, Example 2 shows another way of entering the inequalities given in Example 1.

Example 2: Generating Inequalities with CALIPSO

CALIPSO accepts "for loops" written in a syntax similar to that of the BASIC programming language. More details and other capabilities are given in the help facilities of *LP SOLVER*. You can enter the LP problem in Example 1 as

```
MIN .45b + s + d[0];
for i=1 to 9 do d[i] = 0.075s + 1.047d[i-1];
d[10] + 25000 = 1.075s+ 1.047d[9];          {Start college}
d[11] + 25000 = 1.047d[10];                 {Year 11}
d[12] + 25000 = b + 1.047d[11];             {Bond matures}
25000 = 1.047d12;                           {Graduation year}
```

Note that the variable in the last inequality is written d12 and this same variable is written d[12] in the line above. We did this to show you that CALIPSO accepts either version. When CALIPSO encounters a pair of square brackets, it evaluates the expression between the brackets to get a number and concatenates this number to the variable. You can omit the brackets when the expression between them is a constant; you cannot omit the brackets in d[i] and d[i-1] because they are evaluated within the "for loop."

The formulas for calculating annuities assume that deposits (or withdrawals) will all be of the same amount. When the deposits (or withdrawals) vary in size, annuity problems are usually solved with a spreadsheet program. A *spreadsheet* is a program that allows you to enter data and formulas to be calculated in an array somewhat like an accounting ledger. The difference between a spreadsheet and an accounting ledger is that the computer makes the calculations for you. We will not discuss spreadsheets in this text because they are usually covered in accounting courses and not in finite mathematics courses. Example 3 indicates how to use both a spreadsheet and a linear programming problem to analyze an investment problem.

Example 3: Variable Withdrawals From an Annuity

In deciding a lawsuit brought by Bill Cunningham, who was injured in an industrial accident, the jury has determined that he should receive a cash payment for each of the next 10 years. The jury awarded these payments on an increasing schedule because his medical expenses are expected to grow over the years. Table 1 gives the schedule of payments.

Table 1: Annual Payment Schedule

Year	1	2	3	4	5	6	7	8	9	10
Cash (in $1000)	30	32	35	38	42	45	48	50	55	60

The attorneys for both parties have agreed to allow Bill to accept a lump sum of money in lieu of the annual payments. This sum is to be paid immediately and is of such a size as to allow Bill to invest it and realize at least the annual amounts in Table 1. Bill's attorney wants the lump sum to be large, and the industry's attorney wants it to be small. The judge must determine the size of the lump sum payment.

Bill's attorney argues that the money should be deposited in a government insured savings account because this is the safest investment. He assumes that such an account will pay an effective annual interest of 4.65% for each of the next 10 years. To calculate the lump sum payment, he lets L be the amount of the payment and $B_1, B_2, \ldots, B_9$ be the amounts deposited in savings at the beginning of each of the next 9 years. Since it will be most advantageous to his client, he assumes that the payment will be withdrawn at the beginning of each year. He also assumes that the bank interest will be added to the account at the time the withdrawal is made. Thus, for $2 \leq j \leq 9$, the account balance in year j is

$$B_j = B_{j-1} + 0.0465B_{j-1} - \text{(cash withdrawn)}$$

To find the value of L, he uses a spreadsheet to solve the system of linear equations

$$\begin{aligned}
B_1 &= L - 30000 & B_6 &= 1.0465B_5 - 45000 \\
B_2 &= 1.0465B_1 - 32000 & B_7 &= 1.0465B_6 - 48000 \\
B_3 &= 1.0465B_2 - 35000 & B_8 &= 1.0465B_7 - 50000 \\
B_4 &= 1.0465B_3 - 38000 & B_9 &= 1.0465B_8 - 55000 \\
B_5 &= 1.0465B_4 - 42000 & 0 &= 1.0465B_9 - 60000
\end{aligned}$$

to find that $L = 347,571.29$.

The industry's lawyer argues that Bill can invest his money just as safely and more profitably in a mixed portfolio consisting of a security from Ready Cash, Inc., a City of Moscow bond, a U.S. savings bond, and a bank account to hold loose cash. Table 2 describes the investments he selects as an example.

Table 2: Sample Investment Portfolio

Investment	Initial cost/share ($)	Annual return/share	Years to maturity	Repayment at maturity/share ($)
Ready Cash	1200	85	6	1200
Moscow bond	750	0	7	1200
U.S. bond	1500	0	8	2600
Savings	-	4.5%	-	-

The lawyer interprets Table 2 as follows. Bill will use the lump sum payment L to purchase R shares of Ready Cash, Inc. for $1200R$ dollars. He will receive $85R$ dollars each year and a repayment of $1200R$ dollars after 6 years. Bill will purchase M shares of Moscow bonds for $750M$ dollars and will redeem the bonds for $1200M$ dollars after 7 years. Bill will purchase U shares of U.S. bonds for $1500U$ dollars and will receive $2600U$ dollars after 8 years. He will place all extra cash in savings after taking his annual payment. The lawyer lets $B_i, 1 \leq i \leq 9$ be the amount deposited in savings at the beginning of year i. Thus, the lump sum of L dollars would be used to provide the first year's cash payment of $30,000, with the rest being divided among the investments according to the equation

$$L = 1200R + 750M + 1500U + B_1 + 30,000$$

The lawyer calculates that at the beginning of the second year the amount to be paid Bill and the amount to be left in savings would be determined by the equation

$$\text{income } = 85R + B_1 + 0.045B_1 = B_2 + 32,000 = \text{ disbursements}$$

Arguing in a similar way for the other years, he formulates the LP problem

Minimize: L

Subject to:
$$
\begin{array}{lllll}
L = & 1200R + 750M + 1500U + B_1 + 30,000 & & & \\
& 85R \quad + & 1.045B_1 & = B_2 + 32,000 & \{\text{Start of yr 2}\} \\
& 85R \quad + & 1.045B_2 & = B_3 + 35,000 & \{\text{Start of yr 3}\} \\
& 85R \quad + & 1.045B_3 & = B_4 + 38,000 & \{\text{Start of yr 4}\} \\
& 85R \quad + & 1.045B_4 & = B_5 + 42,000 & \{\text{Start of yr 5}\} \\
& 85R \quad + & 1.045B_5 & = B_6 + 45,000 & \{\text{Start of yr 6}\} \\
& 1285R \quad + & 1.045B_6 & = B_7 + 48,000 & \{\text{Start of yr 7}\} \\
& 1200M \quad + & 1.045B_7 & = B_8 + 50,000 & \{\text{Start of yr 8}\} \\
& 2600U \quad + & 1.045B_8 & = B_9 + 55,000 & \{\text{Start of yr 9}\} \\
& & 1.045B_9 & = 60,000 & \{\text{Start of yr 10}\}
\end{array}
$$

The lawyer uses CALIPSO to find the solution $L = 323,831.32$. We do not know what the judge decided.

The phrases *trust fund, sinking fund*, and *line of credit* all describe financial transactions in which money is delivered to one or more enterprises from one or more sources over a period of time. Many such problems can be modeled as *transportation problems*. We introduced transportation problems in Section 5.7, and you should review that section to recall the idea and the notation we use to solve these problems. Example 4 is a typical problem that can be modeled as a transportation problem.

Example 4: Managing a Line of Credit

To start up a small business George has arranged a \$400,000 line of credit from his bank. Since he does not need all of the money immediately, he plans to access what funds he needs at the start of each quarter. Thus, he will only pay interest on the money from the time he actually borrows it until he repays it. The conditions of the loan are that he pay a 3% loan initiation fee each time he borrows if the amount is \$40,000 or less plus 5% on the amount exceeding \$40,000. (Thus, if he borrows \$30,000, then the fee is $.03 \cdot 30,000 = \$900$. If he borrows \$60,000, the fee is $.03 \cdot 40,000 + .05 \cdot 20,000 = \$2,200$.) He can only borrow money at the start of a quarter for a period of 1 year. He must pay cash for loan fees and interest on the amount of the outstanding debt over the previous quarter at the rate of 4.5% per quarter. He plans to repay the debt at end of 2 years. He is not required to borrow the entire \$400,000.

George estimates that he will need \$60,000, \$90,000, \$80,000, and \$150,000 for the first, second, third, and fourth quarters, respectively. He wants to schedule the quarterly loans to minimize the cost of the loan over the 2-year period. Because the initiation fee depends on the amount of each quarterly loan, he regards each quarterly loan as two loans, one at a 3% initiation fee and one at a 5% initiation fee. Thus, if he borrows x dollars at the 3% and y dollars at the 5% initiation fee, the total fee will be $.03x + .05y$. Because he pays more for an amount over \$40,000, this same formula will work correctly when the total quarterly loan is smaller or larger than \$40,000.

To model the problem as a transportation problem, George introduces 8 *supply points* $S_{ia}, S_{ib}, 1 \le i \le 4$ where S_{ia} corresponds to a loan amount less than or equal to \$40,000 and S_{ib} corresponds to a loan amount in excess of \$40,000. He introduces 4 *demand points* $D_i, 1 \le i \le 4$ that correspond to the amounts he needs for each quarter. George sees that he can ignore the interest accrued during the second year since it is independent of the quarter in which he borrows the money. A solution that minimizes his cost over the first year will also minimize his cost over the 2 years. George also notices that it may be to his advantage to borrow more money in a quarter than he needs for that quarter and save it for use in a later quarter. He makes the Supply/Demand table:

Supply points	Demand points				
	D_1	D_2	D_3	D_4	Supply
S_{1a}	0.21	0.21	0.21	0.21	40,000
S_{1b}	0.23	0.23	0.23	0.23	400,000
S_{2a}	∞	0.165	0.165	0.165	40,000
S_{2b}	∞	0.185	0.185	0.185	400,000
S_{3a}	∞	∞	0.12	0.12	40,000
S_{3b}	∞	∞	0.15	0.15	400,000
S_{4a}	∞	∞	∞	0.075	40,000
S_{4b}	∞	∞	∞	0.095	400,000
Demand	60,000	90,000	80,000	150,000	

Here is how he calculated some of the rows.

Row S_{1a} A dollar borrowed in the first quarter costs the initiation fee of 3% plus 4 quarters at 4.5% for a total of $.03+4(.045) = 0.21$ regardless of which demand point (which quarter) utilizes this dollar.

Row S_{1b} This row is similar to S_{1a} except the initiation fee is 2% higher. He gave the amount of supply as \$400,000 because that is a convenient limit larger than any amount he could draw. The actual amount of the loan will be limited by the quantities in the demand row and his desire to minimize interest.

Row S_{2a} George put ∞ in column D_1 to reflect the fact that he cannot borrow money in the second quarter to use in the first quarter. He eliminates that impossibility from the model by assessing an exhorbitant cost for the transaction. The other entries are the initialization cost of 3% plus 3 interest payments of 4.5% each, that is, $.03+3(.045) = 0.165$.

George numbers the supply points $1, 2, \ldots, 8$ to correspond to $S_{1a}, S_{1b}, \ldots, S_{4a}, S_{4b}$ and introduces 32 variables $x_{ij}, 1 \le i \le 8, 1 \le j \le 4$ where x_{ij} is the amount of money borrowed from supply point i and utilized in demand point D_j. For instance, $x_{43} = 4,200$ corresponds to borrowing $4,200 from supply point 4 (S_{2b}) for use in the third quarter (D_3). With this notation and ∞ replaced by 100, the objective function for the LP model is

$$
\begin{array}{llllllll}
\multicolumn{8}{l}{\text{Minimize: } z =} \\
0.21x_{11} & + & 0.21x_{12} & + & 0.21x_{13} & + & 0.21x_{14} \\
+0.23x_{21} & + & 0.23x_{22} & + & 0.23x_{23} & + & 0.23x_{24} \\
+100 & + & 0.165x_{32} & + & 0.165x_{33} & + & 0.165x_{34} \\
+100 & + & 0.185x_{42} & + & 0.185x_{43} & + & 0.185x_{44} \\
+100 & + & 100 & + & 0.12x_{53} & + & 0.12x_{54} \\
+100 & + & 100 & + & 0.15x_{63} & + & 0.15x_{64} \\
+100 & + & 100 & + & 100 & + & .075x_{74} \\
+100 & + & 100 & + & 100 & + & .095x_{84}
\end{array}
$$

There are 8 constraints on supply of money. The one for supply point 3 (S_{2a}) is

$$x_{31} + x_{32} + x_{33} + x_{34} \le 40,000$$

There are 4 constraints on the quarterly demands for money. The one for the second quarter (D_2) is

$$x_{12} + x_{22} + x_{32} + x_{42} + x_{52} + x_{62} + x_{72} + x_{82} = 90,000$$

In Problem 5 of the exercises you are asked to write the rest of the constraints and use CALIPSO to show that the minimum fees for the first year are $53,100.

10.4A Exercises: Solving Financial Problems with LP

1. The grandmother of Example 1 has discovered another security that may be attractive. It pays 7.2% annual interest and is redeemable at 110% of its face value in 7 years.

   (a) Modify the LP problem to include this security.

   (b) § Solve the new problem with CALIPSO to see if the new security is worth purchasing.

2. A couple wins $500,000 in the lottery. The money will be paid to them on January 1, 1993. They decide to invest their windfall in a way that should realize the following goals.

   (a) On January 1, 1993 they will

      i. Buy a Mercedes-Benz automobile for $32,000.

      ii. Make a down payment of $30,000 on a house and agree to pay $12,000 per year on the first of January 1994 and each January 1 thereafter until they make 20 payments.

      iii. Buy x dollars worth of shares of a security that provides a return of 8% per year and is redeemable for its face value on January 1, 2008.

      iv. Buy y shares of government bonds at 48% of face value that are redeemable at face value on January 1, 2005.

      v. Deposit the remainder of the fund's dollars in a savings account that pays 4.8% effective annual interest. On each January 1 from 1993 to 2018, they will deposit all cash realized from these investments in the savings account.

   (b) Pay taxes of $120,000 on January 1, 1994.

   (c) On January 1, 1998, they will spend $30,000 on a trip to Europe.

   (d) On January 1 in the years 2000, 2001, 2002, and 2003, they will give their son $45,000 to pay for a year of college.

   (e) On January 1, 2005, they will spend $40,000 on a trip to Australia.

   (f) On January 20, 2010 they will give their son $50,000.

   Find the values of x, y, and the initial bank deposit that maximize the amount of the inheritance that remains on January 1, 2019.

3. The lawyer in Example 3 is not well satisfied with the advantages over a savings account displayed by his sample portfolio. He wants to try to improve it by including the same security being considered by the grandmother in Problem 1.
   (a) Modify the LP problem to include this security.
   (b) § Solve the new problem with CALIPSO to see if the new security is worth purchasing.

4. Complete the model in Example 4 and use CALIPSO to solve it.

5. MK, a construction company, has obtained a contract to build an extension to the library at the University of Idaho. To finance the construction MK arranges a \$3.5 million line of credit from a bank at an annual rate of 13.5% compounded quarterly. MK can draw up to \$500,000 to finance the next 3 months of construction at the start of each quarter. All of the money it borrows is to be repaid at the end of a 2-year (8 quarters) period that commences when the first quarterly draw is made. Interest must be paid on each quarterly draw from the time that draw is made. This interest is added to the amount of the loan, and so interest must be paid on the interest in subsequent quarters.

   In addition to the interest, an initiation fee of 3% of the amount of each draw must be paid. This initiation fee is added to the amount of the debt, and, of course, the quarterly interest must also be paid on the fee. In order to avoid initialization fees, MK can elect to draw more money than it needs in a particular quarter and use it in future quarters. Of course, MK could place excess money in an interest bearing account until it needs it for construction; however, to keep things simple, we will not trouble you with that possibility now.

   Table 1 gives MK's estimate of the amounts it will need in each quarter.

   Table 3: Quarterly Estimates in Thousands of Dollars

Quarter	1	2	3	4	5	6	7	8
Cost	300	400	500	700	500	600	200	300

   (a) Make a supply/demand table to model this problem as a transportation problem.
   (b) § Solve the problem with CALIPSO's *LP SOLVER* to minimize the interest MK must pay.

11 Introduction to Logic

In this chapter we examine some basic notions underlying mathematical logic.

11.1 STATEMENTS AND TRUTH TABLES

For our purposes a *statement* is a sentence or expression to which we can assign a truth value: true (T) or false (F). For instance, the sentence "Goldfish can't fly" is a statement since we can assign the truth value, T, to it. Similarly the expression "$5 + 6 = 13$" is a statement since we can assign the truth value, F, to it. On the other hand, the sentence "She is beautiful" is not a statement since, without additional information, we don't know who she is and therefore can't assign a truth value to the sentence. Similarly, it is difficult to assign truth values to the expressions: "Run faster" and "How high is the moon?".

In what follows we will often let p, q, r, etc., represent statements. We can use *connectives* "and," "or," and "not" to form new statements (called *compound statements*) from given statements. In this context we use the symbol $\wedge$ to denote *and*, the symbol $\vee$ to denote *or*, and the symbol $\sim$ to denote *not*. If p and q are statements, then the *disjunction* of p and q is the compound statement $p \vee q$, the *conjunction* of p and q is the statement $p \wedge q$, and the *negation* of p is the statement $\sim p$. At times we use the term *simple statement* to refer to statements that are not compound statements.

Example 1: Forming Compound Statements

If p is the simple statement: "It rained yesterday" and q is the simple statement: $2+6 = 9$, then the disjunction, $p \vee q$, of p and q is the compound statement: "It rained yesterday or $2 + 6 = 9$." (The word *or* is used in the inclusive sense: it includes the possibilities that it rained yesterday, or $2+6 = 9$, or both statements are true.) The conjunction, $p \wedge q$, of p and q is the compound statement: It rained yesterday and $2 + 6 = 9$. The negation, $\sim p$, of the statement p is the statement: It did not rain yesterday; the negation, $\sim q$, of the statement q is the statement $2 + 6 \neq 9$.

It is not surprising that the truth values of statements comprising a compound statement determine the truth value of the compound statement. Tables 1, 2, and 3 give the truth values of the conjunction, disjunction, and negation of statements p and q corresponding to all possible assignments of truth values to p and q.

Table 1: Conjunction

p	q	$p \wedge q$
T	T	T
T	F	F
F	T	F
F	F	F

Table 2: Disjunction

p	q	$p \vee q$
T	T	T
T	F	T
F	T	T
F	F	F

Table 3: Negation

p	$\sim p$
T	F
F	T

Example 2: Applying Truth Tables

If p is the statement "Jim Calvert is a millionaire" and q is the statement "Bill Voxman can run a 10-minute mile," then p is false and q is true. From Table 1 we see that the statement "Jim Calvert is a millionaire *and* Bill Voxman can run a 10-minute mile" is false. Table 2 shows that the statement "Jim Calvert is a millionaire *or* Bill Voxman can run a 10-minute mile" is true. By Table 3, the statement "Jim Calvert is not a millionaire" is true, and the statement "Bill Voxman cannot run a 10-minute mile" is false.

Exclusive Or: As indicated above we have used the word *or* in the sense of an *inclusive or*; thus, the assertion that $p \vee q$ is true allows the possibility that both p and q are true. Occasionally it is useful to utilize an *exclusive or*, which we denote $\oplus$. The statement $p \oplus q$ is true if p is true or if q is true but is false if both p and q are true. Table 4 defines the truth values of the exclusive or.

Table 4

p	q	$p \oplus q$
T	T	F
T	F	T
F	T	T
F	F	F

You can construct truth tables for compound statements that use various connectives between two or more statements. Examples of such statements are $\sim ((p \wedge q) \vee (\sim q))$ and $((p \wedge q) \vee r) \wedge (\sim p \vee r)$.

Note: When using a number of connectives you must be careful that the resulting compound statement is not ambiguous. For instance, $\sim p \vee q$ might be interpreted as the statement $(\sim p) \vee q$ or as the statement $\sim (p \vee q)$. As you can verify, these two statements have different truth values. To avoid this confusion you can insert enough parentheses so that there is no ambiguity. However, to avoid using too many parentheses it is helpful to adopt an order of precedence of the various symbols. The commonly used order is

$$\sim,\ \wedge,\ \vee,\ \oplus \tag{1}$$

Thus, if you use this order of precedence, you know that the statement $\sim p \wedge q$ is to be interpreted as $(\sim p) \wedge q$. In what follows we will adopt the order (1).

Example 3: Assigning Truth Values to Compound Statements

Table 5 gives the truth values of the compound statement

$$(\sim p \wedge q) \vee (\sim q)$$

You can obtain these values by using Tables 1, 2, and 3.

Table 5

p	q	$\sim p$	$\sim q$	$\sim p \wedge q$	$(\sim p \wedge q) \vee (\sim q)$
T	T	F	F	F	F
T	F	F	T	F	T
F	T	T	F	T	T
F	F	T	T	F	T

Table 6 gives the truth values of the compound statement

$$((p \wedge q) \vee r) \wedge (\sim p \vee r)$$

for the various truth assignments to p, q, and r.

Table 6

p	q	r	$p \wedge q$	$(p \wedge q) \vee r$	$\sim p$	$\sim p \vee r$	$((p \wedge q) \vee r) \wedge (\sim p \vee r)$
T	T	T	T	T	F	T	T
T	T	F	T	T	F	F	F
T	F	T	F	T	F	T	T
T	F	F	F	F	F	F	F
F	T	T	F	T	T	T	T
F	T	F	F	F	T	T	F
F	F	T	F	T	T	T	T
F	F	F	F	F	T	T	F

CALIPSO

You can use CALIPSO to evaluate logical expressions by choosing *LOGIC* from the CALIPSO menu. *LOGIC* accepts expressions that use the letters $m, n, \ldots, z$ to represent simple statements. *LOGIC* will generate a truth table by assigning to a set of simple statements in a given expression all sequences that can be made with T and F. For instance, in Table 6 the simple statements are p, q, and r. Table 6 contains a row for each sequence of assignments of T or F to the statements p, q, r. (CALIPSO cannot deal with more than 4 simple statements in any expression because 4 simple statements require 16 rows in a truth table and that is as many as can be displayed on the computer screen.) You form an expression to be evaluated in a column of the truth table by combining the simple statements with the logical connectives *and*, *inclusive or*, *exclusive or*, and *negation* (as well as *implication* and *biconditional* that will be introduced in Section 11.2). *LOGIC* does not accept the symbols $\wedge, \vee$, and $\oplus$ because there is no key on the keyboard for these characters. There is a key for negation ($\sim$). The keys that correspond to the connectives are

Connective	Language Equivalent	Language Symbol	CALIPSO Symbol
Conjunction	And	$\wedge$	&
Disjunction	Or	$\vee$	|
Exclusion	Or	$\oplus$	!
Negation	Not	$\sim$	~

To make a table (like Table 6) that contains columns for several expressions, the entry panel for *LOGIC* allows you to assign an expression to the letters $a, b, c, \ldots, i$. You can then use these letters to build longer expressions. (To help you distinguish letters that represent simple statements from letters that represent expressions, *LOGIC* displays the letters $a, b, \ldots, i$ as capital letters and the letters $m, n, \ldots, z$ as lower-case letters regardless of the case you use to type them.) For instance, to make a table that has columns for $p\&q$, $p| \sim r$ and for the composite $(p\&q)|(p| \sim r)$ of these statements, you can enter $A = p\&q$, $B = p| \sim r$, and $C = A|B$. CALIPSO will evaluate each expression (A, B, C) and construct a truth table with 6 columns headed by

$$p \quad q \quad r \quad A \quad B \quad C$$

The columns headed by A, B, C will contain the truth values of the compound expressions for each assignment. Columns for the simple statements p, q, and r are automatically generated to give all permutations of assignments of T and F to the simple statements. Examples 4 and 5 below illustrate the procedure.

Tautologies and Contradictions: A compound statement A is a *tautology* if A is true regardless of the truth values assigned to the simple statements comprising A. A compound statement B is a *contradiction* if B is false regardless of the truth values assigned to the simple statements comprising B.

Example 4: A Tautology

You can use CALIPSO to construct Table 7 to show that the statement $(\sim p \vee q) \vee (\sim p \wedge q) \vee p$ is a tautology. The second row of Table 7 shows how to enter the terms of the statement in CALIPSO to create this table.

Table 7

p	q	$\sim p \vee q$ $A = \sim p\|q$	$\sim p \wedge q$ $B = \sim p\&q$	$(\sim p \vee q) \vee (\sim p \wedge q)$ $C = A\|B$	$(\sim p \vee q) \vee (\sim p \wedge q) \vee p$ $D = C\|p$
T	T	T	F	T	T
T	F	F	F	F	T
F	T	T	T	T	T
F	F	T	F	T	T

Example 5: A Contradiction

You can use CALIPSO to construct Table 8 to show that the statement $(p \wedge q) \wedge (\sim p)$ is a contradiction. The second row of Table 8 shows how to enter the terms of the statement in CALIPSO to create this table.

Table 8

p	q	$p \wedge q$ $A = p\&q$	$\sim p$ $B = \sim p$	$(p \wedge q) \wedge (\sim p)$ $C = A\&B$
T	T	T	F	F
T	F	F	F	F
F	T	F	T	F
F	F	F	T	F

11.1A Exercises: Statements and Truth Tables

Let p be the statement "The sun rises in the West" and let q be the statement $2 + 2 = 4$. In Problems 1 – 4 write out the English equivalents of the given compound statements.

1.§ (a) $p \wedge q$ (b) $\sim q \vee p$
2. (a) $\sim (p \vee q)$ (b) $p \vee \sim q$
3.§ (a) $\sim p \vee \sim q$ (b) $\sim p \wedge \sim q$
4. (a) $\sim (\sim p \wedge \sim q)$ (b) $\sim (\sim p \vee \sim q)$

5. Let p be the statement "Fred flew to Paris" and let q be the statement "Fred flew to Rome." Write the following statements in symbolic form in terms of p and q.

   (a) § Fred flew to Paris but not to Rome.

   (b) Fred did not fly to either Paris or Rome.

   (c) § Fred either flew to Rome or did not fly to Paris.

   (d) Fred flew to both Paris and Rome.

6. Let p be the statement "Alice is married" and let q be the statement "Alice won the lottery." Write the following statements in symbolic form in terms of p and q.

   (a) Alice is married but did not win the lottery.

   (b) § Alice is unmarried and won the lottery.

(c) Alice is either unmarried or she won the lottery.

7. Construct truth tables for the following compound statements. Check your answers with CALIPSO.

(a) $p \vee (\sim q)$	(b) $\sim p \wedge (\sim q)$
(c) $(p \wedge q) \oplus q$	(d) $(\sim p \vee (\sim q)) \vee (\sim p \wedge q)$
(e) $(\sim p \oplus q) \vee (p \vee q)$	(f) $(p \wedge q) \vee (p \vee (\sim q))$

8. Construct truth tables for the following compound statements. Check your answers with CALIPSO.

(a) $p \wedge (q \vee r)$	(b) $(\sim p \wedge r) \vee (q \vee r)$
(c) $(p \wedge q) \vee (p \wedge r)$	(d) $(p \oplus q) \wedge (\sim p \vee q)$

9. Use CALIPSO to construct truth tables for the following compound statements:

(a) $p \wedge \sim q$	(b) $(p \vee q) \vee (p \vee \sim q)$
(c) $(p \vee q) \oplus (\sim q)$	(d) $\sim ((p \vee q) \wedge r)$
(e) $(\sim p \vee q) \vee (\sim p \wedge \sim q)$	(f) $(p \wedge q) \vee \sim p$
(g) $(\sim p \oplus q) \wedge (\sim p \vee \sim q)$	(h) $(p \oplus q) \vee \sim\sim p$

10. Determine which of the following are tautologies or contradictions. Check your answers with CALIPSO.

(a) § $p \wedge \sim p$	(b) $p \oplus \sim p$
(c) § $(p \wedge q) \wedge \sim q$	(d) $(\sim (p \vee q)) \vee ((\sim p \wedge q) \vee p$
(e) § $\sim p \vee \sim q$	(f) $(p \vee q) \wedge ((\sim p) \wedge (\sim q))$
(g) § $(\sim p \vee \sim q) \wedge (p \wedge q)$	(h) $(p \wedge q) \wedge \sim q$

11.2 IMPLICATIONS AND LOGICAL EQUIVALENCE

The statements:

- If $x = 4$, then $3x = 12$.
- If the temperature exceeds 100, then I will go swimming.

- If the moon is blue, then the Bears will be NFL champions.

are examples of *implications.* In general, if p and q are statements, then the statement: "If p, then q" (denoted $p \to q$) is an implication. The symbol $\to$ is often referred to as a *conditional connective.* If $p \to q$, then the statement p is called the *hypothesis* (or *premise*) and q is called the *conclusion.* Table 1 is the truth table for $p \to q$.

Table 1

p	q	$p \to q$
T	T	T
T	F	F
F	T	T
F	F	T

To see why this table is reasonable, consider the implication: If I win \$1,000,000 in the lottery, then I will buy everyone a round of drinks. This statement is of the form $p \to q$, where p is the statement: I win \$1,000,000 in the lottery; and q is the statement: I will buy everyone a round of drinks.

We check each row of Table 1.

Row 1: If indeed I win the lottery (p is true) and if I buy everyone a round of drinks (q is true), then clearly my offer ($p \to q$) is true.

Row 2: If I win the lottery (p is true) and if I don't buy everyone a round of drinks (q is false), then my offer ($p \to q$) is false.

Rows 3 and 4: If I don't win the lottery (p is false), then my offer cannot be false, whether I buy or I don't buy the drinks. Thus, in either case the statement $p \to q$ must be true.

Example 1: Some Implications

(a) The implication "If $3 + 5 = 9$, then $2 + 2 = 5$" is true since the hypothesis is false.

(b) The implication "If the sun is larger than the earth, then the moon is larger than the earth" is false since although the hypothesis is true, the conclusion is false.

(c) The implication "If $5^2 = 25$, then $6^2 = 36$" is true since both the hypothesis and the conclusion are true.

(d) The implication

$$\text{"If } x = -3, \text{then } x^2 = 9\text{"} \tag{1}$$

is true even though we cannot say that the statement $x = -3$ is necessarily true. However, if $x = -3$ is true, then so is the statement $x^2 = 9$ and therefore the statement (1) is true. If the statement $x = -3$ is false, then it follows from Table 1 that the statement (1) is true. Therefore, the implication (1) is true in all instances.

CALIPSO

You enter the connnective $\rightarrow$ in CALIPSO's *LOGIC* program by typing ">". The ">" displays on the screen as "$\rightarrow$".

As the next example illustrates, the connective $\rightarrow$ can be used with other connectives introduced in the previous section. The order of precedence is $\sim, \wedge, \vee, \oplus, \rightarrow$.

Example 2: An Implication

Table 2 gives the truth values for the statement $(\sim p \wedge q) \rightarrow (p \vee q)$.

Table 2

p	q	$\sim p$	$\sim p \wedge q$	$p \vee q$	$(\sim p \wedge q) \rightarrow (p \vee q)$
T	T	F	F	T	T
T	F	F	F	T	T
F	T	T	T	T	T
F	F	T	F	F	T

Biconditional Connective Frequent use is made of the *biconditional connective* $\leftrightarrow$. The biconditional statement $p \leftrightarrow q$ indicates that if p is true, then so is q; and if p is false, then so is q. Table 3 is the truth table for $\leftrightarrow$.

Table 3

p	q	$p \leftrightarrow q$
T	T	T
T	F	F
F	T	F
F	F	T

The statement $p \leftrightarrow q$ is often read "p if and only if q."

CALIPSO

You enter the biconditional connective in CALIPSO with the key "–" (this is the minus sign). The symbol $\leftrightarrow$ is displayed on the screen.

Implications ($\rightarrow$) and biconditional ($\leftrightarrow$) have a lower precedence than the other connectives. The order of precedence of all of the connectives is

$$\sim, \wedge, \vee, \oplus, \rightarrow, \leftrightarrow$$

or, given in terms of CALIPSO symbols, the order is

$$\sim, \&, |, !, >, -$$

Example 3: A Biconditional Statement

The statement "$x = 4$ if and only if $3x = 12$" is a biconditional statement. You can write this statement in the form $p \leftrightarrow q$, where p is the statement "$x = 4$" and q is the statement "$3x = 12$." Observe that if p is true, then so is q; and if p is false, so is q.

Logical Equivalence. We say that two statements are *logically equivalent* if they have identical truth values. It follows that statement A is logically equivalent to statement B when $A \leftrightarrow B$ is true for all assignments of T and F to the simple statements that comprise A and B, that is, when $A \leftrightarrow B$ is a tautology.

Example 4: Logically Equivalent Statements

Table 4 shows (in two ways) that the statements $A = p \wedge (q \vee r)$ and $B = (p \wedge q) \vee (p \wedge r)$ are logically equivalent. One way is to verify that A and B have the same truth values and the other is to check that $A \leftrightarrow B$ is a tautology.

Table 4

p	q	r	$A = p \wedge (q \vee r)$	$B = (p \wedge q) \vee (p \wedge r)$	$A \leftrightarrow B$
T	T	T	T	T	T
T	T	F	T	T	T
T	F	T	T	T	T
F	T	T	F	F	T
T	F	F	F	F	T
F	T	F	F	F	T
F	F	T	F	F	T
F	F	F	F	F	T

11.2A Exercises: Implications and Logical Equivalences

In Problems 1 – 8 decide whether the statement is true or false.

1. § If $2 + 3 = 5$, then $2 + 4 = 7$.
2. If $2 + 4 = 7$, then $2 + 3 = 5$.
3. § If $x = -5$, then $x^2 = 25$.
4. If $x = 5$, then $3x = 15$.
5. § If $x = 0$, then $x^2 = x$.
6. $x = 6$ if and only if $4x - 1 = 23$.
7. § $x = x^2$ if and only if $x = 1$.
8. If the earth weighs one pound, then the sun weighs 4 pounds.

In Problems 9 – 19 construct truth tables for the given compound statements. Check your answers with CALIPSO. Also determine which of the statements are tautologies or contradictions.

9. § $\sim p \rightarrow \sim q$
10. $(p \leftrightarrow q) \wedge (q \rightarrow p)$
11. § $p \rightarrow (p \vee q)$
12. $\sim (p \vee q) \rightarrow (q \wedge \sim r)$

13. § $((p \wedge r) \vee (q \wedge (\sim r))) \leftrightarrow ((\sim p \wedge r) \vee (\sim q \wedge \sim r))$

14. $((\sim q) \wedge (p \rightarrow q)) \rightarrow \sim p$

15. § $((p \rightarrow q) \wedge (q \rightarrow r)) \rightarrow (p \rightarrow r)$

16. $\sim (p \wedge q) \leftrightarrow (p \vee \ q)$

17. § $((p \wedge q) \rightarrow r) \leftrightarrow (p \wedge q \wedge r)$

18. $(\sim p \rightarrow q) \leftrightarrow (p \rightarrow q)$

19. § $(\sim p \wedge q) \leftrightarrow (\sim q \vee p)$

In Problems 20 – 24, determine which pairs of statements are logically equivalent. Check your answers with CALIPSO.

20. $\sim (p \rightarrow \sim q)$; $p \wedge q$

21. § $p \rightarrow \sim q$; $p \vee \sim q$

22. $p \wedge (q \vee r)$; $(p \wedge q) \vee (p \wedge r)$

23. § $p \rightarrow (q \wedge r)$; $(p \rightarrow q) \wedge (p \rightarrow r)$

24. $(\sim ((\sim p) \vee (\sim q))) \vee r$; $p \vee q \vee r$

11.3 CONDITIONAL STATEMENTS AND VALID ARGUMENTS

The implication $\rightarrow$ is an example of a conditional statement. In this section we consider additional conditional statements and how they relate to each other.

Definition. Suppose p and q are statements. The *converse* of the implication $p \rightarrow q$ is the statement $q \rightarrow p$. The *contrapositive* of the statement $p \rightarrow q$ is the statement $\sim p \rightarrow \sim q$.

Example 1: Converse and Contrapositive of $p \rightarrow q$

Let A be the implication: If the entries of M are all 0, then M is not invertible. (In this statement M is a matrix.)

- The converse of A is the statement: If M is not invertible, then the entries of M are all 0.

- The contrapositive of A is the statement: If M is invertible, then not all of the entries of M are 0.

Observe that in Example 1 the statement A is true, but the converse of A is not true for all matrices M. Observe, however, that the contrapositive of A is true. In fact, as Table 1 shows, an implication $p \to q$ and its contrapositive $\sim q \to \sim p$ are logically equivalent. This is of considerable importance in proving many mathematical theorems, since it is often easier to establish the contrapositive of a theorem than to give a direct proof of the theorem itself.

Table 1 gives the truth tables for the implication $p \to q$ and its converse and contrapositive.

Table 1

p	q	$p \to q$	$q \to p$	$\sim q \to \sim p$
T	T	T	T	T
T	F	F	T	F
F	T	T	F	T
F	F	T	T	T

Observe from Table 1 that the converse of $p \to q$ does not have the same truth values as $p \to q$, and so is not logically equivalent to this implication.

The next example shows that if an implication is true, its converse may or may not be true.

Example 2: More Converses and Contrapositives

(a) Let A be the implication "If I live in Chicago, then I live in Illinois." The converse of A "If I live in Illinois, then I live in Chicago" is false. However, the contrapositive of A "If I don't live in Illinois, then I don't live in Chicago" is true.

(b) Let B be the implication "If $x = 6$, then $3x + 4 = 22$." You can verify that the converse and contrapositive of B are true.

(c) Let C be the implication "If $x = 3$, then $x^2 = 9$." You can verify that the contrapositive of C is true, but that the converse of C "If $x^2 = 9$, then $x = 3$" is not true (since, if $x^2 = 9$, then x can equal -3).

11.3A Exercises: Conditional Connectives

Determine the converse and contrapositive of the statements given in Problems 1 – 5 and give the truth value of each statement.

1. § If I live in Iowa, then I live in Dubuque.

2. If $4 + 5 = 9$, then $2 + 2 = 5$.

3. § If $x = 10$, then $x^2 = 100$.

4. If ab is an odd integer, then a and b are odd integers. (Assume that a and b are integers.)

5. § If the product AB of two matrices is a square matrix, then the matrices A and B are square matrices.

6. Let A be the statement $p \to q$. Show that the conjunction of A and its converse is logically equivalent to $p \leftrightarrow q$.

7. § Let B be the statement $p \to q$. Is the negation of the converse of B logically equivalent to the contrapositive of B? Is the negation of the contrapositive of B logically equivalent to the negation of B?

Valid Arguments

An *argument* consists of series of statements called *premises* (or *hypotheses*) and a statement called a *conclusion*. We say the argument is *valid* if, given that all of the premises are true (or, equivalently, given that the conjunction of all of the premises is true), then the conclusion is true. Somewhat more formally we say that an argument with premises $p_1, p_2, \ldots, p_n$ and conclusion c is valid if whenever $p_1, p_2, \ldots, p_n$ are true (or, equivalently, $p_1 \wedge p_2 \wedge \ldots \wedge p_n$ is true), then the implication

$$p_1 \wedge p_2 \wedge \ldots \wedge p_n \to c \qquad (1)$$

is true.

An argument is *invalid* if the conclusion does not follow from the premises; that is, given that the premises are true, the conclusion [or equivalently, (1)] is not true.

Example 3: A Valid Argument

Assume that the premises

If you do homework, you will get an A.
You do homework.

of an argument are true. Then the conclusion "You will get an A" is true, and therefore this is a valid argument.

More formally, let p and q be the statements

p: You do homework.
q: You will get an A.

You need to show that under the assumption that $p \rightarrow q$ and q are true, the implication

$$((p \rightarrow q) \wedge p) \rightarrow q \tag{2}$$

is also true. It follows from the first row of Table 2 (which is the only row we consider since we assume $p \rightarrow q$ and p are true) that (2) is true.

Table 2

p	q	$p \rightarrow q$	$(p \rightarrow q) \wedge p$	$((p \rightarrow q) \wedge p) \rightarrow q$
T	T	T	T	T
T	F	F	F	T
F	T	T	F	T
F	F	T	F	T

Note: Since in Table 2, p and q can be any statements, it follows from the first row of this table that any argument of the form

$$\begin{array}{ll} \text{Premises:} & p \rightarrow q \\ & p \\ \text{Conclusion:} & q \end{array} \tag{3}$$

is valid (where, of course, we assume that the premises are true). The argument (3) is often referred to as the *direct method* for establishing the validity of an assertion. Other names associated with this kind of argument are *law of detachment*, *modus ponens*, and *assuming the antecedent*.

Example 4: A Valid Argument

We check the validity of the argument: All clever persons like math. John is a person. John does not like math. Therefore, John is not a clever person. To analyze this argument, we consider the statements:

p: A person is clever.
q: A person likes math.
r: John is a person.

The premises (which we assume to be true) are

$$A = p \rightarrow q$$
$$B = \sim q \wedge r$$

and the conclusion is

$$C = \sim p \wedge r \text{ (The person is not clever and is John.)}$$

We need to show that under the assumption that A and B are true, the implication

$$(A \wedge B) \rightarrow C$$

or equivalently, the implication

$$((p \rightarrow q) \wedge (\sim q \wedge r)) \rightarrow (\sim p \wedge r)$$

is true. Table 3 shows that this is the case

Table 3

p	q	r	A	B	$A \wedge B$	C	$(A \wedge B) \rightarrow C$
T	T	T	T	F	F	F	T
T	T	F	T	F	F	F	T
T	F	T	F	T	F	F	T
T	F	F	F	F	F	F	T
F	T	T	T	F	F	T	T
F	T	F	T	F	F	F	T
F	F	T	T	T	T	T	T
F	F	F	T	F	F	F	T

Note that to determine the validity of the argument we use only row 7 of Table 3, since this is the only row corresponding to the assumption that the premises $p \to q$ and $\sim q \wedge r$ are both true.

Example 5: An Invalid Argument

Consider the argument:

Premises: If I win the lottery, I go to Hawaii.
I go to Hawaii.
Conclusion: I win the lottery.

Let p be the statement "I win the lottery" and q be the statement "I go to Hawaii." Then the argument is

Premises: $p \to q$
q
Conclusion: p

We assume that the premises are true and under this assumption check the validity of the implication

$$((p \to q) \wedge q) \to p \tag{4}$$

Table 4

p	q	$p \to q$	$(p \to q) \wedge q$	$((p \to q) \wedge q) \to p$
T	T	T	T	T
T	F	F	F	T
F	T	T	T	F
F	F	T	F	T

It follows from row 3 of Table 4 that the argument is not valid, since even though the conjunction of the premises $p \to q$ and q is true, the implication (4) is false.

Example 6: The Law of Syllogism

The *Law of Syllogism* states that the argument

$$\begin{array}{ll} \text{Premises:} & p \to q \\ & q \to r \\ \text{Conclusion:} & p \to r \end{array}$$

is valid. Rows 1, 4, 7, 8 of Table 5 show why.

Table 5

p	q	r	$p \to q$	$q \to r$	$p \to r$	$((p \to q) \wedge (q \to r)) \to (p \to r)$
T	T	T	T	T	T	T
T	T	F	T	F	F	T
T	F	T	F	T	T	T
T	F	F	F	T	F	T
F	T	T	T	T	T	T
F	T	F	T	F	T	T
F	F	T	T	T	T	T
F	F	F	T	T	T	T

11.3B Exercises: Valid and Invalid Arguments

In Problems 1 – 7 determine the validity of the given argument.

1. § If Fred smokes, he will get lung cancer. Fred smokes. Therefore, Fred will get lung cancer.

2. If Idaho wins twenty games and is lucky, then Idaho will win the conference championship. Idaho wins twenty games. Therefore, Idaho wins the conference championship.

3. § If I eat broccoli, I will live to age 80. I do not eat broccoli. Therefore, I will not live to age 80.

4. If I don't drive over 60 mph, I will not have an accident. I drive 70 mph. Therefore, I will have an accident.

5. § If I do my homework, I will pass this course. If I don't play too much frisbee, I will do my homework. I did not pass this course. Therefore, I play too much frisbee.

6. If Fido escapes, she will eat Mitzi. If Mitzi is eaten, I will be in trouble. Therefore, if Fido escapes, I will be in trouble.

7. § If it is too cold, the pool will not be open. If I bring a rubber duck, the 3-year-old will be ecstatic. The 3-year-old is not ecstatic and the pool is open. Therefore, I forgot the rubber duck.

In Problems 8 – 15, determine the validity of the given argument. Check your answers with CALIPSO.

8. Premises: $p \vee \sim q$; $r \vee \sim q$
   Conclusion: $q \rightarrow r$

9. Premises: $q \rightarrow (p \wedge r)$; $\sim p$
   Conclusion: $\sim q$

10. Premises: $p \rightarrow \sim q$; $p \wedge r$
    Conclusion: $\sim q \leftrightarrow \sim r$

11. Premises: $\sim p \rightarrow q$; $q \rightarrow \sim r$
    Conclusion: $\sim p \rightarrow \sim r$

12. Premises: $p \oplus r$; $\sim p \rightarrow q$; $r \rightarrow \sim s$; $q \rightarrow s$
    Conclusion: p

13. Premises: $p \rightarrow q$; $\sim p \leftrightarrow q$
    Conclusion: p

14. Premises: $q \rightarrow r$; $q \wedge \sim p$; $\sim q \rightarrow \sim p$
    Conclusion: t

15. Premises: $p \rightarrow \sim q$; $p \wedge q$
    Conclusion: $p \wedge \sim q$

Appendix: Using CALIPSO

This appendix contains technical and other useful information on using CALIPSO. The CALIPSO distribution disk contains a file named README. You should read this file by typing TYPE README|MORE to see if it contains any updates or suggestions not included here.

If you are an experienced user of computer software and impatient to get started, you can skip reading these instructions, put the CALIPSO disk in a disk drive, and type CALIPSO to start the program. If your computer has a hard disk, use the INSTALL batch file to put the CALIPSO programs on your hard disk.

Note that the text contains many instructions and examples for using CALIPSO to solve problems. Page vii contains a list of page references for these instructions.

A.1 HARDWARE ENVIRONMENT

Hardware

CALIPSO runs on computers that are IBM bios-compatible and have at least 256K of RAM. You must use DOS version 2.1 or higher; version 4.0 or higher is recommended if you have a VGA or EGA monitor.

A graphics monitor (VGA, EGA, CGA, or MCGA) is needed to run the *GRAPH* program and the part of the *STATISTICS* program that draws graphs on the screen. The remaining programs will run on any monitor.

If your computer has a math coprocessor, the programs will use it; otherwise, the programs emulate the coprocessor, and, consequently, run a little slower.

Startup Files

The CONFIG.SYS and AUTOEXEC.BAT files contain information that is read by your computer's operating system each time the computer is started. Your DOS manual gives instructions on the use of these files. The startup file CONFIG.SYS should contain statements setting the DOS variable FILES to a minimum value of 5. You may want to include the command GRAPHICS in your AUTOEXEC.BAT file. This command runs the DOS program GRAPHICS.COM that is needed in order to print graphics from the *GRAPH* and *STATISTICS* programs of CALIPSO (more is said about this below). On the other hand, you may not want to run GRAPHICS every time you start your computer because the program increases the memory used by DOS. This will not impede CALIPSO, but it may cause trouble with other programs that use almost all available memory.

Printers

CALIPSO can use most dot matrix, laser, and ink-jet printers. To print graphs, your printer must have graphics capability and you must run the DOS program GRAPHICS before starting CALIPSO. You may discover that versions of DOS earlier than 4.0 will not print graphics on some printers. In particular with laser printers, you may need to create a printer profile file in order to print graphics; consult your printer manual to do this. DOS versions 5.0 and higher have an option that establishes the profile for most currently manufactured printers.

If you use an EGA or VGA monitor with DOS 3.3 or earlier, you will not be able to print graphics. These monitors are only supported by DOS 4.0 and higher.

Configurations

You can configure CALIPSO in many ways. Here are some examples.

A. HARD DISK. The INSTALL batch file on the CALIPSO disk will create a directory named FINITE on your hard disk and copy all of the necessary files into it. You can run CALIPSO from the directory FINITE and store all files in that directory. This is the easiest way to manage data files. To run the INSTALL batch file, put the distribution disk in a disk drive, change to that drive, and type INSTALL.

   Alternatively, you can create your own directory and/or subdirectories and store the files on the CALIPSO disk wherever you want. If you use more than one directory, be sure to identify the location of all executable and data files in the SETUP file as described in A.2. Also be sure to place CALIPSO.HLP, MODELER.ERR, UTILITY.H, SETUP, and LOGO in the directory that contains CALIPSO.EXE . You start CALIPSO by changing to its directory and typing CALIPSO.

B. ONE 3.5″ DISK. You can run CALIPSO directly from the distribution disk, but you may prefer to make a working copy and save the original as a backup. CALIPSO will fit on a 1 Mb disk formatted for 720 Kb. If you want to start your computer from the working disk, format it with the /S option. Set all of the paths in the SETUP file to blank. Start the program by typing CALIPSO.

C. TWO FLOPPY 5.25″ DISKS. Format a new disk as a system disk by using the DOS format command with the option /S (you only need to use /S if you want to start your computer from the working disk). Copy the contents of the first distribution disk into this disk; we will refer to this disk as Disk A. Format another disk (without the /S command) and copy the contents of the second distribution disk into it; we will refer to this disk as Disk B. Store the distribution disks in a safe place so that you can use them as a backup in case you damage one of the operating disks.

Place Disk A in drive A and Disk B in drive B. If the computer is off, turn it on and the computer will boot from the copy of COMMAND.COM in Disk A. Start the program by typing CALIPSO.

The SETUP file that is on Disk A should work satisfactorily with most computers. If you want to change SETUP, follow the directions in Section A.2 below to create a SETUP file that identifies the location of the various programs. Use the following paths:

Application	File	Path
MATRIX	MATRIX.H	a:
GRAPH	GRAPH.H	a:
PIVOTER	PIVOTER.H	b:
LP SOLVER	SMALL_LP.H	b:
STAT	STAT.H	b:
LOGIC	TRUTH.H	b:
data files		b:

D. RAM DISK. If you have a computer with at least 640K of memory, you can use a simulated (RAM) disk to hold those programs you execute most frequently. Doing this will improve loading time of the programs on the RAM disk. You can put CALIPSO.EXE, UTILITY.H,

CALIPSO.HLP, MODELER.ERR, LOGO, and as many application programs (those with extension .H) as will fit on the RAM disk. The other program files and your data files can be placed in a floppy or hard disk. You use the SETUP file (see Section A.2) to inform CALIPSO of the location of the files. You should not put valuable data files on the RAM disk since you may lose them when you turn the computer off.

You can use the VDISK facility contained in DOS versions 3.0 to 4.0 to make a RAM disk. In higher versions of DOS the name VDISK was changed to RAMDRIVE. If you have extended or expanded memory, you can place the RAM drive in that memory. Consult your DOS manual for details.

Here is a sample command to make a 220 KB RAM drive on a computer with 640K of memory.

```
device=ramdrive.sys 220 512 16
```

This command must be placed in the CONFIG.SYS file and will be executed when the computer starts. You should also create an AUTOEXEC.BAT file that copies CALIPSO.EXE, UTILITY.H, CALIPSO.HLP, MODELER.ERR, LOGO, SETUP, and one application program (with extension .H) into the RAM disk. If you have more than 640KB of memory, you can make a RAM disk of 640KB and copy all of the CALIPSO programs to it.

Here is a sample AUTOEXEC.BAT file that puts the *MATRIX* program on the RAM disk D, switches to the RAM disk, and starts CALIPSO.

```
copy calipso.exe d:
copy utility.h d:
copy modeler.err d:
copy calipso.hlp d:
copy setup d:
copy logo d:
copy matrix.h d:
d:
calipso
```

Be sure the SETUP file directs CALIPSO to look for the programs *MATRIX* and *LP SOLVER* on the D drive.

A.2 SOFTWARE ENVIRONMENT

The first time you run CALIPSO on a computer you should check that the SETUP file suits your configuration. The SETUP file contains the color selection, paths to the program and data files, and information about the hardware.

You change the SETUP file from the main CALIPSO menu by pressing the [↓] key to move the cursor to the box next to SETUP. When the cursor is in place, press [*Enter*]. You will be instructed to make the following specifications.

A. COLOR SELECTION. The first SETUP screen allows you to change colors. The three lines at the top of the screen are displayed in the current colors against the current background. On the lower portion of the screen you will see four color bars that allow you to change the background color and the foreground color used for ordinary, message, and warning text. You change the colors by moving the cursor to the color of your choice with the arrow keys. Place the cursor above the color of your choice and press [*Enter*]. The screen at the top will change to the new color selection.

   If you have a monochrome monitor, you can press the [*F1*] key (as directed on the screen) to automatically obtain a good choice. If you prefer to make your own selections, experiment with color selections until all four lines at the top of the screen are easy to read. All possible color combinations are available. Some combinations look better on amber screens and others look better on green screens.

   When you have the colors set to your satisfaction, press [*Esc*]. The color choice will be saved in the SETUP file and used for later executions of CALIPSO (until you change SETUP again).

   If you want to restore the color selections to those that were originally on the distribution disk, set background to blue, Line 1 to light gray, Line 2 to yellow, Line 3 to light blue, and Line 4 to light blue.

B. VIDEO OUTPUT. The next screen allows you to choose between having video output sent directly to the video buffer or having it directed to the ROM bios. Directing output to the bios is slightly slower, and this would only be desirable if you were using a machine that was IBM bios compatible, but not 100% IBM compatible.

C. PRINTER CODES. You can enter an escape sequence and font code allowing you to print with a condensed font (if your printer has this capability). You can select a preset code that will work on Epson, IBM, and most other line printers, a preset code that will work on HP and many other laser printers, or you can enter your own code. You can also choose not to use a condensed font; however, it is often easier to interpret large matrices when they are compactly printed.

D. PATH SELECTION. The next screen allows you to select a path that will be concatenated before the name of an executable file or data file. This path can be a drive letter followed by a colon (e.g., A:) or a full directory path. If you use a single 3.5″ diskette or a hard disk with all files in the same directory, you can leave all paths blank. Here is an example of how to enter a path to place data files in the subdirectory DATA of the directory FINITE on hard drive C.

c:\finite\data\

Note that it is not necessary to use path names with a hard disk. If you put all of the CALIPSO programs and data in a single directory and start CALIPSO from that directory, you can set all paths to blanks. This is the easiest way and you should use it unless you plan to store many data files.

You enter path names by moving the cursor up or down with arrow keys and typing the path names. When you have made all your entries, press *Enter* to save the new configuration in the SETUP file.

A.3 OPERATING CALIPSO

You will find instructions on using CALIPSO's programs to solve problems throughout the text. Page vii contains a list of page references for these instructions. Actually you will see that few written instructions are needed since most features of CALIPSO are described on the screens from which you operate the programs.

Help

With few exceptions you can find all the help you need to operate CALIPSO by reading instructions on the screen. These instructions often take the form of a menu describing operations you can perform and directions for leaving the screen. When appropriate the screen instructions offer more help by pressing a *function key.* Function keys are labeled [F1], [F2], ... When a function key is offered on the screen, pressing it causes the program to take the action described. The particular meaning of a function key may vary between programs of CALIPSO. *Help* is most often associated with the [F9] key.

Menus

You control CALIPSO from menus that appear on every screen from which you can take action. You usually choose an item from a menu by moving the cursor to it with arrow keys and pressing [Enter]. Sometimes the cursor is being used for another purpose, such as entering data in a matrix, and cannot be used to select from the menu. In such cases directions on the screen tell you how to use function keys.

Sometimes a menu allows you to make several choices. For instance, the screen for saving files in *LP SOLVER* asks you to pick one of several choices of files to save *and* allows you to type a file name. On such a screen the cursor moves through a series of boxes. You select an item in the usual way by moving to it and pressing [Enter], or you select more than one item by typing a character (any character, say X) in the boxes you want to select and waiting to press [Enter] until all your selections have been made.

Data Files

The programs *MATRIX*, *GRAPH*, *PIVOTER*, and *LP SOLVER* allow you to save files. A *file* is information that is stored on disk for future use. For instance, you may run short of time when working on a problem with *MATRIX* and want to stop. If you save the file, you can return later, load the file, and continue working. The file is retained on the disk until it is erased, even when the computer is turned off. The *STATISTICS* and *LOGIC* programs create worksheets that are not appropriate for recall from files, and so have no facility for handling files.

Files are stored under *file names.* A file name is basically a word of 8 or fewer alphabetic (A-Z) or numeric (0-9) characters. The file name is often augmented with a prefix or an extension, usually added by the programs and sometimes by the user. *Extensions* are 1-3 letter words, preceded by a period, and appended to the file name. CALIPSO's programs use the extensions

PROGRAM	EXTENSION
MATRIX	.mat
GRAPH	.grf
PIVOTER	.tab
LP SOLVER	.inq and .tab

You should *never* type an extension when operating CALIPSO. You only need to use an extension to handle the files with DOS (say to erase or copy one).

A file name can be preceded by a *path*. The most common path you use with CALIPSO is the drive designation letter. For instance, if you are using a computer with two disk drives, the A drive and the B drive, and no hard disk, the program needs to know which drive contains a particular file. You indicate the drive by attaching the drive letter followed by a colon to the start of the file name. For instance, if the file name is POTATO and it is located on the A drive, you can refer to the file as A:POTATO. If you are using a computer with a hard disk, the path can take more complicated forms to describe the *directories* of the hard disk. You can learn about directories in your DOS manual.

Generally you do not need to use paths with CALIPSO. The SETUP file described in Section A.2 automatically adds paths before the names you type while running CALIPSO. If you type a path, it will take precedence over the path ordinarily added from the SETUP file. Thus, you can specify a path when you want to place a file on a disk other than the one you usually use.

How to Save a File

Each program that uses files offers a menu selection to save the file. You can save the file under its current name or enter a new name. Note that saving a file overwrites any file with the same name on your disk.

How to Recall a File

When you start a program that uses files from the main CALIPSO menu, a screen appears asking for the name of the file to be processed. To recall a file that you saved earlier, type its name. If there is no file on the disk with this name, a new file is created.

If you cannot recall the name of the file you want, CALIPSO will give you a list of all data files on your disk that can be used by the program you are running. You obtain this list by pressing [↓] when the screen for entering file names is displayed. The cursor moves to a typing box that contains one of the expressions *.mat, *.tab, *.inq, *.grf. These expressions are DOS "wildcard" search directives. They search out all files on your disk that have the given extension; for instance, *.mat will find all files with extension .mat. You can use other DOS "wildcard" expressions. See your DOS manual for more information about them. When the list of files appears on your screen, you make your selection by moving the cursor to the file name you want and pressing *Enter*.

How to Use a Default Path

In addition to establishing paths to CALIPSO's application programs, the SETUP option allows you to set a default path to data files. This default path is concatenated before any file name you type on the file entry screen that does not itself contain a path. For example, suppose you enter the default path name b:\M326\ with SETUP.

- If you type the file name Prob2c-3, the file will be sought under the name b:\ M326\ Prob2c-3.
- If you type the file name a:M326, the file will be sought on the A drive under the name M326. Thus, any path that you give overrides the default path name.
- If you type *.mat in the second entry field of the file entry panel, you will see a list of all files that match b:\ M326*.mat. If there are no such files, you will see a message that says "*.mat not found on the default path".
- If you type a:*.mat, you will see a list of all files on the A drive that match *.mat.

If you are using a hard disk, the CALIPSO program files will be stored in a directory, say one named FINITE. If you make a subdirectory of FINITE, say one named DATA, to store data files, the appropriate path name would be FINITE, not \FINITE. This is the usual DOS convention. The path name \FINITE would direct DOS to search in the root directory for a directory named FINITE, whereas the path name FINITE would direct DOS to search in the current directory for a subdirectory named FINITE.

Worksheets

MATRIX and *STATISTICS* create *worksheets* that contain solutions to several problems. The *MATRIX* worksheet can contain up to 26 matrices. Some of these matrices represent problems you have entered and others are solutions computed by *MATRIX*. The *STATISTICS* worksheet contains the commands you enter and the responses of the program to the commands. You scan through a worksheet with [*PgUp*], [*PgDn*], and arrow keys (specific instructions are given on the screen).

Printing

All of CALIPSO's programs allow you to print a worksheet or a report of the solution. The CALIPSO programs make use of condensed printing mode if your printer has this printing style. This is useful since many of the matrices and tableaus you will print are wider than the printed page and fewer continued lines occur with condensed printing. To use condensed mode the program must know the appropriate escape sequence for your printer. If you are using an IBM, EPSON, almost any other kind of dot-matrix printer, or an HP laser printer, the needed codes can be set automatically from the SETUP utility found on the main CALIPSO menu. If none of the codes that are built in to SETUP work, consult your printer manual and use the SETUP facility that describes how to insert the codes. If you have trouble or do not want to use condensed printing, you can turn this facility off with SETUP.

The *GRAPH* and *STATISTICS* programs allow you to print graphics. Most of the printers on the market will print graphics; however, some models may not. As described earlier, to print graphics you must run the DOS program GRAPHICS *before* you start CALIPSO. You must run it before CALIPSO because it is a *resident* program; that is, it runs concurrently with CALIPSO.

A.4 APPLICATION PROGRAMS

Matrix

The *MATRIX* program allows you to perform operations on matrices with as many as 12 rows and 12 columns. Matrices are named by a single letter of the alphabet and are displayed in a worksheet that can contain up to 26 matrices. You scroll through the worksheet with the *PgUp* or *PgDn* key. Alternatively, you can move directly to a matrix by holding down the *Ctrl* key and pressing the letter corresponding to the name of the matrix you want to see.

The available functions are given in a menu at the bottom of the screen. You move the cursor about in the menu with arrow keys. A short description of the function currently highlighted by the cursor is given below the menu. This description and other directions that appear on the screen provide all the instructions you need to operate *MATRIX*.

The *MATRIX* functions are:

- *By Command*: You can perform row operations on the current matrix.
- *Operations*: You can evaluate expressions involving addition, subtraction, scalar multiplication, matrix multiplication, transposition, and inversion. You can solve systems of equations that have one or infinitely many solutions. You can generate higher and higher powers of a square matrix A in the form A^n or $A^n D$ where D is a column matrix. Complete instructions for entering expressions appear on the screen after you select *Operations*.
- *Echelon Forms*: You can reduce the current matrix to a row echelon form with leading entries set to 1 or left at a nonzero value.
- LU *Factorization*: You can produce the LU factorization of a matrix.
- *Determinant*: You can find the determinant of the current matrix.

GRAPH

The *GRAPH* program allows you to solve geometrically a two-variable linear programming problem. As you enter inequalities the feasible region is drawn on the screen. *GRAPH* will draw level lines of the objective function that move in the direction of increasing values, allowing you to spot the corner point yielding the minimum or maximum value. *GRAPH* will also evaluate the objective function at each corner point, allowing you to select the optimal value.

You can view a list of the available functions by pressing *F1*. You can obtain help with any function by pressing *F9*.

PIVOTER

The *PIVOTER* program allows you to enter the tableau for a linear programming problem and perform pivots on it. The available functions are given in a menu at the bottom of the screen. You move the cursor about in the menu with arrow keys. A short description of the function currently highlighted by the cursor is given below the menu. You can obtain help with any step of the calculation by pressing [*F9*].

LP SOLVER

The *LP SOLVER* program allows you to enter linear programming problems as systems of equations. It also solves the problem and provides postoptimality analysis. *LP SOLVER* is part of another version of CALIPSO that supports linear programming books. It has capabilities that exceed the linear programming covered in most finite mathematics books. If you experiment with *LP SOLVER* and read its help files, you will find that it has extensive capabilities to model LP problems. Among these features is the ability to generate large systems of inequalities using commands that are similar to those in the basic programming language.

You enter an LP problem as an objective function and a set of inequalities. Three pull-down menus are advertised at the top of the screen. [*F2*] describes the editing (word-processing) features, [*F3*] allows you to read help messages about various features, and [*F4*] allows you to set some display options. The help messages give full instructions for using *LP SOLVER*.

After entering the inequalities you press [*F1*] to obtain the solution. When the solution is on the screen, the menu at the top advertises that [*F4*] gives postoptimality analyses. If you press [*F4*], you are told how to obtain a sensitivity analysis, a display of the initial or final tableau, and the inverse basis matrix.

The inequalities, the solution, and postoptimality analyses are displayed in a worksheet. You can move about in this worksheet with [*PgUp*], [*PgDn*], [↑], or [↓]. You can enter remarks or make changes anywhere in the worksheet; however, to maintain integrity of the solution, if you change the inequalities, the solution disappears.

You may find that *LP SOLVER* is useful in many courses in operations research or business management that you may take. It is more fully described in [CV].

STATISTICS

The *STATISTICS* program will perform all of the statistics operations usually found in finite mathematics books. It can draw descriptive or probability histograms, calculate means and standard deviations, calculate binomial distributions, display and calculate the area under a normal curve, and calculate and graph linear regressions. It also contains a calculator that evaluates expressions involving numbers, $C(a,b), P(a,b)$, and $n!$. The calculator also accepts the functions $a_{\overline{n}|i}$ and $s_{\overline{n}|i}$ that are used in financial calculations.

The commands and the resulting calculations are held in a worksheet. You can scroll through the worksheet with [*PgUp*], [*PgDn*], and arrow keys. You can edit the worksheet and print a copy of it.

LOGIC

The *LOGIC* program allows you to enter logical expressions and construct truth tables.

Answers to Selected Exercises

0.1A (p. 3)

1. $B \cup C = \{-15, -12, -9, -6, -3, 0, 3, 5, 6, 7, 8, 9, 10, 11, 12, 13, 15\}$
   $B \cap C = \{6, 9, 12\}$
   $B - C = \{5, 7, 8, 10, 11, 13\}$

3. $(A \cup B) \cap D = \{0, 12\}$ 5. $D \cap (B \cup C) = \{-3, 0, 12\}$.

0.1B (p. 5)

1.
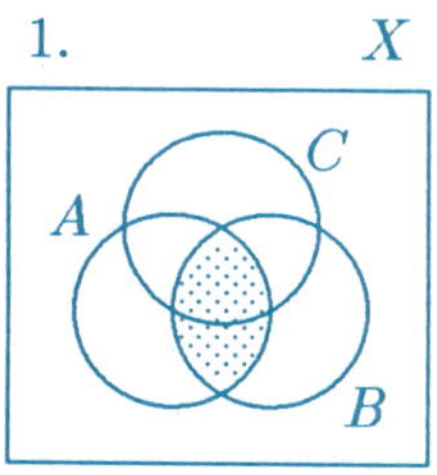

3.
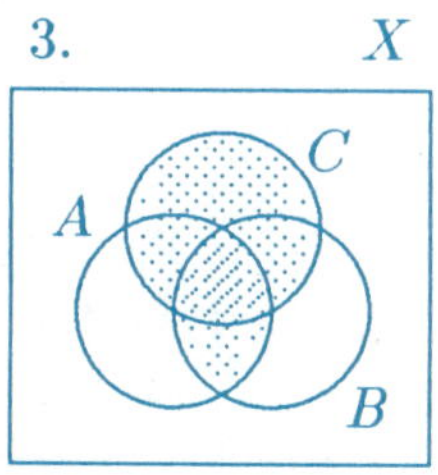

5.
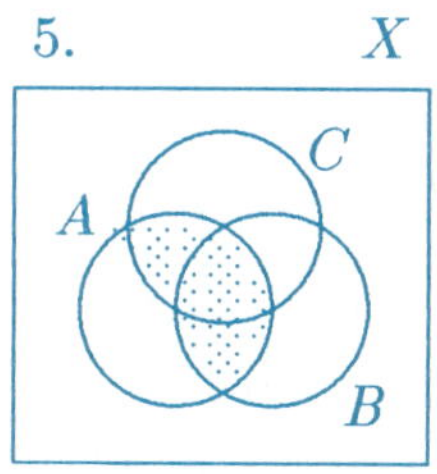

0.2A (p. 10)

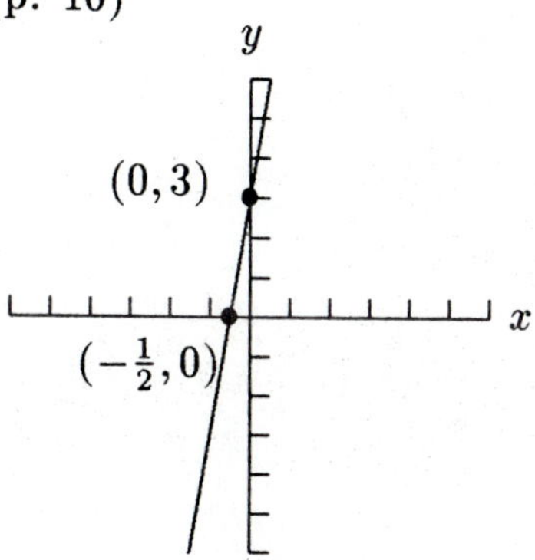

Problem 2

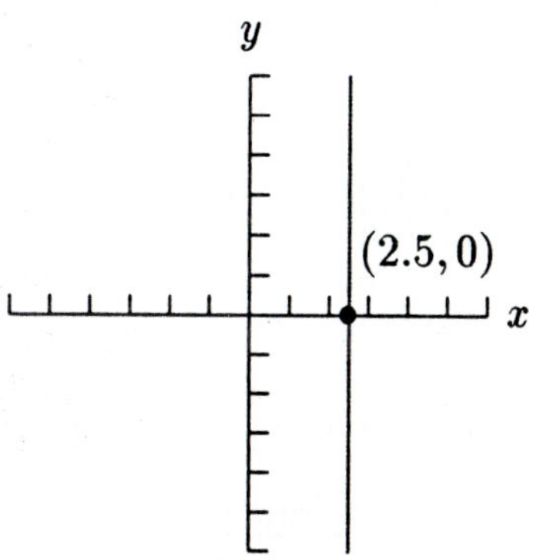

Problem 4

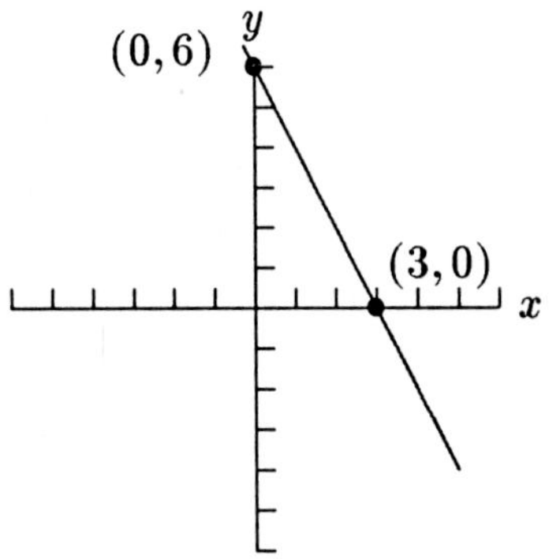

Problem 6

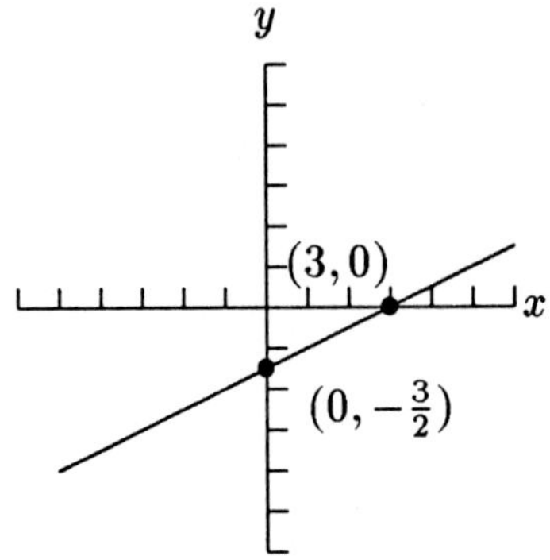

Problem 8

10. $y = -\frac{1}{4}x + \frac{3}{2}$; $m = -\frac{1}{4}$, $b = \frac{3}{2}$ 12. $y = 3x - 12$; $m = 3$, $b = -12$

14. $y = 4x + 6$ 16. $y = -\frac{1}{2}x + 2$ 18. $2y - 5x = -6$

20. (a) $C = \frac{5}{9}(F - 32)$ (b) $\frac{140}{9}$ (c) 14

1.2A (p. 20)

1. (a) $\begin{bmatrix} 6 & 6 \\ 1 & 4 \end{bmatrix}$ (c) $\begin{bmatrix} 3 & 6 \\ -3 & 9 \end{bmatrix}$ 2. (a) $\begin{bmatrix} -3 & 6 & 12 \\ 9 & -12 & 6 \end{bmatrix}$ (c) $\begin{bmatrix} 7 & 4 \\ 12 & 7 \end{bmatrix}$

3. (a) $\begin{bmatrix} 3 & -1 & 2 \\ 5 & 2 & 4 \end{bmatrix}$ (c) $\begin{bmatrix} 0 & 5 & -7 \\ -1 & -1 & -2 \end{bmatrix}$

4. (a) $\begin{bmatrix} -3 & 6 & 6 \\ 5 & 5 & 8 \end{bmatrix}$ (c) $\begin{bmatrix} -1 & -2 & 3 \\ 6 & 0 & 2 \\ 11 & 1 & 2 \end{bmatrix}$

1.2C (p. 23)

1. $\begin{bmatrix} 1 & -14 \\ -6 & -12 \end{bmatrix}$ 3. $\begin{bmatrix} -4 & -52 \\ -3 & 3 \\ -9 & -54 \end{bmatrix}$ 5. $\begin{bmatrix} 4 \\ -56 \end{bmatrix}$ 7. Wrong sizes.

1.3A (p. 30)

2. Instructor 1: 88.5%   Instructor 2: 87.75%.

4. (a) Th: 63,500; Fri: 57,450; Sat: 68,600; Sun: 39,600   (b) 229,150
   (c) A:1300, B:1450, C:1800, D:1600, E:2650, F:1600

6. NYC \$1,359; Moscow \$1,458; Boston \$1,651; Chicago \$1,338; Atlanta \$1,229; Seattle \$1,953.

8. (a) 93,816 (b) 135,660

10. (a) $\begin{bmatrix} 427 & 502 & 444 & 428 \end{bmatrix}$ (b) $\begin{bmatrix} 104.25 \\ 25.75 \\ 235 \\ 67.75 \\ 17.5 \end{bmatrix}$

2.1A (p. 37)

1. (a)
$$\begin{array}{rcrcl} x & - & y & = & 1 \\ 3x & + & 2y & = & 3 \end{array}$$

(c)
$$\begin{array}{rcrcrcrcrcr} x_1 & - & x_2 & + & 2x_3 & & & + & 3x_5 & = & 1 \\ 2x_1 & - & x_2 & + & 2x_3 & + & 4x_4 & + & 3x_5 & = & -2 \\ -2x_1 & + & x_2 & + & 2x_3 & + & x_4 & & & = & 3 \\ 3x_1 & - & 3x_2 & + & 2x_3 & + & x_4 & - & 2x_5 & = & 3 \end{array}$$

2. (a)
$$\begin{bmatrix} 1 & 1 & 2 \\ 2 & -3 & 1 \\ 3 & 1 & -5 \end{bmatrix} \begin{bmatrix} x_1 \\ x_2 \\ x_3 \end{bmatrix} = \begin{bmatrix} 2 \\ -2 \\ 4 \end{bmatrix}$$

(c)
$$\begin{bmatrix} 2 & 1 & 3 & -4 & -2 \\ 1 & -2 & 0 & 3 & -1 \\ 2 & -2 & 4 & 2 & -3 \\ 4 & -2 & 4 & 1 & -2 \end{bmatrix} \begin{bmatrix} x_1 \\ x_2 \\ x_3 \\ x_4 \\ x_5 \end{bmatrix} = \begin{bmatrix} 6 \\ 2 \\ 8 \\ 9 \end{bmatrix}$$

2.2A (p. 45)

2. $\begin{bmatrix} 2 & 6 & -4 & 22 \\ 0 & -12 & 10 & -34 \\ 0 & 0 & 7 & -1] \end{bmatrix}$ 4. $\begin{bmatrix} 2 & 2 & 1 & 4 \\ 0 & -2 & \frac{1}{2} & 0 \\ 0 & 0 & 1 & -2 \end{bmatrix}$

6. $\begin{bmatrix} \frac{1}{2} & \frac{2}{3} & \frac{3}{4} & -\frac{1}{2} \\ 0 & -\frac{7}{18} & -\frac{5}{4} & \frac{4}{3} \\ 0 & 0 & -\frac{1}{56} & -\frac{25}{84} \end{bmatrix}$ 8. $\begin{bmatrix} 2 & 0 & -2 & 4 & 2 & 3 \\ 0 & 4 & -3 & 5 & 5 & \frac{19}{2} \\ 0 & 0 & 8 & -14 & -10 & -18 \\ 0 & 0 & 0 & \frac{165}{16} & -\frac{41}{16} & \frac{57}{16} \\ 0 & 0 & 0 & 0 & \frac{47}{5} & \frac{66}{5} \end{bmatrix}$

10. $x = \frac{33}{13}, y = -\frac{4}{13}$ 12. $x_1 = \frac{4}{7}, x_2 = \frac{16}{7}, x_3 = \frac{4}{7}$

14. $x_1 = -3.4194, x_2 = 5.3839, x_3 = .093548, x_4 = -1.3516, x_5 = 3.2935$

2.2B (p. 48)

3. $\begin{bmatrix} 1 & 2 & 4 & 3 \\ 0 & -2 & -7 & -1 \\ 0 & 0 & 1 & 2 \end{bmatrix}$ 5. $\begin{bmatrix} 2 & 1 & 3 & 4 & 5 \\ 0 & -1 & 5 & 0 & -1 \\ 0 & 0 & 4 & 1 & 6 \\ 0 & 0 & 0 & 1 & 4 \end{bmatrix}$

2.3A (p. 50)

2. $\begin{bmatrix} 1 & 0 & 0 & 0 & \frac{25}{18} \\ 0 & 1 & 0 & 0 & -\frac{13}{80} \\ 0 & 0 & 1 & 0 & -\frac{137}{120} \\ 0 & 0 & 0 & 1 & \frac{17}{45} \end{bmatrix}$ 4. $\begin{bmatrix} 1 & 0 & 0 & 0 & -1.0673 \\ 0 & 1 & 0 & 0 & 2.0924 \\ 0 & 0 & 1 & 0 & 4.0833 \\ 0 & 0 & 0 & 1 & 1.9977 \end{bmatrix}$

6. $\begin{bmatrix} 1 & 0 & 0 & 0 & 0 & 3.2793 \\ 0 & 1 & 0 & 0 & 0 & .34423 \\ 0 & 0 & 1 & 0 & 0 & .63532 \\ 0 & 0 & 0 & 1 & 0 & -.32845 \\ 0 & 0 & 0 & 0 & 1 & -.43686 \end{bmatrix}$

2.3B (p. 53)

2. 22 and 13 4. Fleet A: 8 vans, Fleet B: 6 vans, Fleet C: 4 vans.

6. Chintz 75%; Scrumch 11.25%; Tallow 27.5%.

8. Tomatoes: 2.0005, Beans: 3.1534, Peas: 0.65249, Corn: 4.1665, Carrots: 5.0272.

10. Metals: -55556; Stocks: 240,740; Bonds: 148,150; Bank: 166,670. A negative investment in metals is impractical.

2.4A (p. 57)

2. $\begin{bmatrix} 1 & 6 & 0 \\ 0 & 1 & -\frac{3}{22} \\ 0 & 0 & 1 \\ 0 & 0 & 0 \end{bmatrix}$ 4. $\begin{bmatrix} 1 & -\frac{1}{2} & -\frac{3}{2} \\ 0 & 1 & 4 \\ 0 & 0 & 1 \\ 0 & 0 & 0 \\ 0 & 0 & 0 \end{bmatrix}$

6. $\begin{bmatrix} 1 & 4 & 3 & 2 \\ 0 & 1 & \frac{1}{3} & \frac{1}{6} \\ 0 & 0 & 0 & 0 \\ 0 & 0 & 0 & 0 \end{bmatrix}$ 8. $\begin{bmatrix} 1 & \frac{1}{2} & -\frac{3}{2} \\ 0 & 1 & -9 \\ 0 & 0 & 1 \\ 0 & 0 & 0 \end{bmatrix}$

2.4B (p. 58)

2. $\begin{bmatrix} 1 & 0 & 0 \\ 0 & 1 & 0 \\ 0 & 0 & 1 \\ 0 & 0 & 0 \end{bmatrix}$ 4. $\begin{bmatrix} 1 & 0 & 0 \\ 0 & 1 & 0 \\ 0 & 0 & 1 \\ 0 & 0 & 0 \\ 0 & 0 & 0 \end{bmatrix}$

6. $\begin{bmatrix} 1 & 0 & -\frac{1}{6} & -\frac{1}{3} \\ 0 & 1 & -\frac{3}{2} & 2 \\ 0 & 0 & 0 & 0 \\ 0 & 0 & 0 & 0 \end{bmatrix}$ 8. $\begin{bmatrix} 1 & 0 & 0 \\ 0 & 1 & 0 \\ 0 & 0 & 1 \\ 0 & 0 & 0 \\ 0 & 0 & 0 \end{bmatrix}$

2.5A (p. 62)

1. $\left[\begin{array}{ccc|c} 1 & 0 & -1 & 0 \\ 0 & 1 & -5 & 0 \\ 0 & 0 & 0 & 1 \end{array}\right]$ 3. $(x_1, x_2, x_3) = (2t - \frac{11}{3}, -3t + \frac{13}{3}, t)$

6. $(x_1, x_2, x_3) = (\frac{1}{4}t + \frac{15}{4}, \frac{11}{4}t - \frac{3}{4}, t)$.

2.5B (p. 63)

1.

$$\begin{bmatrix} x_1 \\ x_2 \\ x_3 \\ x_4 \\ x_5 \end{bmatrix} = t\begin{bmatrix} -\frac{11}{7} \\ -\frac{12}{7} \\ \frac{13}{7} \\ -\frac{16}{7} \\ 1 \end{bmatrix} + \begin{bmatrix} -\frac{19}{15} \\ -\frac{29}{15} \\ \frac{53}{15} \\ -\frac{5}{3} \\ 0 \end{bmatrix}$$

2.

$$\begin{bmatrix} x_1 \\ x_2 \\ x_3 \\ x_4 \\ x_5 \end{bmatrix} = t\begin{bmatrix} -\frac{79}{38} \\ -\frac{93}{76} \\ -\frac{63}{76} \\ 1 \\ 0 \end{bmatrix} + s\begin{bmatrix} \frac{2}{19} \\ \frac{5}{38} \\ -\frac{15}{38} \\ 0 \\ 1 \end{bmatrix}\begin{bmatrix} \frac{121}{38} \\ \frac{179}{76} \\ \frac{109}{76} \\ 0 \\ 0 \end{bmatrix}$$

2.5C (p. 66)

2. $x_1 = -\frac{12}{5}$, $x_2 = \frac{17}{5}$, $x_3 = 10$, $x_4 = -3$, $x_5 = 4$

4. $x_1 = 9$, $x_2 = -\frac{37}{4}$, $x_3 = \frac{163}{12}$, $x_4 = -5$, $x_5 = 7$

6.
$$\begin{bmatrix} x_1 \\ x_2 \\ x_3 \end{bmatrix} = \begin{bmatrix} 3 \\ 0 \\ 2 \end{bmatrix} + r\begin{bmatrix} -2 \\ -4 \\ 3 \end{bmatrix} + s\begin{bmatrix} -3 \\ 1 \\ 0 \end{bmatrix} + t\begin{bmatrix} 4 \\ 0 \\ -1 \end{bmatrix}$$

8.
$$\begin{bmatrix} x_1 \\ x_2 \\ x_3 \\ x_4 \end{bmatrix} = \begin{bmatrix} -1 \\ 0 \\ -2 \\ 0 \end{bmatrix} + r\begin{bmatrix} 7 \\ 3 \\ -2 \\ 2 \end{bmatrix} + s\begin{bmatrix} -4 \\ -4 \\ 0 \\ 3 \end{bmatrix} + t\begin{bmatrix} 1 \\ 0 \\ 1 \\ 0 \end{bmatrix} + u\begin{bmatrix} 2 \\ 6 \\ -1 \\ 4 \end{bmatrix}$$

3.1A (p. 72) 1. $\begin{bmatrix} \frac{3}{14} & \frac{1}{7} \\ -\frac{1}{7} & \frac{1}{14} \end{bmatrix}$ 3. $\begin{bmatrix} \frac{1}{4} & -\frac{1}{2} \\ \frac{1}{4} & 0 \end{bmatrix}$ 5. $\begin{bmatrix} \frac{2}{5} & \frac{1}{5} & -\frac{2}{5} \\ -\frac{1}{5} & \frac{2}{5} & \frac{1}{5} \\ 1 & 0 & 0 \end{bmatrix}$

7. $\begin{bmatrix} -.61152 & -.09999 & -.08669 & .51970 \\ .25826 & .03741 & .05769 & .09391 \\ .12021 & .00601 & .07677 & -.04756 \\ .26715 & .09925 & .15193 & -.18038 \end{bmatrix}$

3.1B (p. 75)

1. $\begin{bmatrix} 51 & 5 & 85 & 56 & 15 & 83 & 137 \\ 124 & 20 & 150 & 84 & 60 & 172 & 268 \end{bmatrix}$

3.1C (p. 75)

1. Finite Mathematics is Fun.

4. $\begin{bmatrix} 71 & 158 & 190 & 131 & 57 & 121 & -140 \\ 26 & -58 & -53 & -22 & 21 & -14 & 112 \\ -4 & -133 & -149 & -79 & -6 & -119 & 196 \\ 26 & 11 & 4 & 2 & 33 & 4 & 28 \end{bmatrix}$

3.1D (p. 78) 2. $x = \frac{6}{5}, y = -\frac{2}{5}$ 4. $x = \frac{12}{5}, y = \frac{14}{5}$

3.1E (p. 79)

2. $x = \frac{135}{28}$, $y = -\frac{59}{28}$, $z = \frac{7}{4}$.

4. $x_1 = -\frac{95}{63}$, $x_2 = -\frac{116}{63}$, $x_3 = -\frac{8}{9}$, $x_4 = -\frac{8}{63}$.

6. $x_1 = 0.27248$, $x_2 = 0.14605$, $x_3 = 0.63034$, $x_4 = -0.97789$.

8. (a) $\begin{bmatrix} -8 & 4 \\ 9 & -3 \end{bmatrix}$ (b) 9,996 were sick last week.

3.1F (p. 82)

2. $x_1 = 100, x_2 = 70$ 4. 24.324 units of lumber and 35.135 units of steel.

3.1G (p. 83)

2. $X = [73,167 \ \ 79,116 \ \ 46,359 \ \ 60,542 \ \ 50,772]^T$.

4. $X = [17,568 \ \ 27,274 \ \ 18,234 \ \ 15,729 \ \ 25,455 \ \ 12,745 \ \ 28,212 \ \ 23,307]^T$.

3.2A (p. 88) 1. $x_1 = -\frac{9}{4}$, $x_2 = -\frac{15}{2}$ 3. $x_1 = -16$, $x_2 = \frac{13}{2}$, $x_3 = 6$.

3.2B (p. 91)

1. $\begin{bmatrix} 2 & 1 \\ 2 & 4 \end{bmatrix} = \begin{bmatrix} 1 & 0 \\ 1 & 1 \end{bmatrix} \begin{bmatrix} 2 & 1 \\ 0 & 3 \end{bmatrix}$

3. $\begin{bmatrix} 2 & 3 & -1 \\ -2 & 2 & 4 \\ 1 & -1 & 2 \end{bmatrix} = \begin{bmatrix} 1 & 0 & 0 \\ -1 & 1 & 0 \\ \frac{1}{2} & -\frac{1}{2} & 1 \end{bmatrix} \begin{bmatrix} 2 & 3 & -1 \\ 0 & 5 & 3 \\ 0 & 0 & 4 \end{bmatrix}$

5. $L = \begin{bmatrix} 1 & 0 & 0 & 0 \\ 2 & 1 & 0 & 0 \\ -1 & -1 & 1 & 0 \\ -3 & -2 & -4 & 1 \end{bmatrix}$

3.2C (p. 92)

2. (b) $Y = [18.2 \ \ 16.233 \ \ -9.4413 \ \ -20.594]^T$,
$X = [1.4803 \ \ 11.934 \ \ -1.1391 \ \ -6.8707]^T$

4. (b) $Y = [8.2 \ \ .66154 \ \ -4.0714 \ \ 4.6243]^T$,
$X = .6917 \ \ .71924 \ \ 1.0192 \ \ .73577]^T$

8. (b) $x_1 = \frac{35}{16}$, $x_2 = \frac{5}{8}$, $x_3 = \frac{3}{2}$.

4.1A (p. 98)

Problem 1

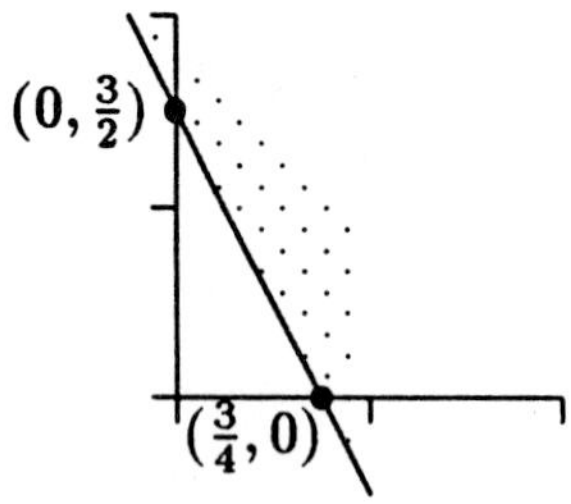

Problem 3

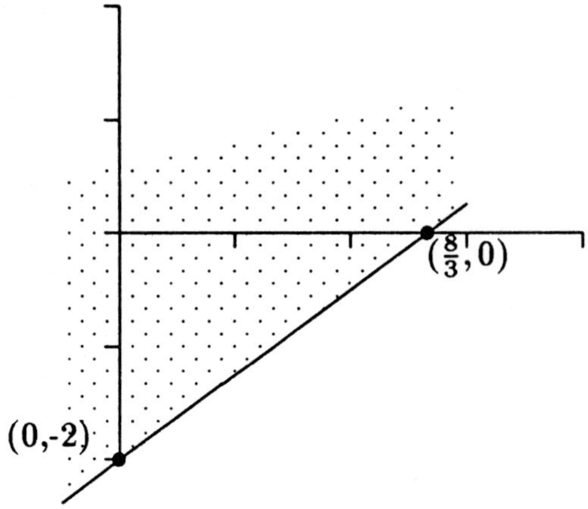

Problem 5

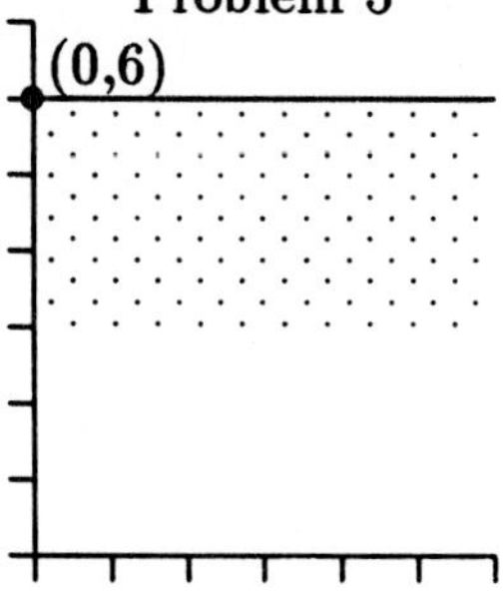

Problem 7

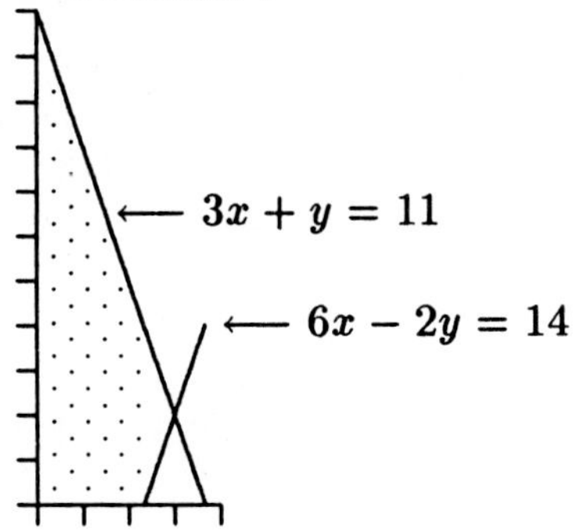

Problem 9

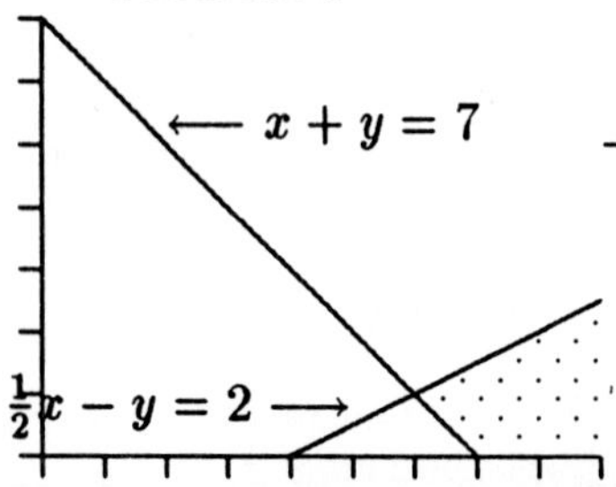

Problem 11

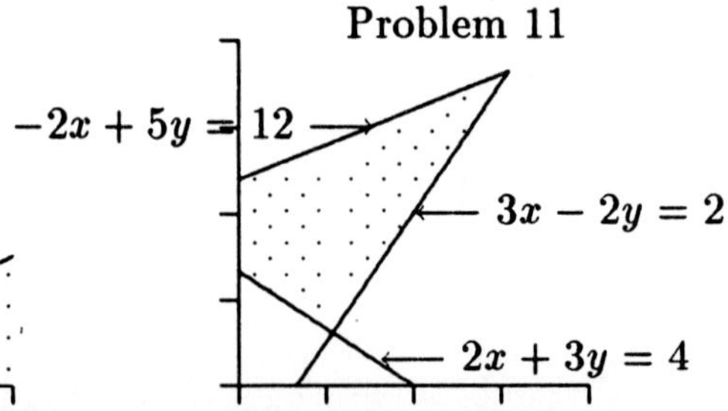

4.2A (p. 107)

3. (a)

Granola	Protein	Carbohydrate	Fat	Cost per Pound
Greaseball	4	4	24	.50
Sugarpop	12	4	8	.80
Units needed	6	4	12	

(b) Let x and y be the number of pounds of Greaseball and Sugarpop.

$$\begin{array}{ll} \text{Minimize:} & z = .50x + .80y \\ \text{Subject to:} & 4x + 12y \geq 6 \\ & 4x + 4y \geq 4 \\ & 24x + 8y \geq 12 \\ & x, y \geq 0 \end{array}$$

5. (a)

Product	R	S	T	Profit per Unit ($)
A	5	4	1	6
B	3	4	3	6
Supply (g)	9000	9000	4000	

(b) Let x and y be the number of units of A and B.

$$\begin{array}{ll} \text{Maximize:} & z = 6x + 6y \\ \text{Subject to:} & 5x + 3y \leq 9000 \\ & 4x + 4y \leq 9000 \\ & x + 3y \leq 4000 \\ & x, y \geq 0 \end{array}$$

4.3A (p. 112)

1. $x = 160, y = 480; z = 48000$ 3. $x = .75, y = .25; z = .575$
5. $x = 1.5, y = 2.5; z = 430$ 7. $x = 3, y = 5; z = 51$
9. $x = 3.6, y = 1.6; z = 33.2$ 11. $x = 12, y = 0; z = 1440$

4.4A (p. 121)

1. $x_1 = 0, x_2 = 4; z = 12$ 3. $x_1 = \frac{5}{2}, x_2 = \frac{3}{2}, z = 430$

5. $x_1 = 6, x_2 = 0, x_3 = 8, z = 50$ 7. $x_1 = 0, x_2 = \frac{6}{5}, x_3 = \frac{48}{5}, z = \frac{576}{5}$

4.4B (p. 126)

1. $x = \frac{3}{2}, y = \frac{5}{2}; z = 430$ 3. $x_1 = 1, x_2 = 0, x_3 = 0; x_4 = 3, z = \frac{11}{3}$

5. $x_1 = 0, x_2 = \frac{44}{15}, x_3 = \frac{7}{3}, x_4 = \frac{53}{15}; z = \frac{1190}{3}$

7. $x_1 = 2.041, x_2 = 0, x_3 = 0, x_4 = 1.7541; z = 357.18$

4.4C (p. 129) Same answers as 4.4B.

5.1A (p. 133)

1. $x_1 = 30, x_2 = 0; z = 30;$ 3. $x_1 = 0, x_2 = 3, z = 15.$

5.1B (p. 135)

1. $x_1 = 2, x_2 = 0; z = 4$

3. $x_1 = 0, x_2 = \frac{1}{3}, x_3 = \frac{29}{9}, z = \frac{28}{3}$
   There are alternate solutions, e.g., $x_1 = \frac{29}{6}, x_2 = \frac{1}{3}, x_3 = 0.$

5. $x_1 = 3, x_2 = 0, x_3 = 0, x_4 = 0; z = 3$ 7. $x_1 = \frac{11}{12}, x_2 = \frac{65}{12}, x_3 = \frac{19}{4};$ $z = \frac{747}{2}$

5.2A (p. 140)

The answers are the same as those for Exercises 5.1B. We give the initial tableaus with $M = 100$.

1.

	x_1	x_2	s_1	s_2	y	RHS
s_1	2	3	1	0	0	5
y	1	−3	0	−1	1	2
OBJ	−98	296	0	100	0	−200

3.

	x_1	x_2	x_3	s_1	s_2	s_3	s_4	y_1	y_2	RHS
s_1	2	1	3	1	0	0	0	0	0	10
s_2	2	−2	3	0	1	0	0	0	0	9
y_1	1	3	2	0	0	−1	0	1	0	4
y_2	2	−1	3	0	0	0	−1	0	1	5
OBJ	−302	−199	−503	0	0	100	100	0	0	−900

5.

	x_1	x_2	x_3	x_4	s_1	s_2	s_3	y_1	y_2	RHS
y_1	1	0	1	−4	−1	0	0	1	0	2
s_2	1	−1	3	1	0	1	0	0	0	9
y_2	1	1	−1	1	0	0	−1	0	1	3
OBJ	−199	−98	3	304	100	0	100	0	0	−500

7.

	x_1	x_2	x_3	s_1	y_1	y_2	s_3	RHS
y_1	5	−2	3	−1	1	0	0	8
y_2	1	−1	2	0	0	1	0	5
s_3	2	1	1	0	0	0	1	12
OBJ	−620	266	−536	100	0	0	0	−1300

5.3A (p. 144)

We give a tableau that shows infeasibility because it contains a row corresponding to an equation with a negative number on the right-hand side and all nonnegative numbers on the left-hand side. Other answers are possible.

2.

	x	y	s_1	s_2	s_3	RHS
s_1	2	0	1	0	2	−10
s_2	7	0	0	1	2	30
y	2	1	0	0	1	5
	−5	0	0	0	−1	−5

4.

	x	y	s_1	s_2	s_3	RHS
y	0	1	$\frac{2}{5}$	$\frac{3}{5}$	0	$\frac{4}{5}$
x	1	0	$-\frac{1}{5}$	$-\frac{4}{5}$	0	$\frac{3}{5}$
s_3	0	0	$\frac{3}{5}$	$\frac{7}{5}$	1	$-\frac{4}{5}$
	0	0	$\frac{1}{5}$	$-\frac{6}{5}$	0	$\frac{17}{5}$

6.

	x_1	x_2	x_3	s_1	s_2	s_3	RHS
x_1	1	$-\frac{3}{2}$	0	$-\frac{1}{2}$	0	$-\frac{1}{2}$	$\frac{7}{2}$
s_2	0	−34	0	−9	−1	−5	8
x_3	0	$\frac{17}{2}$	1	$\frac{5}{2}$	0	$\frac{3}{2}$	$-\frac{11}{2}$
	0	$\frac{85}{2}$	0	$\frac{27}{2}$	0	$\frac{17}{2}$	$-\frac{69}{2}$

5.3B (p. 148)

2.

Initial tableau for MSM method

	x	y	s_1	s_2	RHS
s_1	1	−1	1	0	1
s_2	−1	−1	0	−1	2
OBJ	−2	−3	0	0	0

Tableau showing unboundedness

	x	y	s_1	s_2	RHS
s_1	2	0	1	−1	3
y	1	1	0	−1	2
OBJ	1	0	0	−3	6

4.

Initial tableau for MSM method

	x	y	s_1	s_2	RHS
s_1	−2	−2	1	0	−20
s_2	3	−2	0	1	20
OBJ	−3	1	0	0	0

Tableau showing unboundedness

	x	y	s_1	s_2	RHS
y	0	1	$-\frac{3}{10}$	$-\frac{1}{5}$	2
x	1	0	$-\frac{1}{5}$	$\frac{1}{5}$	8
OBJ	0	0	$-\frac{3}{10}$	$\frac{4}{5}$	22

5. The initial tableau for the Big-M method is

	x_1	x_2	x_3	s_1	s_2	s_3	y	RHS
s_1	3	−2	−2	1	0	0	0	4
y	−2	1	0	0	−1	0	1	1
s_3	−2	5	0	0	0	1	0	8
OBJ	206	−102	−2	0	100	0	0	−100

The following tableau shows that the problem is unbounded because all entries in the x_3 column are ≤ 0.

	x_1	x_2	x_3	s_1	s_2	s_3	y	RHS
s_1	−1	0	−2	1	−2	0	2	6
x_2	−2	1	0	0	−1	0	1	1
s_3	8	0	0	0	5	1	−5	3
OBJ	2	0	−2	0	−2	0	102	2

5.3C (p. 153)

2. Several optimal tableaus are possible because there are alternative

optimal solutions. A final tableau is

	x_1	x_2	s_1	s_2	s_3	RHS
s_1	0	0	1	$-\frac{7}{2}$	$\frac{1}{2}$	5
x_2	0	1	0	$\frac{5}{2}$	$-\frac{1}{2}$	5
x_1	1	0	0	$-\frac{3}{2}$	$\frac{3}{2}$	9
OBJ	0	0	0	1	0	8

You can get two other optimal tableaus by pivoting in the s_3 column and either the s_1 or x_1 row.

4. Several optimal tableaus are possible because there are alternative optimal solutions. A final tableau is

	x_1	x_2	x_3	s_1	s_2	s_3	RHS
x_3	0	-13	1	3	0	-1	16
s_2	0	32	0	-8	1	$\frac{5}{2}$	3
x_1	1	5	0	-1	0	$\frac{1}{2}$	2
OBJ	0	0	0	5	0	0	100

There are alternative optimal solutions because there is a 0 entry in the column corresponding to the nonbasic variable s_3.

8. The initial Big-M tableau (with $M = 100$) is

	x_{11}	x_{12}	x_{13}	x_{21}	x_{22}	x_{23}	s_1	s_2	y_1	y_2	y_3	RHS
s_1	1	1	1	0	0	0	1	0	0	0	0	60
s_2	0	0	0	1	1	1	0	1	0	0	0	20
y_1	1	0	0	1	0	0	0	0	1	0	0	15
y_2	0	1	0	0	1	0	0	0	0	1	0	30
y_3	0	0	1	0	0	1	0	0	0	0	1	15
OBJ	-96	-98	-97	-98	-96	-97	0	0	0	0	0	-6000

This final tableau has alternate optimal solutions because there is a 0 in the objective row in the column corresponding to the nonbasic variable x_{23}.

	x_{11}	x_{12}	x_{13}	x_{21}	x_{22}	x_{23}	s_1	s_2	y_1	y_2	y_3	RHS
s_1	1	0	0	0	-1	-1	1	0	0	-1	-1	15
s_2	-1	0	0	0	1	1	0	1	-1	0	0	5
x_{21}	1	0	0	1	0	0	0	1	1	0	0	15
x_{12}	0	1	0	0	1	0	0	0	0	1	0	30
x_{13}	0	0	1	0	0	1	0	0	0	0	1	15
OBJ	2	0	0	0	2	0	0	0	98	98	97	-135

5.4A (p. 162) The answers are given in a format that can be used by LP SOLVER.

1. {Let S, C, A, BAN, BEEF be the number of units of Squash, Carrots, Apples, Bananas, and Beef, resp.}
   Min .05S + .09C + .10A + .12BAN + .90BEEF;
   .5S + .8C + .4A + 1.2BAN + 6.0BEEF >= 80; {Protein}
   3S + 4C + 7A + 10BAN >= 50; {Vitamin C}
   .4S + .4C + .5A + .04BAN + 1.5BEEF >= 10; {Iron}

Answer: Eat 160 units of Squash at a cost of $8.00.

3. {Let x1,x2 be the number of hours plants P1,P2 operate, resp.}
   min 8000x1 + 11000x2;
   200x1 + 100x2 >= 1000; {Demand for A}
   60x1 + 200x2 >= 800; {Demand for B}
   90x1 + 150x2 >= 900; {Demand for C}
   130x1 + 80x2 >= 1500; {Demand for D}

Answer: $x1 = 11.132, x2 = 0.66038; z = 96,320.755$.

5. {Let S, B, R, SAV be the amounts Jennifer invests in stocks, bonds, real estate, and savings, respectively.}
   .12S + .08B +.10R + .06SAV; {Maximize Jennifer's return}
   S + R <= SAV; {No more in stocks and real estate than in savings}
   B >= 2/3S; {Amount in bonds at least 2/3 amount in stocks}
   S + B + R >= 50000; {At least $50,000 in stocks, bonds, and real estate}
   S + B + R + SAV = 200000; {Invest it all}

Answer: S=75,000, B=50,000, SAV= 75,000; z = 17,500.

7. {We start by making a table of the data:

	Apple	Orange	Lime	Cranberry	Peach	Profit/quart
Drink 1	5	8	12	3	4	0.60
Drink 2	8	0	0	16	8	0.40
Drink 3	4	12	3	13	0	0.80
Drink 4	5	0	6	8	13	0.50
Avail	7500	5000	1200	4000	1500	

Let D1, D2, D3, D4 be the number of quarts of each drink made. }

$$\begin{array}{lllllllll}
\text{Max} & .6D1 & + & .4D2 & + & .8D3 & + & .5D4; & \\
& 5D1 & + & 8D2 & + & 4D3 & + & 5D4 & <= 32*7500; \quad \{\text{Apple}\} \\
& 8D1 & & & + & 12D3 & & & <= 32*5000; \quad \{\text{Orange}\} \\
& 12D1 & & & + & 3D3 & + & 6D4 & <= 32*1200; \quad \{\text{Lime}\} \\
& 3D1 & + & 16D2 & + & 13D3 & + & 8D4 & <= 32*4000; \quad \{\text{Cranberry}\} \\
& 4D1 & + & 8D2 & & & + & 13D4 & <= 32*1500; \quad \{\text{Peach}\}
\end{array}$$

$D1 = 783.67, D2 = 9665.31; z = 8,202.449.$

9. {Let xi be the number of times pattern i is used, for 1 <= i <= 5}

$$\begin{array}{lllllllllll}
\text{Min} & 2x1 & + & 4x2 & + & 0x3 & + & 6x4 & + & 2x5; & \\
& 4x1 & + & 2x2 & + & 2x3 & & & & & >= 150; \\
& & & x2 & & & + & 2x4 & + & x5 & >= 200; \\
& & & & & x3 & & & + & x5 & >= 300;
\end{array}$$

Answer: $x3 = 100, x5 = 200; z = 400.$

11. {Let xij be the amount shipped from Supply i to Demand j.}

$$\begin{array}{ll}
\text{min } 8x11 + 6x12 + 13x13 + 11x14 + & \\
7x15 + 3x21 + 1x22 + 7x23 + 14x24 + & \\
9x25 + 5x31 + 8x32 + 9x33 + 6x34 + 15x35; & \\
x11 + x12 + x13 + x14 + x15 <= 420; & \{\text{Out of Supply 1}\} \\
x21 + x22 + x23 + x24 + x25 <= 160; & \{\text{Out of Supply 2}\} \\
x31 + x32 + x33 + x34 + x35 <= 300; & \{\text{Out of Supply 3}\} \\
x11 + x21 + x31 >= 150; & \{\text{Into Demand 1}\} \\
x12 + x22 + x32 >= 100; & \{\text{Into Demand 2}\} \\
x13 + x23 + x33 >= 250; & \{\text{Into Demand 3}\} \\
x14 + x24 + x34 >= 90; & \{\text{Into Demand 4}\} \\
x15 + x25 + x35 >= 230; & \{\text{Into Demand 5}\}
\end{array}$$

Answer: $x11 = 30, x34 = 90, x15 = 230, x12 = 100, x23 = 160, x31 = 120, x33 = 90; z = 5520.$

13. { The cost table is

	D1	D2	D3	D4	Supply
S1	6	10	8	14	80
S2	7	13	5	9	120
S*	100	100	100	100	40
Demand	55	85	40	60	

Let xij be the amount shipped from Supply i to Demand j }

$$\begin{aligned}
&\text{Min } 6x11 + 10x12 + 8x13 + \\
&14x14 + 7x21 + 13x22 + 5x23 + \\
&9x24 + 100x31 + 100x32 + 100x33 + 100x34; \\
&x11 + x12 + x13 + x14 <= 80; \quad \{\text{Out of Supply 1}\} \\
&x21 + x22 + x23 + x24 <= 120; \quad \{\text{Out of Supply 2}\} \\
&x31 + x32 + x33 + x34 <= 40; \quad \{\text{Out of Supply 3}\} \\
&x11 + x21 + x31 >= 55; \quad \{\text{Into Demand 1}\} \\
&x12 + x22 + x32 >= 85; \quad \{\text{Into Demand 2}\} \\
&x13 + x23 + x33 >= 40; \quad \{\text{Into Demand 3}\} \\
&x14 + x24 + x34 >= 60; \quad \{\text{Into Demand 4}\}
\end{aligned}$$

Answer: $x12 = 45, x24 = 60, x11 = 35, x21 = 20, x32 = 40, x23 = 40; z = 5540$.

6.1A (p. 174)

1. $x_1 = 70, x_2 = 10; z = 400$ 3. $x_1 = 0, x_2 = \frac{28}{5}, x_3 = \frac{3}{5}; z = \frac{34}{5}$

6. Infeasible. 8. The row corresponds to an equation of the form

$$a_1x_1 + a_2x_2 + \cdots + a_nx_n = b$$

where all of the coefficients a_i are nonnegative and b is negative. It is impossible to satisfy such an equation with nonnegative x_i.

6.2A (p. 177)

1.

$$\begin{array}{lrcrcrcrcr}
\text{Minimize: } w = & 8y_1 & + & 12y_2 & + & 18y_3 & + & 17y_4 & & \\
\text{Subject to:} & 4y_1 & + & 4y_2 & + & 3y_3 & + & 6y_4 & \geq & 9 \\
& 2y_1 & + & y_2 & + & 2y_3 & + & y_4 & \geq & 3 \\
& -2y_1 & - & 2y_2 & - & 2y_3 & + & 2y_4 & \geq & -3 \\
& \multicolumn{9}{l}{y_i \geq 0, 1 \leq i \leq 4}
\end{array}$$

3.

$$\begin{array}{lrcrcrcr}
\text{Maximize: } w = & 3y_1 & + & 2y_2 & + & 3y_3 & & \\
\text{Subject to:} & 4y_1 & + & 2y_2 & + & y_3 & \leq & 1 \\
& 2y_1 & + & 3y_2 & + & y_3 & \leq & 1 \\
& \multicolumn{7}{l}{y_i \geq 0, 1 \leq i \leq 3}
\end{array}$$

6.2C (p. 185)

1. (a) Produce 50 gallons of juice B for a profit of 1000 cents. The final tableau is

	A	B	s_1	s_2	RHS
s_1	2	0	1	$-.5$	150
B	1	1	0	.5	50
OBJ	5	0	0	10	1000

(b) Let y_1 be the marginal value of an ounce of apple juice and y_2 be the marginal value of an ounce of cherry juice. The dual problem is

$$\begin{array}{lrcrcr} \text{Minimize: } w = & 200y_1 & + & 100y_2 & & \\ \text{Subject to:} & 3y_1 & + & 2y_2 & \geq & 15 \\ & y_1 & + & 2y_2 & \geq & 20 \\ & y_1, y_2 \geq 0 & & & & \end{array}$$

(c) To find the solution of the dual problem look in the objective row of the final tableau given in (a). The dual price corresponding to the first constraint is in the S_1 column (the slack variable column corresponding to the first constraint) and the dual price for the second constraint is found in the S_2 column. Thus, the solution to the dual problem is $y_1 = 0, y_2 = 10$. The minimum value of the objective function for the dual problem is 1000, the same as the maximum value of the objective function for the primal problem.

3. (a) Produce 6.52 cars of type B and 52.17 cars of type C for a profit of \$355.43.

(b) Let y_1 be the marginal value of a unit of raw material and y_2 be the marginal value of a unit of time. The dual problem is

$$\begin{array}{lrcrcr} \text{Minimize: } w = & 450y_1 & + & 300y_2 & & \\ \text{Subject to:} & 6y_1 & + & 4y_2 & \geq & 4.5 \\ & 5y_1 & + & 6y_2 & \geq & 6.5 \\ & 8y_1 & + & 5y_2 & \geq & 7 \\ & 4y_1 & + & 8y_2 & \geq & 6 \\ & y_1, y_2 \geq 0 & & & & \end{array}$$

(c) The dual prices are in the objective row and slack variable columns S_1 and S_2. The prices are $y_1 = 0.152174, y_2 = 0.956522$. The objective function of the primal problem is $450 \cdot 0.152174 + 300 \cdot$

0.956522. Thus, a unit change in the amount of raw materials will change profit by \$0.152174 and a unit change in the available time will change profit by \$0.956522.

5. (a) Make 9.434 units of Grate A and 1.887 units of Grate B for a profit of \$924.53.

   (b) The dual prices are 0.820755, 1.03774, and 0. They give profit per unit of steel, labor, and machine time, respectively. The objective function can be computed as $924.53 = .820755 \cdot 1000 + 1.03774 \cdot 100 + 0 \cdot 120$. Thus, a unit change in the amount of steel available will change profit by \$0.82755, a unit change in the availability of labor will change profit by \$1.03774, and a unit change in machine time will not change the profit because all of the available time is not used.

   (c) The dual problem is

$$\begin{array}{llcccccccc}
\text{Minimize: } w = & 1000y_1 & + & 100y_2 & + & 120y_3 & & \\
\text{Subject to:} & 80y_1 & + & 9y_2 & + & 5y_3 & \geq & 75 \\
 & 130y_1 & + & 8y_2 & + & 7y_3 & \geq & 115 \\
 & y_1, y_2, y_3 \geq 0 & & & & & &
\end{array}$$

7. The dual problem is

$$\begin{array}{llcccccccccc}
\multicolumn{12}{l}{\text{Minimize: } w = 32(7500y_1 + 5000y_2 + 1200y_3 + 4000y_4 + 1500y_5)} \\
\text{Subject to:} & 5y_1 & + & 8y_2 & + & 12y_3 & + & 3y_4 & + & 4y_5 & \geq & 0.6 \\
 & 8y_1 & & & & & + & 16y_4 & + & 8y_5 & \geq & 0.4 \\
 & 4y_1 & + & 12y_2 & + & 3y_3 & + & 13y_4 & & & \geq & 0.8 \\
 & 5y_1 & & & + & 6y_3 & + & 8y_4 & + & 13y_5 & \geq & 0.5 \\
 & y_i \geq 0, 1 \leq i \leq 5 & & & & & & & & & &
\end{array}$$

The units of the dual variables are the amounts of profit per quart that each drink contributes to the total profit. The solution is $y_1 = 0, y_2 = 0, y_3 = 0.03673, y_4 = 0.05306; z = 8,202.449$.

9. Let R, F, D, V be the respective marginal values of a refrigerator,

freezer, dishwasher, and oven in dollars per unit.

$$\begin{array}{llllllllll}
\text{Maximize:} & z = 1500R + 1000F + 1000D + 750V \\
\text{Subject to:} & 30R & + & 25F & + & 60D & + & 5V & \leq & 10000 \\
& 35R & + & 10F & + & 40D & & & \leq & 15000 \\
& 50R & + & 40F & + & 10D & & & \leq & 8000 \\
& 20R & + & 15F & + & 35D & + & 60V & \leq & 15000 \\
& 15R & + & 10F & + & 60D & + & 30V & \leq & 12000 \\
& R, F, D, V \geq 0
\end{array}$$

The answer is $R = 143.58, F = 0, D = 82.0097, V = 154.295$.

6.3A (p. 190)

2. $-\infty < c_1 \leq 0.4,\ 0 \leq c_2 < \infty,\ -\infty < c_3 \leq 9.26667,\ -11 \leq c_4 \leq 0$.

4. (a) No. The allowable increase is 1.375
   (b) Yes. The allowable decrease is 0.273077
   (c) Yes. You must solve the problem again to check this.

6.3B (p. 192)

2. $8.333 \leq b_1 \leq 58.2,\ -6.47059 \leq b_2 < \infty,\ -4.5 \leq b_3 \leq 10.8$

4. No. The change in ounces is from $5000 \cdot 32$ to $6000 \cdot 32$. The amount of increase actually does not matter since b_2 can vary over the range $122,253 \leq b_2 < \infty$. Another reason that increasing b_2 does not matter is that the slack variable for the second constraint has the value $s_2 = 37,746.939$, which means that not all of this resource was used.

6. None, b_3 can vary over the range $240 \leq b_3 \leq 330$.

6.3C (p. 196)

2. Yes, the set of basic variables changes. Substitute $\Delta b_1 = 1, \Delta b_2 = 3, \Delta b_3 = 1$ in the system of inequalities

$$\begin{array}{rcrcrcrl}
.059\Delta b_1 & & & + & .118\Delta b_3 & + & 1.706 & \geq 0 \\
.176\Delta b_1 & & & - & .147\Delta b_3 & + & .118 & \geq 0 \\
-.294\Delta b_1 & + & \Delta b_2 & + & .912\Delta b_3 & + & 14.471 & \geq 0
\end{array}$$

and observe that the second inequality is not satisfied.

5. In CALIPSO's notation the changes are $\Delta b_1 = -30, \Delta b_2 = 0, \Delta b_3 = 40, \Delta b_4 = 10, \Delta b_5 = -20, \Delta b_6 = 0, \Delta b_7 = 5, \Delta b_8 = -20$. The set of basic variables changes.

6.3D (p. 198)

2.

$$\begin{array}{rrrrrrrrl} 1(-\Delta c_1) & + & .8\Delta c_2 & & & - & .4(-\Delta c_4) & + & 2.4 \geq 0 \\ & & .2\Delta c_2 & & & + & .0667(-\Delta c_4) & + & .9333 \geq 0 \\ & & .2\Delta c_2 & - & \Delta c_3 & + & 1.7333(-\Delta c_4) & + & 13.2667 \geq 0 \\ & & & & & & .333(-\Delta c_4) & + & 1.6667 \geq 0 \end{array}$$

6.4A (p. 208)

2. No. b_5 can increase no more than 533. The problem becomes infeasible after the change.

4. We substitute the new costs in equations (1) – (4) and calculate the new cost coefficients to be $c_1 = 863, c_2 = 729, c_3 = 1019, c_4 = 718.95$. Subtracting the new cost coefficients from the old ones gives $\Delta c_1 = 74.2, \Delta c_2 = 73, \Delta c_3 = 82, \Delta c_4 = 85.2$. We find that these changes satisfy the system of inequalities (16) and conclude that the solution does not change. The new value of the objective function is \$173,357.15.

6. We substitute the new costs in equations (1) – (4) and calculate the new cost coefficients to be $c_1 = 896.8, c_2 = 760, c_3 = 1048, c_4 = 716.1$. Subtracting the new cost coefficients from the old ones gives $\Delta c_1 = 108, \Delta c_2 = 104, \Delta c_3 = 111, \Delta c_4 = 82.35$. We find that these changes satisfy the system of inequalities (16) and conclude that the solution does not change. The new value of the objective function is \$178,023.39.

8. Resource value b_4 can increase by 271.76 or decrease by 2462.26 .

7.1A (p. 214)

2. 14 4. 54 6. (a) 6 (b) 12 (c) 6 8. 32 10. 11

12. 70 14. 83

16.

(a) 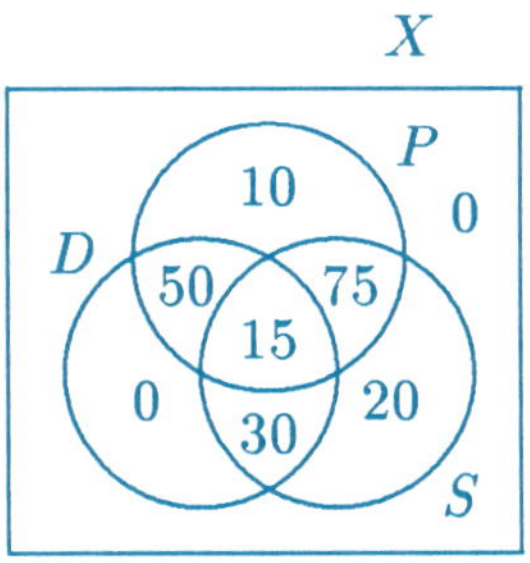

(b) 95 (c) 65 (d) 10 (e) 50

7.2A (p. 219)

2. Without repetitions $24 \cdot 23 \cdot 22 = 12{,}144$;
with repetitions $24^3 = 13{,}824$.

4. $12 \cdot 11 \cdot 10 \cdot 9 \cdot 8 \cdot 7 = 665{,}280$ 6. $5 \cdot 12 \cdot 8 \cdot 16 = 7{,}680$.

8. $19 \cdot 18 \cdot 19 = 6498$

10. (a) 26 (b) $26 + 26 \cdot 37 = 988$
(c) $26(1 + 37 + 37^2 + \ldots + 37^6) = 68{,}561{,}911{,}262$.

7.3A (p. 222)

2. $52 \cdot 51 \cdot 50 \cdot 49 \cdot 48 \cdot 47 \cdot 46 \cdot 45 \doteq 3.0 \cdot 10^{13}$.

4. $14 \cdot 13 \cdot 12 \cdot 11 \cdot 10 \cdot 9 = 2{,}162{,}160$.

6. In any order $18!$; in alternating order $(9!)^2$. 8. $12!$

7.4A (p. 227)

2. $C(34,11) = 286{,}097{,}760$ 4. $C(11,4) = 330$

6. $C(52,13) = 635{,}013{,}559{,}600$

8. $6 \cdot 8 \cdot 2 = 96$. Even though the word "combination" appears in the statement, this is not a problem that uses $C(m,n)$.

10. $C(12,4) + C(12,6) = 1419$. 12. $C(7,3) \cdot C(5,3) = 350$

7.5A (p. 230)

2. $\frac{8!}{3!1!1!1!2!} = 3360$ 4. $\frac{20!}{6!5!4!3!2!} = 97{,}772{,}875{,}200$ 6. $\frac{10!}{4!3!2!1!} = 12{,}600$

7.5B (p. 232)

2. $\frac{30!}{11!11!5!3!} = 231,215,072,961,600$ 4. $\frac{20!}{8!7!5!} = 99,768,240$

6. $\frac{100!}{40!25!35!} \doteq 7.1364 \cdot 10^{44}$

7.5C (p. 235)

2. $(\frac{1}{3} + \frac{2}{3})^4 = 1$ 4. $2^{10} = 1024;\ 1024 - 11 = 1013$ 6. $2^5 = 32$

8. $2^{10} \cdot 2^5 - 1 = 2^{15} - 1$ (assuming you do not have an empty dish).

10. $0 = (1-1)^n = C(n,0)1^0(-1)^n + C(n,1)1^1(-1)^{n-1} + \ldots + C(n,n)1^n(-1)^0$

8.1A (p. 242)

1. (a) $\frac{C(13,5)}{C(52,5)} \doteq .000495$ (b) 4 times the answer to (a)
   (c) $\frac{2 \cdot C(13,2) \cdot C(4,3) \cdot C(4,2)}{C(52,5)} \doteq .00144$ (d) $1 - \frac{C(39,5)}{C(52,5)} \doteq 0.7785$

3. (a) $\frac{C(12,3)C(15,4)}{C(27,7)} \doteq .338$ (b) $\frac{C(12,7)}{C(27,7)} + \frac{C(15,7)}{C(27,7)} \doteq .00814$
   (c) $1 - \frac{C(15,7)}{C(27,7)} \doteq .993$

5. Let S be the set of people who drink Pepsi, C be the set of people who drink Coke, and N be the set of people who drink neither.
   (a) $P(S \cup C) = P(S) + P(C) - P(S \cap C) = \frac{85+64-32}{140} \doteq 0.8357$
   (b) $\frac{85-32}{140} \doteq 0.3786$ (c) $1 - \frac{85+64-32}{140} \doteq 0.1643$.

7. $1 - \frac{C(93,10)+C(93,9)C(7,1)+C(93,8)C(7,2)}{C(100,10)} \doteq .0208$ 9. $\frac{1}{19}$

11. $\frac{C(30,6) \cdot C(50,6)}{C(80,12)} \doteq 0.157$ 13. $\frac{C(8,4)+C(16,4)}{C(24,4)} \doteq .178$ 15. $\frac{1}{C(7,3)} = \frac{1}{35}$

8.2A (p. 250)

1. Let E_3 be the event that 3 women are chosen and E_1 be the event that at least 1 woman is chosen.

$$\begin{aligned} P(E_3|E_1) &= \frac{P(E_3 \cap E_1)}{P(E_1)} = \frac{P(E_3)}{P(E_1)} \\ &= \frac{C(12,3)}{C(12,1)C(8,2)+C(12,2)C(8,1)+C(12,3)C(8,0)} \doteq .2030 \end{aligned}$$

3. $P(2 \text{ girls}|\text{at least 1 girl}) = \frac{P(2 \text{ girls})}{P(\text{at least 1 girl})} = \frac{1/4}{3/4} = \frac{1}{3}$ 5. $\frac{1}{2}$

7. $P(3 \text{ clubs}|4 \text{ hearts}) = \frac{P(3 \text{ clubs and } 4 \text{ hearts})}{P(4 \text{ hearts})} = \frac{C(13,3)C(13,4)}{C(39,3)C(13,4)} \doteq .0313.$

9. The tree is

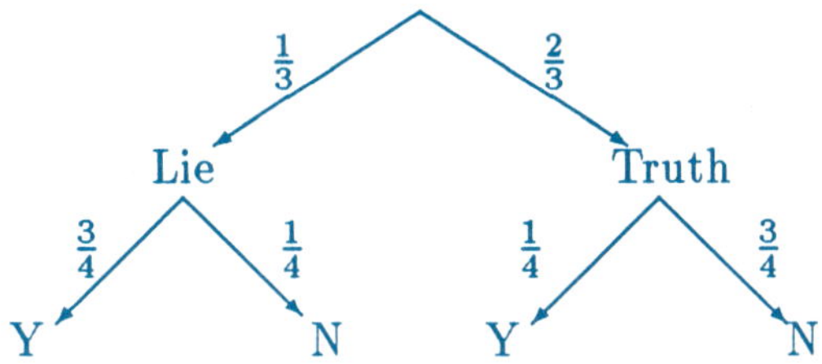

The probability that a respondent answers "yes" is

$$\frac{1}{3}\cdot\frac{3}{4}+\frac{2}{3}\cdot\frac{1}{4}=\frac{5}{12}$$

11. In the tree in the answer to Problem 9, replace $\frac{1}{4}$ by p and $\frac{3}{4}$ by $1-p$. Then $\frac{1}{3}(1-p)+\frac{2}{3}p=\frac{3}{5}$ from which we conclude that $p=\frac{4}{5}$.

13. (a) $(.95)(.88) = .836$ (b) $(.05)(.12) = .006$

15. Of all the 2^3 three-tuples of H's and T's, only one consists of all T's; consequently, $P(E_1 \cap E_2 \cap E_3) = \frac{1}{8}$. Also, $P(E_1) = P(E_2) = P(E_3) = \frac{1}{2}$. Hence, the events are independent.

17. (a) $(.2)(.6) = .12$ (b) $(.2)(.4) + (.8)(.6) = .56$ (c) $(.8)(.4) = .32$

8.3A (p. 258)

1. Let E_i be the event that the marble comes from Jar i, R be the event that the marble is red, and G be the event that the marble is green. To apply Bayes' formula we need to find the probabilities:

$$P(R) = \frac{3+6+8+4}{37} = \frac{21}{37} \quad P(G) = \frac{5+4+0+7}{37} = \frac{16}{37}$$
$$P(E_1) = \frac{8}{37} \quad P(E_2) = \frac{10}{37} \quad P(E_3) = \frac{8}{37} \quad P(E_4) = \frac{11}{37}$$
$$P(R|E_1) = \frac{3}{8} \quad P(R|E_2) = \frac{6}{10} \quad P(R|E_3) = 1 \quad P(R|E_4) = \frac{4}{11}$$
$$P(G|E_1) = \frac{5}{8} \quad P(G|E_2) = \frac{4}{10} \quad P(G|E_3) = 0 \quad P(G|E_4) = \frac{7}{11}$$

(a)

$$P(E_2|G) = \frac{\frac{10}{37} \cdot \frac{4}{10}}{\frac{8}{37} \cdot \frac{5}{8} + \frac{10}{37} \cdot \frac{4}{10} + \frac{8}{37} \cdot 0 + \frac{11}{37} \cdot \frac{7}{11}} = \frac{1}{4}$$

(b) Note that $P(\text{Not from Jar 1 or 4}) = P(E_2) + P(E_3)$, and, therefore, $P(\text{Not from Jar 1 or 4}|R) = P(E_2|R) + P(E_3|R)$.

$$P(E_2|R) = \frac{\frac{10}{37} \cdot \frac{6}{10}}{\frac{8}{37} \cdot \frac{3}{8} + \frac{10}{37} \cdot \frac{6}{10} + \frac{8}{37} \cdot 1 + \frac{11}{37} \cdot \frac{4}{11}} \doteq .286$$

$$P(E_3|R) = \frac{\frac{8}{37} \cdot 1}{\frac{8}{37} \cdot \frac{3}{8} + \frac{10}{37} \cdot \frac{6}{10} + \frac{8}{37} \cdot 1 + \frac{11}{37} \cdot \frac{4}{11}} \doteq .381$$

$$P(E_2|R) + P(E_3|R) \doteq .286 + .381 = .667$$

3. Let D be the event the person has the disease, F be the event the person does not have the disease, and Y be the event the test is positive.

$$P(D|Y) = \frac{P(D)P(Y|D)}{P(D)P(Y|D) + P(F)P(Y|F)} = \frac{(.2)(.85)}{(.2)(.85) + (.8)(.05)} \doteq .8095$$

5. We denote the event that we catch a certain type of fish by the first letter of its name, the event the fish is under 12 in long by U, and organize the data into the table

Type	$P(\text{Type})$	$P(U|\text{Type})$
C	.45	.8
R	.35	.75
S	.15	.90
O	.05	.95

$$P(C|U) = \frac{(.45)(.8)}{(.45)(.8) + (.35)(.75) + (.15)(.90) + (.05)(.95)} \doteq .447$$

7.

$$\frac{(.25)(.98)}{(.30)(.97) + (.20)(.95) + (.25)(.98) + (.25)(.96)} \doteq .254$$

9.

$$\frac{(.01)(.90)}{(.01)(.90) + (.99)(.10)} \doteq .0833$$

You will have to form your own opinion as to whether or not this test is sufficiently reliable.

8.4A (p. 262)

1. 0.12683 3. $1 - 0.99036 = .00964$ 5. 0.25082 7. 0.47441

9. $P(5 \text{ heads}) = 0.24609$ $P(\text{at least 5 heads}) = 0.62305$

11. $3.87 \cdot 10^{-7}$. Yes. 13. 0.25172

8.5A (p. 272)

1. (a) No. Add the first column. (b) No. A negative entry. (c) No.

3. Only d.

5. (a) The transition matrix is $T = \begin{bmatrix} .8 & .3 \\ .2 & .7 \end{bmatrix}$.

   (b) The distribution matrix is $D = \begin{bmatrix} .6 \\ .4 \end{bmatrix}$. Since $TD = D$ the D is a stable distribution matrix. So, 60% are predicted to donate every year.

   (c) 45%

7. (a) $\begin{bmatrix} .5 & .2 & .3 \\ .3 & .6 & .3 \\ .2 & .2 & .4 \end{bmatrix}$ (b) $\begin{bmatrix} .320 \\ .426 \\ .254 \end{bmatrix}$ (c) 0.24

9. "$T^n D$ approaches E as n gets large" means that the entries in $T^n D$ can be made as close as we desire to the entries in E for all n large enough, say $n \geq N$. Then, the entries in T^{n+1} are also close to those in E because $n + 1 > N$; that is, $T^{n+1}D \doteq T^n D \doteq E$. If we replace $T^n D$ and $T^{n+1}D$ by E in the expression $T^{n+1}D = T(T^n D)$, we get $E = T(E)$.

11. You can produce successive values of $T^n D$ by entering $E = T\#D$ in MATRIX *Operations* (you may use other letters to name your matrices.) Both choices for D give the stable distribution matrix

$$E = \begin{bmatrix} .13376 \\ .21497 \\ .44108 \\ .21019 \end{bmatrix}$$

13. After 3 months, 36.25%. In the long run, 40%.

15. (a) 36.572% (b) 50.072% (c) 16.637%
(d) Ford: 34.884%, GM: 47.907%, Bike: 17.209%.

17. (*b*) $\begin{bmatrix} .1872 \\ .1393 \\ .1767 \\ .2133 \\ .1686 \\ .1149 \end{bmatrix}$ (*c*) $\begin{bmatrix} .17256 \\ .14673 \\ .20967 \\ .1888 \\ .15207 \\ .13016 \end{bmatrix}$ (*d*) $\begin{bmatrix} .17256 \\ .14677 \\ .20991 \\ .18862 \\ .1519 \\ .13024 \end{bmatrix}$

8.6A (p. 280) 1. (a) $e_1 = \frac{5}{6}, e_2 = \frac{1}{6}$ (b) $e_1 = \frac{1}{5}, e_2 = \frac{4}{5}$ (c) Not regular.
(d) Not regular. (e) $e_1 = \frac{15}{31}, e_2 = \frac{16}{31}$ (f) Not regular.

8.6B (p. 282)

1. Matrices a, c, and d are regular.

4. 2/3

6. Clear: 0.37727, Rain: 0.22315, Snowy: 0.2622, Foggy: 0.13738

8.7A (p. 290)

1. Not absorbing.

3. The standard form matrix is

	1	0	2
1	1	$\frac{1}{2}$	0
0	0	$\frac{1}{2}$	$\frac{1}{2}$
0	0	0	$\frac{1}{2}$

$A = \begin{bmatrix} \frac{1}{2} & 0 \end{bmatrix}$

$B = \begin{bmatrix} \frac{1}{2} & \frac{1}{2} \\ 0 & \frac{1}{2} \end{bmatrix}$ $I - B = \begin{bmatrix} \frac{1}{2} & -\frac{1}{2} \\ 0 & \frac{1}{2} \end{bmatrix}$ $(I - B)^{-1} = \begin{bmatrix} 2 & 2 \\ 0 & 2 \end{bmatrix}$

(a)

	1	0	2
1	1	1	1
0	0	0	0
0	0	0	0

(b) Both 1. (c) Both 1. (d) Both $\frac{1}{2}$

8.7B (p. 295)

1. (a) No. (b) No way to reach the absorbing state from every state.

3. (a) Yes. (b) $\begin{bmatrix} 1 & \frac{2}{5} & \frac{4}{5} & \frac{1}{5} \\ 0 & \frac{1}{10} & \frac{1}{10} & \frac{1}{2} \\ 0 & \frac{1}{10} & 0 & 0 \\ 0 & \frac{2}{5} & \frac{1}{10} & \frac{3}{10} \end{bmatrix}$

5. (a) No.
(b) No way to reach the absorbing state from every state.

7. A standard form matrix with rows and columns in the order of states 2, 3, 1, 4, 5 is

$$(a)\ A = \begin{bmatrix} 1 & 0 & \frac{1}{4} & \frac{1}{3} & 0 \\ 0 & 1 & \frac{1}{4} & 0 & 0 \\ 0 & 0 & 0 & \frac{1}{6} & \frac{1}{3} \\ 0 & 0 & \frac{1}{2} & 0 & \frac{1}{3} \\ 0 & 0 & 0 & \frac{1}{2} & \frac{1}{3} \end{bmatrix} \quad (b)\ A^3 = \begin{bmatrix} 1 & 0 & .4375 & .5 & .3287 \\ 0 & 1 & .27083 & .083333 & .125 \\ 0 & 0 & .083333 & .097222 & .13889 \\ 0 & 0 & .125 & .13889 & .17593 \\ 0 & 0 & .083333 & .18056 & .23148 \end{bmatrix}$$

(c) These probabilities are given by the stable matrix

$$A^{69} = \begin{bmatrix} 1 & 0 & .65385 & .80769 & .73077 \\ 0 & 1 & .34165 & .19231 & .26923 \\ 0 & 0 & 0 & 0 & 0 \\ 0 & 0 & 0 & 0 & 0 \\ 0 & 0 & 0 & 0 & 0 \end{bmatrix}$$

(d) The fundamental matrix (in order of states 1, 4, 5) is

$$(I - B)^{-1} = \begin{bmatrix} 1.3846 & .76923 & 1.0769 \\ .92308 & 1.8462 & 1.3846 \\ .69231 & 1.3846 & 2.5385 \end{bmatrix}$$

The average number of transitions to reach an absorbing state is the sum of the entries in the column corresponding to that state. The sums are State 1: 3; State 4: 4; State 5: 5.

11.

$$(a) \begin{bmatrix} 1 & 0 & .22 & .13 & .09 & .06 \\ 0 & 1 & 0 & 0 & 0 & .94 \\ 0 & 0 & 0 & 0 & 0 & 0 \\ 0 & 0 & .78 & 0 & 0 & 0 \\ 0 & 0 & 0 & .87 & 0 & 0 \\ 0 & 0 & 0 & 0 & .91 & 0 \end{bmatrix}$$

$$(b) \begin{bmatrix} 1 & 0 & .41953 & .2558 & .1446 & .06 \\ 0 & 1 & .58047 & .7442 & .8554 & .94 \\ 0 & 0 & 0 & 0 & 0 & 0 \\ 0 & 0 & 0 & 0 & 0 & 0 \\ 0 & 0 & 0 & 0 & 0 & 0 \\ 0 & 0 & 0 & 0 & 0 & 0 \end{bmatrix}$$

(c) The probabilities that members of each class will graduate are given in the second row of the matrix in (b) in columns 3–6.

13. (a) We give a standard form transition matrix in which the rows and columns are ordered to correspond to chambers 6,8,3,4,5,1,7,2. You could get another standard form matrix ordered as 8,6,3,4,5,2,7,1.

		Current							
		6	8	3	4	5	1	7	2
	6	1	0	$\frac{1}{2}$	$\frac{1}{7}$	0	0	$\frac{1}{4}$	0
	8	0	1	0	0	$\frac{1}{2}$	0	$\frac{1}{4}$	0
	3	0	0	0	$\frac{1}{7}$	0	$\frac{1}{2}$	0	0
Next	4	0	0	$\frac{1}{4}$	0	$\frac{1}{4}$	0	$\frac{1}{2}$	$\frac{1}{2}$
	5	0	0	0	$\frac{1}{7}$	0	0	0	$\frac{1}{4}$
	1	0	0	$\frac{1}{4}$	0	0	0	0	$\frac{1}{4}$
	7	0	0	0	$\frac{2}{7}$	0	0	0	0
	2	0	0	0	$\frac{2}{7}$	$\frac{1}{4}$	$\frac{1}{2}$	0	0

The stable matrix is

	Current							
	6	8	3	4	5	1	7	2
6	1	0	.82975	.62423	.29601	.69479	.56212	.55982
8	0	1	.17025	.37577	.70399	.30521	.43788	.44018
3	0	0	0	0	0	0	0	0
4	0	0	0	0	0	0	0	0
5	0	0	0	0	0	0	0	0
1	0	0	0	0	0	0	0	0
7	0	0	0	0	0	0	0	0
2	0	0	0	0	0	0	0	0

(b) Chamber 6: .69479 Chamber 8: .30521

(c) Chamber 6: .29601 Chamber 8: .70399

(d) The fundamental matrix is

	3	4	5	1	7	2
3	1.33740	0.42945	0.23313	0.92025	0.21472	0.50307
4	0.75153	1.91100	0.83742	1.09510	0.95552	1.43870
5	0.23313	0.47853	1.28830	0.45399	0.23926	0.67485
1	0.46012	0.31288	0.22699	1.52760	0.15644	0.59509
7	0.21472	0.54601	0.23926	0.31288	1.27300	0.41104
2	0.50307	0.82209	0.67485	1.19020	0.41104	1.87730

For each state, the average numbers of doors to pass through to reach a trap are the sums of the columns of the fundamental matrix. The sums are

$$[\ 3.5 \quad 4.5 \quad 3.5 \quad 5.5 \quad 3.25 \quad 5.5\ ]$$

(e) .029695 (This is the fifth power of the matrix in (a)).

9.1B (p. 308)

5. (a) Let X be the amount you win.

X	2	6	10
$P(X)$	$\frac{4}{6}$	$\frac{1}{6}$	$\frac{1}{6}$

(b) $P(X = 2) = \frac{4}{6}$ $\quad P(X \geq 3) = \frac{2}{6}$ $\quad P(2 < X \leq 10) = \frac{2}{6}$

7. (a)

X	2	3	4	5	6	7	8	9	10	11	12
$P(X)$	.01	.04	.10	.16	.19	.18	.14	.10	.05	.02	.01

(b) $P(X = 7 \text{ or } 11) = .20$, $P(X \geq 10) = .08$, $P(4 < X < 9) = .67$, $P(X \leq 5) = .31$

9. (a)

X	1	2	3	4	5	6	7	8	9
Colors	*RRR*	*RRB*	*RRO*	*RBB*	*RBO*	*ROO*	*BBO*	*BOO*	*OOO*
$P(X)$	$\frac{1}{120}$	$\frac{6}{120}$	$\frac{15}{120}$	$\frac{3}{120}$	$\frac{30}{120}$	$\frac{30}{120}$	$\frac{5}{120}$	$\frac{20}{120}$	$\frac{10}{120}$

11. Let X be the number of hearts in the 7-card hand.

X	0	1	2	3	4	5	6	7
$P(X)$	.11497	.31703	.33568	.17583	.048843	.007128	.00050	.0000128

9.2A (p. 314)

1. $\bar{x} = 26.1$, median $= 27.5$ 3. $\bar{x} = 28.7143$, median $= 24$

5. $\bar{x} = 1.76991$, median $= 1.2$ 7. $E(X) = 6.9$ 9. $E(X) = 13.9$

11. $E(X) = \frac{1}{6}(6 + 12 + 18) - \frac{1}{6}(2 + 6 + 10)$. 13. 64

9.3A (p. 320)

1. $(a)\ \bar{x} = 3,\ s^2 = 9.5,\ s = 3.08221$ $(b)\ \mu = 3,\ \sigma^2 = 7.6,\ \sigma = 2.75681$

4. $(a)\ \bar{x} = 9.2,\ s^2 = 16.8444,\ s = 4.1042$ $(b)\ \mu = 9.2,\ \sigma^2 = 15.1601,$
   $\sigma = 3.8936$

7. $P(3 \le X \le 7) \ge 0.4375$ 9. $c = 3\sqrt{3}$ 11. $\mu = 35,\ \sigma = 4.7697$

13. $\mu = 500, \sigma = 19.3649$

14. Problem 1. $E(X^2) - \mu^2 = 16.6 - 9 = 7.6$
    Problem 3. $E(X^2) - \mu^2 = 30.82 - 4.1616 = 26.6584$

9.4A (p. 327)

1. $0.2119, 0.7881, 0.6730$ 3. $0.0054, 0.9946, 0.8795$

5. $0.5126, 0.8295, 0.3170$ 7. $0.1361, 0.8943, 0.5$ 9. $0.1545, 0.2023, 0.9088$

11. (a) .3319 (b) .3938 (c) 3.232% 13. $P(X \ge 3800) = .008198$

15. $P(95 \le X \le 97) = .9545$, $P(X \le 94) = 3.167 \times 10^{-5}$

9.5A (p. 332)

1. Approximation with normal distribution gives .70932; actual probability is .67307.

3. $\mu = 230,\ \sigma = 14.99,\ P(199.5 \le X \le 240.5) \doteq .7372$

5. $\mu = 18,\ \sigma = 4.236,\ P(X \le 3.5) \doteq .0003097,\ P(X \ge 1.5) \doteq 1$

7. $\mu = 1000$, $\sigma = 22.36$, $P(H \leq 990.5) = .3355$,
$P(849.5 \leq H \leq 950.5) = .01342$

9.6A (p. 336)

1. $y = 0.485x + 0.59$ (*a*) $y = 7.38$ (*b*) $x = 11.155$ (*c*) $x = 4.970$

3. $y = 3.91x + .589$ (*a*) $y = 27.959$ (*b*) $y = 33.824$ (*c*) $x = 1.128$

5. $y = -2.54x + 182$ (*a*) 101.72 (*b*) 207.4 (*c*) 29.6

7. $y = 4.12x + 10$ When $x = 25, y = 113$.

9.6B (p. 341)

1,3,5,7 The answers are the same as for Exercises 9.6A. Here are the matrices to solve Problem 1.

$$A = \begin{bmatrix} 1 & 1 \\ 4 & 1 \\ 6 & 1 \\ 8 & 1 \\ 12 & 1 \end{bmatrix} \qquad B = \begin{bmatrix} 3 \\ 0 \\ 3 \\ 5 \\ 7 \end{bmatrix}$$

You can find m and b in $y = mx + b$ by using MATRIX *Operations* and entering $C = A'B/(A'A)$.

9. $m = \frac{31}{35}$, $b = -\frac{24}{35}$ 11. $y = -10x + 290.67$ 13. $y = -2.5357x + 71.821$

10.1A (p. 347)

1. (a) 7000 (b) 6200 (c) 6333.33 3. (a) 3500 (b) 2900 (c) 3000

5. (a) 7449.23 (b) 6351.19 (c) 6970.35

7. (a) 13,498.26 (b) 11,972.00 (c) 12,840.00

9. (a) 4214.36 (b) 3127.89 (c) 3722.04 11. 60,496.47

13. 10,600; 10,616.78; 10,618.00; 10,618.31; 10,618.37

15. (a) 0.1607545 (b) 0.1615834 17. 20,510.97

10.2A (p. 353)

1. 7159.29 3. 4,144.70 5. 488.07 7. 4,151.33

9. 13,180.30 11. 1,081.56

13. (a) After multiplying the second parenthesized expression through by s, all the powers of s except 1 and s^k subtract away. (b) Factor the expression. If you cannot see how, let $x = 1 + s + s^2 + \cdots + s^{k-1}$. Then the left-hand side of the equation in (a) is $x - sx$, which equals $(1-s)x$. (c) Divide both sides by $1 - s$.

10.3A (p. 359)

1. 719.62 3. 7,635.47 5. 5,935.08 7. 6,342.77

9. 7,606.50 11. 129,555.60 13. 218,088.18

15. (a) $W = \frac{81{,}000}{a_{\overline{300}|.125/12}} = 883.19$ (b) $300W = 264{,}957$
   (c) $D = \frac{81000 a_{\overline{180}|.125/12}}{a_{\overline{300}|.125/12}} = 71{,}656.88$
   (d) $i = \frac{.125}{12} D = 746.43$ $P = 883.19 - 746.43 = 136.76$

17. (a) $W = \frac{13{,}500}{a_{\overline{42}|.145/12}} = 411.76$ (b) $42 \cdot 411.76 - 13500 = 3793.92$
   (c) $D = \frac{13500 a_{\overline{18}|.145/12}}{a_{\overline{42}|.145/12}} = 6625.25$ (d) $W = \frac{12{,}000}{a_{\overline{42}|.145/12}} = 366.01$

10.4A (p. 368)

1. (b) The optimal solution does not use the new security.

3. (b) Yes. The new lump sum payment is $313,305.69.

5. (b) MK must pay $685,500 in interest and fees.

11.1A (p. 376)

1. (a) The sun rises in the West and $2 + 2 = 4$.
   (b) $2 + 2 \neq 4$ or the sun rises in the West.

3. (a) The sun does not rise in the West or $2 + 2 \neq 4$.
   (b) The sun does not rise in the West and $2 + 2 \neq 4$.

5. (a) $p \wedge \sim q$ (c) $q \oplus \sim p$ 6. (b) $\sim p \wedge q$

10. (a) Contradiction (c) Contradiction (e) Neither (g) Contradiction

11.2A (p. 381)

1. F 3. T 5. T 7. F

9.

p	q	$\sim p$	$\sim q$	$\sim p \to \sim q$
T	T	F	F	T
T	F	F	T	T
F	T	T	F	F
F	F	T	T	T

11. Tautology

p	q	$p \vert q$	$p \to (p \vert q)$
T	T	T	T
T	F	T	T
F	T	T	T
F	F	F	T

13. Contradiction

p	r	q	$A = p \& r$	$B = q \& \sim r$	$C = \sim p \& r$	$D = \sim q \& \sim r$	$(A \vert B) \leftrightarrow (C \vert D)$
T	T	T	T	F	F	F	F
T	T	F	T	F	F	F	F
T	F	T	F	T	F	F	F
T	F	F	F	F	F	T	F
F	T	T	F	F	T	F	F
F	T	F	F	F	T	F	F
F	F	T	F	T	F	F	F
F	F	F	F	F	F	T	F

15. Tautology

p	q	r	$A = p \to q$	$B = q \to r$	$C = p \to r$	$D = (A \& B) \to C$
T	T	T	T	T	T	T
T	T	F	T	F	F	T
T	F	T	F	T	T	T
T	F	F	F	T	F	T
F	T	T	T	T	T	T
F	T	F	T	F	T	T
F	F	T	T	T	T	T
F	F	F	T	T	T	T

17. Neither a tautology nor a contradiction

p	q	r	$A = p\&q$	$B = A \to r$	$C = p\&q\&r$	$D = B \leftrightarrow C$
T	T	T	T	T	T	T
T	T	F	T	F	F	T
T	F	T	F	T	F	F
T	F	F	F	T	F	F
F	T	T	F	T	F	F
F	T	F	F	T	F	F
F	F	T	F	T	F	F
F	F	F	F	T	F	F

19. Contradiction

p	q	$A = \sim p\&q$	$B = \sim q\|p$	$C = A \leftrightarrow B$
T	T	F	T	F
T	F	F	T	F
F	T	T	F	F
F	F	F	T	F

21. Not logically equivalent 23. Logically equivalent

11.3A (p. 384)

1. Converse: If I live in Dubuque, then I live in Iowa. The converse is true because Dubuque is in Iowa. Contrapositive: If I do not live in Dubuque, then I do not live in Iowa. The contrapositive is false because I could live in Iowa without living in Dubuque.

3. Converse: If $x^2 = 100$, then $x = 10$. This is false since $(-10)^2 = 100$. Contrapositive: If $x^2 \neq 100$, then $x \neq 10$. This is true since x^2 must be 100 when $x = 10$.

5. Converse: If A and B are square matrices, then AB is a square matrix. This is false since A could be 3×3 and B could be 4×4. Contrapositive: If A or B is not a square matrix, then AB is not a square matrix. This is false since A could be 3×4 and B could be 4×3.

7. No to both questions.

11.3B (p. 388)

1. Valid 3. Invalid 5. Valid 7. Valid 9. Valid

11. Valid 13. Invalid 15. Valid

INDEX

L

M

N

O

P

R

S